Lecture Notes in Physics

Edited by H. Araki, Kyoto, J. Ehlers, München, K. Hepp, Zürich
R. Kippenhahn, München, H. A. Weidenmüller, Heidelberg
and J. Zittartz, Köln
Managing Editor: W. Beiglböck

242

Exactly Solvable Problems in Condensed Matter and Relativistic Field Theory

Proceedings of the Winter School and
International Colloquium Held at
Panchgani, January 30–February 12, 1985
and Organized by Tata Institute of
Fundamental Research, Bombay

Edited by B. S. Shastry, S. S. Jha and V. Singh

Springer-Verlag
Berlin Heidelberg New York Tokyo

Editors

B.S. Shastry
S.S. Jha
V. Singh
Tata Institute of Fundamental Research
Homi Bhabha Road, Colaba, Bombay 400005, India

ISBN 3-540-16075-2 Springer-Verlag Berlin Heidelberg New York Tokyo
ISBN 0-387-16075-2 Springer-Verlag New York Heidelberg Berlin Tokyo

<u>FOREWORD</u>

We have pleasure in presenting the Proceedings of the "Winter School and International Colloquium on Exactly Solvable Problems in Condensed Matter and Relativistic Field Theory" organized by the Tata Institute of Fundamental Research, Bombay, and held at Panchgani between January 30 and February 12, 1985. Some of the lectures presented in the School are reproduced in this volume.

The article by B. Sutherland reviews the coordinate space Bethe Ansatz in the context of the 1-d Hubbard model. The article also surveys the area of the asymptotic Bethe Ansatz, an approach pioneered by the author and F. Calogero, leading to the solution of several interesting models, including the quantum - Toda lattice.

T. Miwa in his article provides an introduction to the theory of the τ-function, an approach pioneered by the author and his collaborators M. Jimbo, M. Sato, E. Date and M. Kashiwara, in the context of the 2-d Ising and Kadomtsev-Petviashvili equations. The article by C.K. Majumdar reviews the application of the Bethe Ansatz to the Heisenberg model and discusses broken symmetry in antiferromagnets in general.

The articles by L. Faddeev, L. Takhtadjan and by Bogoliubov, Izergin and Korepin present the Leningrad approach to exactly integrable systems, focusing on the algebraic structure of the Yang Baxter equations, in various contexts.

We expect that this volume, containing several expository articles written by experts in the field, will be of considerable interest to students as well as specialists.

We acknowledge with thanks the help provided by L. Faddeev and B. Sutherland in arranging the conference. Additional financial aid was received from the Department of Science and Technology (DST). Visits of several scientists were financed under the U.S.-India Exchange Programme, the Yamada Foundation and the INSA-Soviet Academy exchange programme.

B.S. Shastry
S.S. Jha
& V. Singh

TABLE OF CONTENTS

B. Sutherland
AN INTRODUCTION TO THE BETHE ANSATZ 1

T. Miwa
AN INTRODUCTION TO THE THEORY OF τ FUNCTIONS 96

C.K. Majumdar
ANTIFERROMAGNETS 142

L.D. Faddeev
CLASSICAL AND QUANTUM L-MATRICES 158

L.A. Takhtajan
INTRODUCTION TO ALGEBRAIC BETHE ANSATZ 175

N.M. Bogoliubov, A.G. Izergin, V.E. Korepin
QUANTUM INVERSE SCATTERING METHOD AND CORRELATION FUNCTIONS ...220

LIST OF PARTICIPANTS 317

AN INTRODUCTION TO THE BETHE ANSATZ

Bill Sutherland
Department of Physics
University of Utah
Salt Lake City, Utah 84112 USA

I. INTRODUCTION

The term "Bethe ansatz" is a code name for a wave function with a particular structure, much as the Hartree, Hartree-Fock and Jastrow wave functions denote other kinds of structure. And since we are talking of wave functions, the physics we use is quantum mechanics, although the systems do have well-defined classical limits. And the structure of the Bethe ansatz wave function does persist in the classical systems.

What then is the physical content of this structure? This is probably best expressed by saying the systems are non-diffractive or integrable. Non-diffraction is a term borrowed from wave-phenomena, and it implies that geometrical optics or ray tracing in fact gives a correct answer, even though we really have wave behavior. Integrability comes from mechanics, and indicates that the system admits a complete set of constants or integrals of motion. We will illustrate these concepts in a few moments.

The surprising fact is that there exist interesting physical systems for which this form provides the exact wave function. By physically interesting I mean both that the system might be approximately realized in the real world or on the theorist's list of favorite models, and that the system is not simply free, but in fact has a non-trivial interaction.

There are numerous models which are solved by a wave function of Bethe ansatz form. They are all, however, essentially one-dimensional. The traditional model to cut one's teeth on has been the delta-function boson gas as first solved by McGuire, Lieb, and Liniger, and Yang and Yang. (Incidently, I don't find the delta-function boson gas as simple as it first appears.) The original ansatz was applied by Bethe to a spin-chain model for magnetism.

In these introductory lectures, however, I will introduce the techniques through a discussion of the Hubbard model. I have chosen this model because it has more structure than the simple models, and therefore is more interesting and illustrates a greater range of techniques. We will not have to switch models as we become more sophisticated. On the other hand, the resulting calculations may be carried out in more detail, or at least more explicitly, since the Hubbard model is a lattice system rather than a continuum system. Finally, the Hubbard model is simpler than the spin chains, both in superficial aspects and intrinsic structure. Also, my understanding is that the Hubbard model is not (yet) susceptible to treatment by the algebraic Bethe ansatz methods of Professors Faddeev, Takhtadjan and Korepin, so in this respect my lecture complements theirs.

However, before attacking the Hubbard model, let us return to the concepts of non-diffraction and integrability, and illustrate them by some simple examples from wave phenomena.

Let us take two mirrors α, β and place them together at right angles as in fig. (1). We shine a beam of light (for instance) into the wedge, making an angle ϕ with mirror α. What happens?

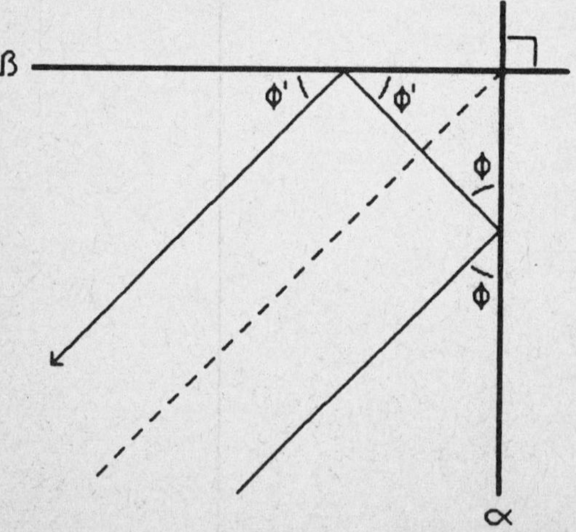

Fig. (1) Reflection from a 90° kaleidoscope. The dashed line divides the wave front into a portion which strikes mirror α first, and a portion which strikes mirror β first.

Let us ray-trace a ray which strikes mirror α first. It is specularly reflected from α with angle ϕ , crosses over to hit mirror β with angle ϕ' is reflected once again from β, and emerges from the wedge. Since $\phi + \phi' = \pi/2$, it emerges in the opposite direction from which it entered.

If the ray had struck β first, then this ray would simply retrace the first ray. Thus both outgoing rays emerge in the same direction. This is good, but it is not enough to insure that there is no diffraction in the problem.

There is one ray making an angle ϕ with respect to mirror α, which strikes the origin where the two mirrors meet. This line divides the two types of rays and thus the two portions of the wave front, one from another. For our ray tracing to represent a solution of the wave problem, the amplitudes of the emerging wave fronts must match across this line.

Let us assume the wave to be $\psi = e^{i\vec{k}\cdot\vec{x}}$, or superpositions thereof. The boundary conditions on the mirrors we take to be $\psi|_\alpha = \psi|_\beta = 0$. Then the solution is given by the method of images as $\psi = \sum_j (\pm 1) e^{i\vec{k}_j \cdot \vec{x}}$. The $\vec{k}_j$'s are $\vec{k}$ after successive reflections. The values are most easily seen in fig. (2). We begin with $\vec{k}_1$ and assume the amplitude to be +1. After bouncing off α, we have $\vec{k}_2$, and amplitude -1. After then bouncing off β we have $\vec{k}_3$ emerging with amplitude +1. On the other hand if we had struck β first, we would have $\vec{k}_2'$ with amplitude -1. This is followed by a collision with α, giving $\vec{k}_3'$ and amplitude +1. We see, as we verified previously, $\vec{k}_3 = \vec{k}_3'$. Further the amplitudes are both, +1, so there is no diffraction, and we have verified the method of images.

Another way of looking at the problem is to recognize that we have two integrals of motion: $L_x = k_x^2$, $L_y = k_y^2$. All four $\vec{k}$'s consistent with these constants of motion occur in the decomposition of ψ .

Does it matter if we have "partially silvered" mirrors? No, this just gives us the class of separable potentials in two-dimensions, $V = V_x(x) + V_y(y)$, and is integrable.

To see how delicate our solution is, however, let us increase the angle between the mirrors a small amount from $\pi/2$ to $\pi/2 + \epsilon$. Now $\phi + \phi' + \pi/2 + \epsilon = \pi$ or $\phi + \phi'$ $= \pi/2 - \epsilon$. Then the angle between the incoming ray and the outgoing ray, or the angle

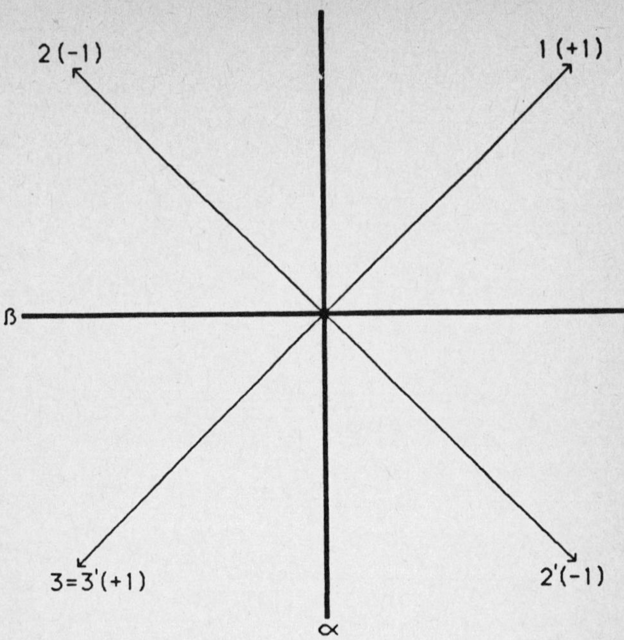

Fig. (2) The reflection of the k-vectors in the mirrors of fig. (1). The numbers in parentheses are the amplitudes after successive reflections.

between the two outgoing rays, is the angle $\pi/2 + \varepsilon - (\phi + \phi') = 2\varepsilon$. Thus if we shine a beam of light (for instance) on the mirror wedge, then by ray tracing we open up a wedge of darkness of angular width 2ε. This of course cannot be a solution to the wave equation, for the amplitude cannot take a jump from 1 to 0 and then back up to 1. In fact there must be diffraction near these two shadow edges. Another way of looking at it: We now have only the single integral of motion $|\vec{k}|^2 = L$.

Separable potentials are in some sense trivial, so let us look at a three particle system in one dimension interacting by pair potentials. The coordinates are y_1, y_2, y_3. The center of mass coordinate $(y_1+y_2+y_3)/\sqrt{3}$ separates out, so let us take it as one of our variables, with the other two coordinates perpendicular and in the plane of the paper. That is, we project down along the diagonal of the $y_1 y_2 y_3$ system and see the three planes $y_j = y_\ell$ as lines making angles of $\pi/3$ with each other. The projection of the y_j axes coincide with these lines. Fig. (15) illustrates the arrangement of mirrors. Such an experimental apparatus is familiar--it is called a kaleidoscope.

Let us assume for convenience that the potential is equivalent to $\psi = 0$ for $y_j = y_\ell$. Then we have specular reflection from the planes, with a change of sign of the amplitude. We show the successive $\vec{k}$'s after the two possible strings of successive reflections in fig. (3). Once again, the k's emerge in the same direction and the amplitudes are the same. Thus there is no diffraction.

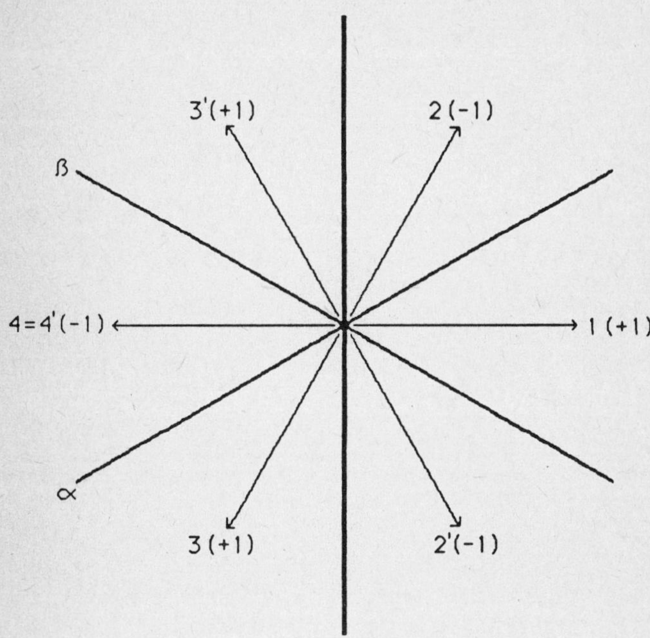

Fig. (3) The reflection of the k-vectors in the mirrors of the 60° kaleidoscope. Again the amplitudes are in parentheses.

A second way of looking at the problem is to realize that there are three integrals of motion: $P = \sum k_j$, $E = \sum k^2_j$, $L = \sum k^3_j$. If we fix these three integrals, the only solutions are the 3! permutations of the incoming momenta.

Our basic question is now the following: Are there "partially silvered mirrors", i.e., two-body potentials, which give diffraction-less kaleidoscopes? If so, the solution to these problems will be given by the Bethe ansatz.

The Hubbard model is one such kaleidoscope. Surprisingly, there are many others.

Before continuing with our discussion of the Bethe ansatz, let me now give some references. Since this lecture is primarily pedagogic --as is appropriate for a school-- and few new results are to be reported, I have kept the references to original works to a minimum. Those that are given will be collected here and the selection is personal instead of comprehensive. One reason for this, is that I haven't worked in the area for several years, and am not aware of the recent advances. However, I know that Professors Faddeev, Takhtadjan and Korepin are resonsible for many of these advances, and I am certain that they will pick up where I leave off, so that by combining my references with theirs, the participants should have a rather complete list.

The name, of course, honors the contribution of Bethe; this is a very beautiful paper and one should occasionally revisit it for inspiration and humility. This first application of the wavefunction was to the antiferromagnetic Heisenberg chain. It was followed by studies of Hulthén[2], Orbach[3], Walker[4], des Cloizeaux and Pearson[5], Griffiths[6,7] and des Cloizeaux and Gaudin[8]. In addition there were beautiful numerical studies on the same system by Bonner and Fisher[9], and Bonner[10] to keep the theorists honest. These investigations concluded in a series of papers by Yang and Yang[11-15] --very elegant to my eye--closing the period prior to the solution of the XYZ chain by Baxter.

At the same time that the properties of the spin chains were being unraveled, there was a parallel investigation of the quantum many-body systems of particles interacting by short range forces --the delta-function potentials. The first application of the Bethe ansatz was by Lieb and Liniger[16], and Lieb[17] to the delta-function boson gas. The scattering states for other statistics were investigated by McGuire[18-20], and Zinn-Justin and Brezin.[21] Finally the attractive fermion problem was solved by Gaudin[22,23], followed by the repulsive fermion problem by Yang[24], and then general statistics by Sutherland[25]. The statistical mechanics of the boson problem was treated by Yang and Yang[26], introducing new techniques.

We apply the Bethe ansatz to the Hubbard model[27], a model for interacting electrons from solid state physics. This was first solved in one dimension by Lieb and Wu[28].

Review articles on the Bethe ansatz are hard to find; workers are always more interested in what is to come rather than where we have been. Hopefully this school will correct this lack. Meanwhile, there are some older articles reprinted in the Lieb and Mattis book.[29] There are excellent lectures by Gaudin[30] in French. I understand that there is a new book by Gaudin[31] only recently published; I eagerly look forward to it. The book of Baxter[32] thoroughly covers that line of development of the Bethe ansatz. I would also add the articles of Yang[33], and Sutherland[34].

II. THE HUBBARD MODEL - AN EXAMPLE OF BETHE ANSATZ AT WORK

A) The Model

The Hubbard model is a lattice model for interacting electrons, i.e. fermions with two components of spin labeled $s = \pm 1$. We will use only the one-dimensional version. The Hamiltonian consists of a kinetic energy term which hops the electrons to nearest-neighbor sites with amplitude -1, plus an interaction term of magnitude $+2c$ between two electrons on the same site, and these must have opposite spins.

We must first determine the symmetries of this system, and to discuss the symmetries, we must write out the Hamiltonian explicitly. For this purpose, it is most convenient to use the language of second quantization. If this language is unfamiliar, don't worry, for we will never again use it after the section B) on symmetries.

We begin with an empty one dimensional lattice of N sites labeled by $j = 1, 2, ..., N$; when convenient, we shall assume N to be even. Let us also impose periodic boundary conditions, forming the line into a ring by identifying the $N + 1$ site with the first, and thus $j + N = j$. On this lattice we place N_{+1} electrons with spin $s = +1$, N_{-1} electrons with spin $s = -1$. Then we allow the electrons to hop about and interact.

Thus, N_s is separately conserved for each s, and clearly we must have $0 < N_s < N$ due to the exclusion principle. Let us write n_{js} as the number of electrons on site j with spin s, and

$$N_s/N = \sum_j n_{js}/N \equiv n_s \quad .$$

Therefore we are to solve the problem--for instance find the ground state energy-- for each sector labeled by the two values $0 < n_s < 1$, $s = \pm 1$.

Now the interaction term of the Hamiltonian is simply written as

$$2c \sum_j n_{j,+1}\, n_{j,-1} \quad .$$

However, to write the hopping term, we must introduce the electron creation and annihilation operators b^+_{js} and b_{js} at site j with spin s. Each is the adjoint of the other, and they obey the canonical anticommutation relations:

$$b_{js}\, b_{\ell t} + b_{\ell t}\, b_{js} = b^+_{js}\, b^+_{\ell t} + b^+_{\ell t}\, b^+_{js} = 0 \quad ,$$

$$b_{js}\, b^+_{\ell t} + b^+_{\ell t}\, b_{js} = \delta_{j\ell}\, \delta_{st} \quad .$$

Further we have the expression for the number operator

$$n_{js} = b^+_{js}\, b_{js} \quad .$$

Let us restate that $b^+_{js} = (b_{js})^+$.

Given these definitions, the Hamiltonian for our Hubbard model is written as:

$$H(c) = -\sum_{js}(b^+_{j+1,s} b_{j,s} + b^+_{j,s}\, b_{j+1,s}) + 2c \sum_j n_{j,+1}\, n_{j,-1} \quad .$$

However, we emphasize that this "second quantization" is just a way to write the matrix elements of a many-body operator, and has no new physical content.

B) <u>Symmetries</u>

For us, the main advantage of the second quantization formalism is that it makes it easy to exhibit a symmetry of the problem. A physical system is defined by the canonical anticommutation relations for the creation and annihilation operators b^+, b, and an expression H[b] for the Hamiltonian in terms of these operators. (b^+ is always the adjoint of b, so we show only the dependence on b.) Now if we exhibit a transformation

$$b \rightarrow b' = b'(b) \quad ,$$

which preserves the canonical anticommutation relations, then

$$H[b] \rightarrow H[b'] = H[b'(b)] \equiv H'[b] \quad .$$

Clearly H[b] and H'[b] have the same spectrum, and thus are related by a unitary transformation.

In what follows, we will exhibit some of these symmetries or transformations of the Hubbard Hamiltonian. They will be given both in words and in second quantization, so it will not be necessary for one to be conversant in the language of second quantization. More or less obvious verifications will be left to the reader, while slightly more complicated results will be sketched.

(a) First, it doesn't matter which direction we call "up". Thus, let $s' \equiv$ and let

$$b_{js} \rightarrow b_{js}' \quad , \quad H(c) \rightarrow H(c) \quad , \quad n_s \rightarrow n_s' \quad .$$

(b) The signs of the kinetic energy terms for the two spin components individually do not matter, for we can multiply the wave function by -1 everytime we have an up spin on an odd site. That is , if

$$b_{js} \rightarrow (-1)^j b_{js} \quad , \quad s = \pm 1 \text{ or both,}$$

then the sign of the kinetic energy term is changed for $s = \pm 1$ or both. n_s is not changed, but the "momentum" of a particular state is obviously changed.

(c) If we place particles with spin up on certain sites, then we may say that the remaining unoccupied sites have spin up "holes" on them. We then use these holes as dynamical variables. Or, we let

$$b_{js} \rightarrow b_{js}^{+} \quad , \quad s = \pm 1 \text{ or both.}$$

Then

$$n_s \rightarrow 1 - n_s \quad ,$$

so

$$H(c) \rightarrow H(-c) + 2cN_s' \quad , \quad s = \pm 1$$

or

$$H(c) \rightarrow H(c) + 2c(N - N_s - N_s') \quad , \quad s = \text{both.}$$

(d) One further observation: The spectrum of -c is that of +c inverted. Thus, if we are willing to study <u>all</u> states, we need only treat one sign of c.

Let us now examine the phase plane of all possible values of $0 \leqslant n_s \leqslant 1$, $s = \pm 1$, labeling all sectors of the Hubbard model. This would be the range of the independent variables for some property like the ground state energy, for example. We shall say two points in the phase plane are equivalent if the spectrum at these two points differ only by an additive constant. Using the symmetries a) and c), we indicate in fig. (4) equivalent points in the phase plane by asterisks. Points equivalent to these by inversion through symmetries c) and d), we also indicate in fig.(4) by circles. Thus, if we restrict ourself to the fundamental region (f.r.) defined by

$$0 < n_{+1} + n_{-1} \equiv n < 1 \quad , \qquad \text{(f.r.)}$$

$$0 < n_{+1} - n_{-1} \equiv s < 1 \quad ,$$

and to a single sign of coupling constant c, then by symmetries of the system, we may obtain the behavior in the entire phase plane for either sign of c. This f.r. we show shaded in fig. (4).

In addition we have labeled certain selected points and lines in the phase plane:

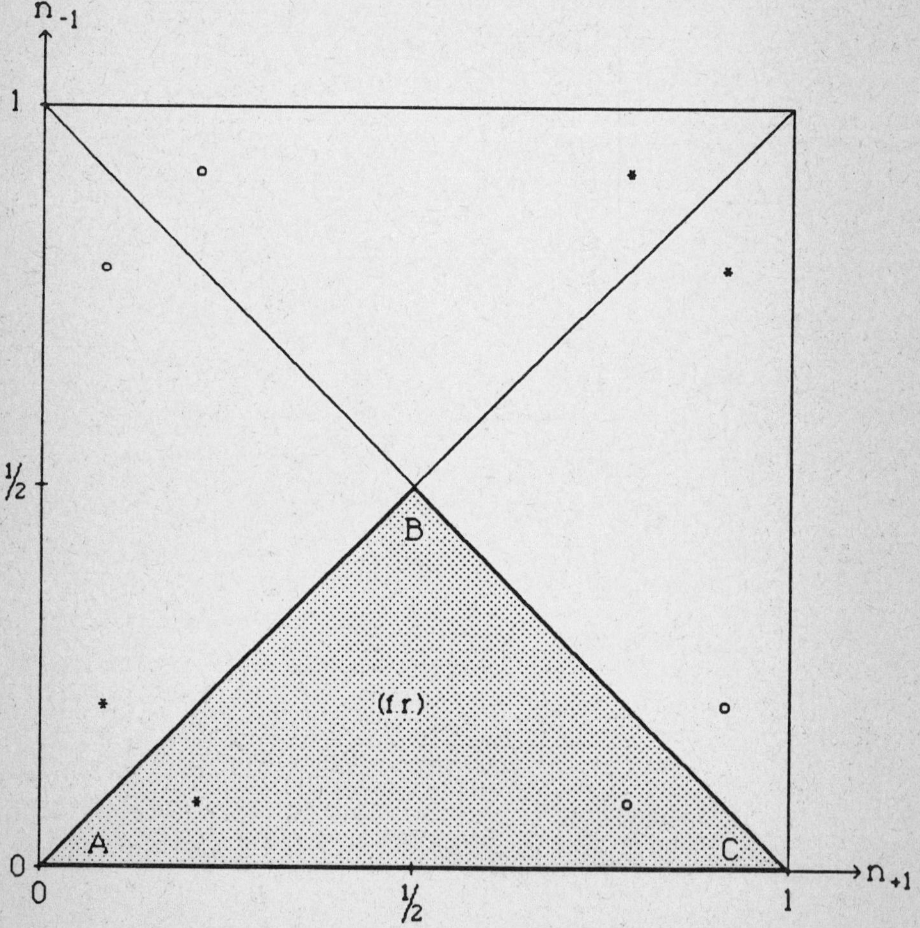

Fig. (4) The phase plane of the Hubbard model. The explanation of the symbols and features is in the text.

A: no particles at all;

B: total spin zero, half-filled band;

C: full lattice, all spin up;

AC: all spin up, free fermions, x-y model;

AB: total spin zero;

BC: half-filled band.

Note that AB and BC are equivalent by a change of sign of the coupling constant. The point B is the point of highest symmetry, and thus the leading candidate for the most singular point. It is invariant under change of sign of c. Finally, we state without derivation, that for c sufficiently large and positive we expect the model to be equivalent on the line BC, by degenerate second order perturbation theory, to the Heisenberg antiferromagnet.

At this point, the reader may well be wondering why so much time and effort has been spent cataloging symmetries, and mapping out a minimal region in parameter space --the fundamental region-- before even discussing how we might solve the problem. If we have a solution through the Bethe ansatz, why do we need to worry about symmetries?

A fact not generally recognized is that the Bethe ansatz scheme is a solution only for the low-density phase. That is, in the solution of a particular problem by Bethe ansatz techniques, we are able to calculate physical properties only in a phase which includes the zero density limit (in an appropriate realization), and thus the properties we calculate will be analytic at zero density. If we do encounter a singularity, it is likely to be an essential one, and we will not be able to continue beyond it except by a symmetry transformation.

Thus if all goes well, we will calculate within a fundamental region, and encounter (essential) singularities only on the boundaries. Then by _symmetry_, we may fill in the entire phase plane, and the fundamental region will coincide with a pure phase. (Of course, the phase may be larger than the f.r. , but if it is smaller, we are in trouble.) Such considerations are basic to all Bethe ansatz studies, except the most trivial.

What might these singularities mean? Consider as an example the ground state energy per site ε_0. If ε_0 were to have a cusp as a function of $s \equiv n_{+1} - n_{-1}$

at $s = 0$, as might happen for attractive c, then a binding energy for pairs of spins would be indicated, and a form of antiferromagnetism. On the other hand, if ε_0 developed a cusp in $n \equiv n_{+1} + n_{-1}$ at $n = 1/2$, as might happen for repulsive c, then this would indicate the opening of a band gap in the middle of the band, and thus an insulating state for the half-filled band.

There is one further set of symmetries we wish to discuss. These indicate that the statistics of the problem are less important than one might expect; such is often the case in one dimension. Suppose we are able to exhibit a change from the b operators obeying the canonical anticommutation relations, to new σ operators obeying a different set of canonical commutation or anticommutation relations. Let the relation be through $b = b(\sigma)$. Then

$$H[b] = H[b(\sigma)] = H'[\sigma] \ .$$

$H[b]$ and $H'[\sigma]$ clearly have the same spectrum; they are the same operator.

(e) Let us first show that we can treat the spin up and spin down electrons as two types of one-component fermions and thus not worry about the spin part of the wave function. Let us write

$$c_{j,+1} = b_{j,+1} \ ,$$
$$c_{j,-1} = b_{j,-1} \ \prod_{\ell}(2n_{j,+1} - 1)$$

Then we verify the following properties of $\sigma_{js}^z \equiv 2n_{js} - 1$:

$$(\sigma_{js}^z)^+ = \sigma_{js}^z \ , \ (\sigma_{js}^z)^2 = I$$
$$[b_{js}, \sigma_{\ell s'}^z] = 0 \ , \qquad s' = -s$$
$$[b_{js}, \sigma_{\ell s}^z] = 0 \ , \qquad \ell \neq j$$
$$b_{js} \sigma_{js}^z + \sigma_{js}^z b_{js} = 0 \ \ .$$

Then the c_{js} obey canonical anticommutation relations for each s, while c_{js} commute for different s. Thus the c_{js} represent two different kinds of fermions for different s. On the other hand bilinears in b_{js} for the same s are invariant under this transformation. Thus, both σ_{js}^z and $H[c]$ are invariant.

(f) Let us go one step further. We define:

$$\bar{\sigma}_{js} = c_{js} \prod_{\ell<j} \sigma_{\ell s}^z$$

Then pairs of σ_{js}^- for different s commute, while for the same js they obey canonical anticommutation relations. This leaves

$$\sigma_{js}^- \sigma_{ks}^- = c_{js} \, \sigma_{ks}^z \, c_{ks} \prod_{j>\ell>k} \sigma_{\ell s}^z \; , \; j > \; = \; -c_{js} \, c_{ks} \, \sigma_{ks}^z \prod_{j>\ell>k} \sigma_{\ell}^z$$

$$= +c_{ks} \, c_{js} \, \sigma_{ks}^z \prod_{j>\ell>k} \sigma_{\ell s}^z = \sigma_{ks}^- \sigma_{js}^-$$

Therefore, σ_{js}^- commute for different j same s. The same holds for the adjoint.

Again σ_{js}^z is invariant under the transformation, justifying the notation. Likewise n_{js} is invariant. However,

$$c_{j+1}^+ \, c_j = \sigma_{j+1}^+ \, \sigma_j^z \, \sigma_j^- = -\sigma_{j+1}^+ \, \sigma_j^- \; .$$

But the signs of the kinetic energy terms were never important, by symmetry (b), so we have H[c] is invariant.

These canonical commutation relations can be interpreted as two sets of spin 1/2 spins, labeled by s, with one of each to a lattice site. Another interpretation, is that σ_{js}^+, σ_{js}^- are creation and annihilation operators for hard-core bosons on site j. These bosons are of two types, labeled by s. (Two hard-core bosons of the same type cannot both occupy a site, due to an infinite on-site repulsion.)

Therefore, if we realize the problem as in (e) or (f), there is no spin-part of the wave function. We simply antisymmetrize in (e) or symmetrize in (f) the coordinates with the same label s. In what follows, we will switch from one picture to another as convenience dictates, but of course all are equivalent. We remind you that in all realizations, the n_s are the same.

C) <u>Simple Cases</u>.

This is enough talk for now; let's do a warm-up calculation for the line AC, corresponding to all M spins up. Then the exclusion principle prevents two particles from being on the same site, and therefore there is no interaction. Thus, the particles are free fermions, and the eigenstates are Slater determinants of plane waves:

$$\Psi(x_1 \ldots x_n) = \sum_{P \in S_M} (-1)^P \exp \left[i \sum_{j=1}^M x_j P_{Pj} \right] \; .$$

This gives us the opportunity to introduce some notation which we will use repeatedly in what follows. The symbol P denotes any one of the M! permutations of the integers 1 to M in the symmetric group S_M. Let the permutation P be written out as the collection of integers (P1, P2, ..., PM); this means P puts P1 in the first place, etc. The signature of P is $(-1)^P = \pm 1$, depending on whether P is even or odd.

Then since each particle behaves as if independent, the Hamiltonian acting on the wavefunction gives

$$H\Psi = -\sum_{j=1}^{M} [\Psi(\ldots x_j +1 \ldots) + \Psi(\ldots x_j -1 \ldots)] = -\sum_{j=1}^{M} (e^{ip_j} + e^{-ip_j})\,\Psi \quad ,$$

so that

$$E = -2 \sum_{j=1}^{M} \cos p_j \quad .$$

The periodic boundary condition on the wavefunction gives us a quantization condition on the p's, namely

$$e^{ip_j N} = 1, \text{ or } p_j = 2\pi I_j/N \quad .$$

The quantum numbers I_j are integers lying between $-N/2$ and $N/2$, so $-\pi < p_j \leqslant \pi$. Then in the thermodynamic limit as $N, M \to \infty$, M/N fixed, the possible p's are spaced $2\pi/N$ apart, and so can be expected to distribute themselves with a density $N\rho(p)$. For the ground state, we expect the p's to be as dense as possible about $p = 0$, or $\rho(p) = 1/2\,\pi, \ |p| \leqslant p_0$. The limit p_0 is determined by the normalization condition

$$\int_{-p_0}^{p_0} \rho(p)dp = M/N \equiv n = p_0/\pi \ ,$$

so $p_0 = \pi n$. For the ground state energy per site then,

$$\varepsilon_0 \equiv E/N = -1/\pi \int_{-\pi n}^{\pi n} \cos p\, dp = -2\sin(\pi n)/\pi \quad .$$

We now make two remarks:

(1) We are ignoring possible even-odd effects, which will not be important in the thermodynamic limit of a large system. Alternatively, we choose the parity of N, M, etc. appropriately. This is consistent with our previous considerations, since certain symmetries such as (b) and (f) will not be consistent with periodic boundary conditions unless N, N_{+1}, N_{-1} have the correct parity.

(2) For a given value of $n = n_{+1} + n_{-1}$, this wavefunction is always an eigenstate of the Hamiltonian. Thus by the variational principle the ground
state energy per site $\varepsilon_0(n,s,c)$ for the complete problem is concave upwards
as a function of s between the end points

$$\varepsilon_0(n,\pm n,c) = -2\sin(\pi n)/\pi \quad .$$

D) The Bethe Ansatz

We seek a wavefunction in the fundamental region (f.r.) in the form of the Bethe
ansatz:

Let

$$N_{+1} + N_{-1} \equiv M \equiv Nn, \; 0 < n < 1; \quad N_{-1} \equiv K;$$

$$N_{+1} - N_{-1} = M-2K \equiv Ns, \; 0 < s < 1 \quad .$$

Then the f.r. is defined by $N \geqslant M \geqslant 2K \geqslant 0$.

Usually we think of the first particle having coordinate y_1, the second coordinate $y_2,\ldots$, the Mth particle having coordinate y_M. Then for a particular ordering--
called a sector--we have

$$y_{Q1} < y_{Q2} < \cdots < y_{QM} \quad ,$$

where Q is a permutation of the labels 1 to M. For the Bethe ansatz, it is more convenient to separate the labels (Q1, Q2, ..., QM) from the coordinates, so in any
sector labeled by $Q = (Q1, \ldots, QM)$, we always have

$$x_1 < x_2 < \cdots < x_M \quad .$$

The label Q means particle Q1 is in the first position, Q2 in the second, etc. The
identity permutation $Q = I$ is the reference ordering before the particles are mixed.

The Bethe ansatz states that we look for a solution of the form:

$$\Psi(x|Q) = \sum_{P \varepsilon S_M} A(Q|P)\exp[i \sum_{j=1}^{M} x_j p_{Pj}] \quad .$$

If the particles are well separated, then we immediately have that the energy eigenvalue must be

$$E = -2 \sum_{j=1}^{M} \cos p_j \quad .$$

One further symmetry of the problem is:

(g) the Hamiltonian is invariant under translations of all the coordinates.

Thus, the Hamiltonian commutes with a translation operator T with eigenvalue e^{iP}, $e^{iPN} = 1$, and T acts on the wavefunction ψ by

$$T\psi = \psi(x_1+1,\ldots,x_j+1,\ldots,x_M+1) = e^{i\sum_{j=1}^{M} p_j}\,\psi \quad .$$

Thus we identify the momentum P--(not to be confused with a permutation)--as

$$P = \sum_{j=1}^{M} p_j = 2\pi I/N \quad ,$$

with I an integer $0 < I < N$.

(h) These of course don't exhaust the symmetries of the problem. For instance there is the continuous SU(2) symmetry to tell us that the spin states must be in SU(2) multiplets of singlet, doublet, triplet, etc. This continuous symmetry can be enlarged to include a mixing of particles and holes; we see no immediate use however, and thus pursue them no further.

We leave the question of the statistics of the problem--whether we have spin-1/2 fermions, two types of fermions, or two types of hard-core bosons--until later. In all three realizations, of course, the exclusion principle is in effect; this restriction to, at most, two particles on a site will be crucial in the following section.

E) <u>Verification - the Amplitudes</u>

Examining the Bethe ansatz form, it is clear that a sector boundary, when two particles are on the same site, can be considered in either of two sectors Q or Q'. For example, if $x_j = x_{j+1} = x$, then this is on the boundary between

$$Q = (Q1\ldots Qj, Qj+1\ldots QM)$$

and

$$Q' = (Q1\ldots Qj+1, Qj\ldots QM).$$

That is, $Q'j = Qj+1$, $Q'j+1 = Qj$, all other labels the same. Then continuity demands

$$\psi(\ldots x,x\ldots|Q) = \psi(\ldots x,x\ldots|Q') \quad .$$

When the particles are well separated, the wave function obeys the free particle Schrödinger equation,

$$E\,\psi(x|Q) = \pm\sum[\psi(\ldots x_j+1\ldots|Q) + \psi(\ldots x_j-1\ldots|Q)] \quad ,$$

with E as given previously. This equation can be used to extend the function $\psi(x|Q)$

beyond the sector $x_1 < \ldots < x_M$, although it will <u>not</u> coincide with the wavefunction. Such an extension will be useful in the following paragraph.

Now suppose two particles are at the same place, $x_j = x_{j+1} = x$; Q, Q' are as before. Then the equation for the wavefunction is

$$- \Psi(\ldots x-1, x \ldots |Q) - \Psi(\ldots x, x+1 \ldots |Q) - \Psi(\ldots x-1, x \ldots |Q') - \Psi(\ldots x, x+1 \ldots |Q')$$
$$+ 2c\Psi(\ldots x, x \ldots |Q) + \text{other terms} = E\Psi(\ldots x, x \ldots |Q) \quad .$$
$$= - \Psi(\ldots x-1, x \ldots |Q) - \Psi(\ldots x+1, x \ldots |Q) - \Psi(\ldots x, x-1 \ldots |Q) - \Psi(\ldots x, x+1 \ldots |Q)$$
$$+ \text{other terms}.$$

The last line follows from extending the wavefunction, as discussed above.

Cancelling terms, we obtain the following equation whenever two particles are on the same site:

$$\Psi(\ldots x-1, x \ldots |Q') + \Psi(\ldots x, x+1 \ldots |Q') - \Psi(\ldots x+1, x \ldots |Q) - \Psi(\ldots x, x-1 \ldots |Q)$$
$$= 2c \, \Psi(\ldots x, x \ldots |Q) \quad .$$

Combined with the continuity equation, this makes a pair of equations for each pair of particles on the same site. Since at most two particles can be at a site, due to the exclusion principle, these are the <u>only</u> constraints on the wavefunction. This exclusion principle will be inforced by the wave function itself; it will automatically vanish whenever two particles of the same type are on the same site.

To really appreciate the above observation, the reader should try for him/herself the three component problem, and see why the Bethe ansatz doesn't work. Then one can avoid being overly optimistic in the future.

We now seek to satisfy the constraint equations by considering pairs of terms in the Bethe expression for Ψ :

Let
$$P = (P1 \ldots Pj, \, Pj+1 \ldots PM) \quad ,$$
$$P' = (P1 \ldots Pj+1, Pj \ldots PM) \quad .$$

Thus P, P' are related, as were Q, Q', by an exchange in the j and $j+1$ places. These are the two terms we consider in satisfying the constraints. Then with the further definitions $p \equiv p_{Pj}$, $p' \equiv p_{Pj+1}$ the constraint equations become

$$A(Q|P) + A(Q|P') = A(Q'|P) + A(Q'|P')$$
$$A(Q|P)(2c + e^{ip} + e^{-ip'}) + A(Q|P')(2c + e^{ip'} + e^{-ip})$$

$$= A(Q'|P)(e^{-ip} + e^{ip'}) + A(Q'|P')(e^{-ip'} + e^{ip}) \quad .$$

Given two equations for four unknowns, we choose to solve for $A(Q|P')$, $A(Q'|P')$ in terms of $A(Q|P)$, $A(Q'|P)$, visualizing this as a scattering from P to P', where the jth particle with momentum p strikes the $j+1^{th}$ particle with momentum p'.

With the auxiliary definitions

$$\phi \equiv \sin p, \quad \phi' \equiv \sin p' \ ,$$

the solution is

$$A(Q|P') = - \frac{i(\phi - \phi')A(Q'|P) - cA(Q|P)}{c - i(\phi - \phi')} \ ,$$

$$A(Q'|P') = - \frac{i(\phi - \phi')A(Q|P) - cA(Q'|P)}{c - i(\phi - \phi')} \ .$$

We may combine all the equations for different Q's by defining an M! component vector A(P) whose elements are $A(Q|P)$ for all M! Q's. We also define a permutation operator P_{jk} which acts on this vector by permuting Q_j and Q_k. Thus, with P, P' as before, we combine all the equations into the M-1 matrix equations

$$A(P') = - \frac{c + i(\phi - \phi')P_{j,j+1}}{c - i(\phi - \phi')} \ A(P) \quad .$$

At this point one should pause to ponder the beauty and simplicity of these equations.

As a simple example, if we pick the totally antisymmetric representation for P_{jk}, $P_{j,j+1} A = - A$, so $A(P') = - A(P)$, and $A(P) = (-1)^P A(I)$, or $A(Q|P) = (-1)^{P+Q}$. This reproduces the section (C).

We remark that picking a representation which is antisymmetric when two particles are interchanged insures that the wave function vanishes when the two particles are on the same site; however, if the wave function vanishes, then there are no constraint equations at all for this boundary. It is easy enough to transform the constraint equations so that fermions become bosons by the unitary transformation on the A(P)'s with matrix elements

$$M_{QQ'} = \delta_{QQ'} (-1)^Q \quad .$$

Then

$$P_{jk} \rightarrow MP_{jk}M^{-1} = -P_{jk} \quad .$$

The transformed connection equations now read

$$A(P') = -\frac{c - i(\phi - \phi')P_{j,j+1}}{c - i(\phi - \phi')} A(P) \quad .$$

$$\equiv T_{j,j+1}(\phi - \phi') A(P) \quad .$$

We shall prefer this representation, and use it henceforth. The reasons for our "taste" in representation is that now we can use the spins s_j themselves as labels for the particles. The space part of the wavefunction now has the permutation symmetry that the spin part would have had in the original Hubbard model for electrons. Thus, if we have two particles with spin up, then the wave function is now necessarily symmetric under interchange, and thus it makes sense to simply label it $[\uparrow\uparrow]$.

F) Consistency

The connection equations are $M!(M-1)$ matrix equations for $M!$ vectors. We now must answer the important question: Are they mutually consistent? These connection formulae for the amplitudes are identical with the formulae for the delta-function problem, when we use the variables

$$\phi_j \equiv \sin p_j \quad ,$$

so we may take over Yang's method completely. Notice that at this point we are considering only scattering states on an infinite line; we have not yet imposed periodic boundary conditions.

The connection formulae allow us to start from the amplitude for a given permutation P, and by applying a sequence of T's between nearest neighbors, arrive at the amplitude for a new permutation P'. Suppose we start from the identity permutation for four particles, and apply some sequence such as T_{12}, T_{34}, T_{12}, T_{23}, T_{34} in this order to arrive at the permutation P' = (1 4 3 2). This operation is shown in a self-evident graphical notation in fig. (5). Time is the vertical axis and position on the line the horizontal axis. The index j on $T_{j,j+1}$ shows that the nearest neighbor pair j,j+1 is scattered. Of course, the arguments of the T's must be picked appropriately so that if ϕ is the variable of the left-most particle of nearest-neighbor pair j,j+1 ϕ' the right-most, we use $T_{j,j+1}(\phi - \phi')$.

Each line or string in fig. (5) is labeled by a variable ϕ, so to match the variables in the sequence of T's, it is easy to read off the variables from the graph.

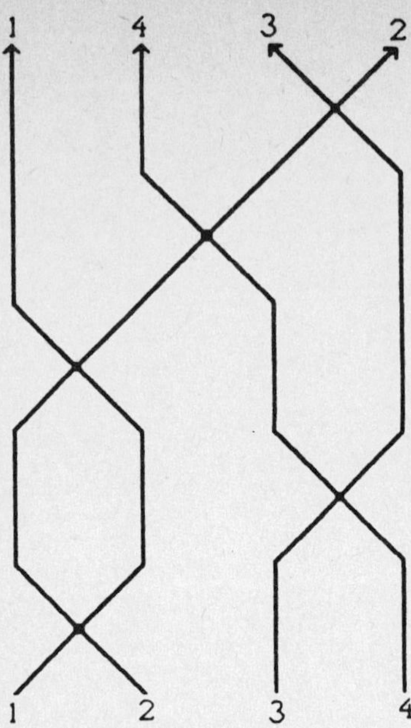

Fig. (5) A sequence of scattering operators applied to an amplitude, as explained in the text.

We see that fig. (5) corresponds to the connection equation:

$$A(1432) = T_{34}(\phi_2 - \phi_3)T_{23}(\phi_2 - \phi_4)\, T_{12}(\phi_2 - \phi_1)T_{34}(\phi_3 - \phi_4)\cdot$$

$$\cdot T_{12}(\phi_1 - \phi_2)A(1234) \quad .$$

It is clear we could have arrived at the same permutation P' from the identity permutation by many other routes. For instance we could have taken the sequence T_{23}, T_{34}, T_{23} shown in fig. (6). For consistency, the connection formulas must give the <u>same</u> amplitude A(P').

Thus, all possible equations which produce A(P') from A(P) may be obtained by fastening the ends of the strings in a graph at the top and bottom, then tangling the strings in a sufficiently devious way. (Strings may pass through one another.) It is clear that tangles on disjoint sets of strings do not interfere, since the T's commute.

After drawing a few diagrams, readers will convince themselves that in fact we need only have the equivalence of the fundamental 2-string and 3-string tangles, as shown in fig. (7), for consistency.

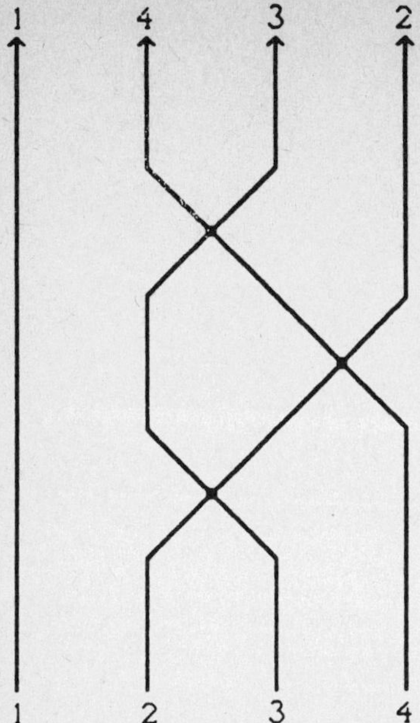

Fig. (6) An alternate sequence of scattering operators, which should lead to the same final amplitude as the previous sequence of fig. (5).

When these elementary tangles are translated back into operator relations, we have two identities that must be satisfied. Let us write the permutation operators as

$$P_{12} = \alpha \ , \ P_{23} = \beta \quad ,$$

so that they satisfy the defining relations for the group S_3,

$$\alpha^2 = \beta^2 = I \ , \ \alpha\beta\alpha = \beta\alpha\beta \quad .$$

For relation I,

$$T\alpha(\phi'- \phi)T\alpha(\phi - \phi') =$$

$$\frac{[c + i(\phi -\phi')\alpha][c - i(\phi - \phi')\alpha]}{c^2 + (\phi - \phi')^2}$$

$$= \frac{c^2+ (\phi - \phi')^2\alpha^2}{c^2 + (\phi - \phi')^2} = I \quad .$$

Thus relation I is an identity.

Relation II requires that

$$T_\beta (\phi - \phi'')T\alpha(\phi - \phi')T\alpha(\phi'' - \phi')$$

$$= T\alpha (\phi'' - \phi')T\beta(\phi - \phi')T\alpha(\phi - \phi'') \quad .$$

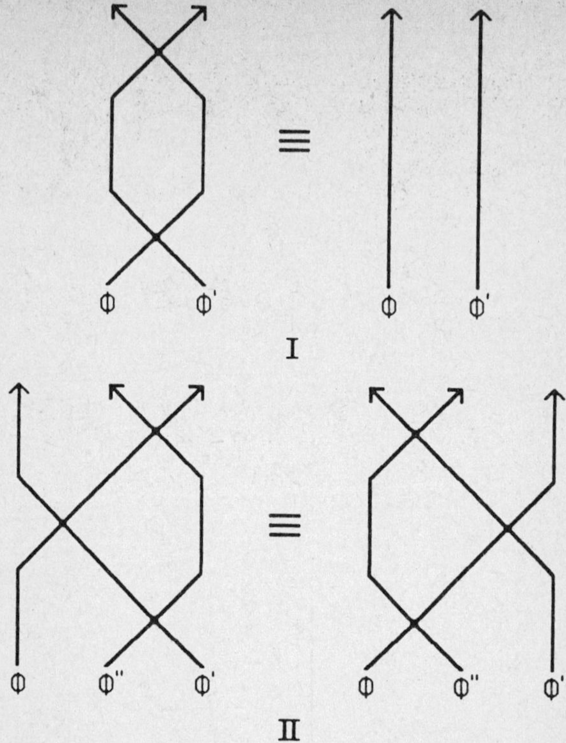

Fig. (7) The fundamental two-tangles and three-tangles needed for consistency of the amplitudes.

The denominator of the left side is the same as the denominator of the right. Thus, the numerator of the left side is

$$- [c - i(\phi - \phi'')\beta][c - i(\phi - \phi')\alpha][c - i(\phi'' - \phi')\beta]$$

$$= - \{c^3 - ic^2(\phi - \phi')(\beta + \alpha)$$

$$- c[\beta^2(\phi - \phi'')(\phi'' - \phi') + \beta\alpha(\phi - \phi'')(\phi - \phi')$$

$$+ \alpha\beta(\phi - \phi')(\phi'' - \phi')] + i(\phi - \phi'')(\phi - \phi')(\phi'' - \phi')\beta\alpha\beta\} .$$

On the right side, the numerator is

$$- \{c^3 - ic^2(\phi - \phi')(\beta + \alpha)$$

$$- c[(\phi - \phi'')(\phi'' - \phi')\alpha^2 + \beta\alpha(\phi - \phi'')(\phi - \phi')$$

$$+ \alpha\beta(\phi - \phi')(\phi'' - \phi')] + i(\phi - \phi'')(\phi - \phi')(\phi'' - \phi')\alpha\beta\alpha\} .$$

Using the defining relations for S_3, we see that the two are the same. Thus, relation II is an identity, and hence all the connection formulae are consistent. We then have a solution to the problem on an infinite line--the scattering problem.

G) Two-Body Scattering and Bound States

All of the dynamics of the scattering states, including the bound states, is included in the connection operators $T_{j,j+1}(\phi - \phi')$. Thus T is more properly called the scattering operator. Let us imagine a two-body incoming state which consists of a spin up particle on the left with momentum p, and a spin down particle on the right with momentum p'. If this is to be an incoming state, then we must have $p > p'$ for the particles to collide. We write the configuration of such an incoming state as $[\uparrow\downarrow]$. We will explain this notation in more detail in section K, when all questions are to be answered. For now, imagine the notation to be simply a labeling scheme for the different scattering channels.

This incoming state has amplitude 1, so the operator $T = T_{12}(\phi - \phi')$ produces the following out-going amplitudes

$$T[\uparrow\downarrow] = - \frac{c - i(\phi - \phi')P_{12}}{c - i(\phi - \phi')} \quad [\uparrow\downarrow]$$

$$= - \frac{c}{c - i(\phi - \phi')} \; [\uparrow\downarrow] + \frac{i(\phi - \phi')}{c - i(\phi - \phi')} \; [\downarrow\uparrow]$$

$$\equiv r(\phi - \phi') \; [\uparrow\downarrow] + t(\phi - \phi')[\downarrow\uparrow] \quad .$$

The last line defines the two-body reflection and transmission amplitudes,

$$r(\phi) = - c/(c - i\phi) \; , \; t(\phi) = i\phi/(c - i\phi) \quad .$$

Note the important relation $r + t = -1$, so that when two up spins scatter, the amplitude changes sign due to the exclusion principle.

Let us look for bound states of two particles with p_1 , p_2. Since the total momentum must be real, $\mathrm{Im}\, p_1 = -\mathrm{Im}\, p_2$. Since the total energy must also be real, $\mathrm{Re}\, p_1 = \mathrm{Re}\, p_2$. We write $p_1 = P/2 + i\kappa$, $p_2 = P/2 - i\kappa$, with κ real, > 0. Then the Bethe ansatz state is

$$\Psi(x_1 x_2 | Q) = A(P|Q)\exp[\kappa(x_2 - x_1) + iP(x_2 + x_1)/2]$$

$$+ A(P'|Q)\exp[-\kappa(x_2 - x_1) + iP(x_2 + x_1)/2] \quad .$$

By definition $x_2 > x_1$, so for a bound state we can only have the second term, and thus

$A(P|Q) = 0$. The only way $A(P')$ can be non-zero when $A(P) = 0$ is for $T(\phi)$ to have poles.

Thus we are to look for bound states as poles of $T(\phi - \phi')$, or as zeros of the denominators of r,t. Since $\phi \equiv \sin p$, $\mathrm{Re}\,\phi = \mathrm{Re}\,\phi'$, $\mathrm{Im}\,\phi = -\mathrm{Im}\,\phi' > 0$, and thus the condition for bound states is

$$c + 2\mathrm{Im}\,\phi = 0 \quad \text{or} \quad \mathrm{Im}\,\phi = -c/2 > 0 \quad .$$

Obviously there is only one bound state, and this exists only for an attractive potential when $c < 0$.

Let us now redefine ϕ as

$$\phi \equiv \mathrm{Re} \sin p \ ,$$

so that the new definition coincides with the old usage when p is real. The only situation when p is complex is for a bound state with

$$\sin p = \phi \pm ic/2 \ ;$$

the $\phi - ic/2$ is always to the right of $\phi + ic/2$. (Remember $c < 0$!) Thus ϕ parameterizes the bound states.

The energy of a bound state is:

$$E(\phi) = -4\mathrm{Re} \cos p = -4\mathrm{Re} \ \sqrt{1- (\phi - ic/2)^2}$$

The momentum of a bound state is

$$P(\phi) = 2\mathrm{Re}\,p = 2\mathrm{Re} \ \sin^{-1} (\phi - ic/2) \quad .$$

The ambiguity in choosing the branches of these functions will be discussed in the following section (H).

At the bound state, $r/t = -1$, so the state is $[\uparrow\downarrow - \uparrow\downarrow]$, or antisymmetric, and thus exists only between particles of different types as demanded by the exclusion principle. Remember that we are using a boson realization. When translated back into the Hubbard model of electrons, it is the <u>spin part</u> of the wave function which will be antisymmetric, not the space part, and so a singlet.

H) The Proper Choice for Variables

By now, it is clear that the ϕ variables are much more useful and natural for this problem than the p variables. We no longer even use the plane wave factors in the Bethe ansatz wave function, but instead work with the amplitudes through the

connection formulae, and the scattering operators. The scattering operators themselves depend on ϕ , ϕ' through the difference $\phi - \phi'$, reflecting a sort of hidden
symmetry much like Galilean or Lorentz invariance, but occuring instead in a discrete
system. Thus, if we boost all p's by

$$p = \sin^{-1} \phi \rightarrow p' = \sin^{-1}(\phi + \delta\phi) \quad ,$$

the scattering amplitudes are unchanged.

To make the change of variables complete, we take a complex Φ . For the parti-
cles or bound states respectively, $\Phi = \phi$, or $\phi \pm ic/2$, ϕ real. Also $p = \sin^{-1} (\Phi)$,
$\omega \equiv - \cos p = - [1 - \Phi^2]^{1/2}$. The energy of a particle is 2ω, and of a bound state is
$4\mathrm{Re}\ \omega$. The momentum of a particle is p, and of a bound state is 2Rep.

The range of complex p is taken to be $-\pi/2 < \mathrm{Re}p \le 3\pi/2$, as shown in the complex
p-plane of fig. (8). This leads to a double covering of the entire complex Φ -plane,

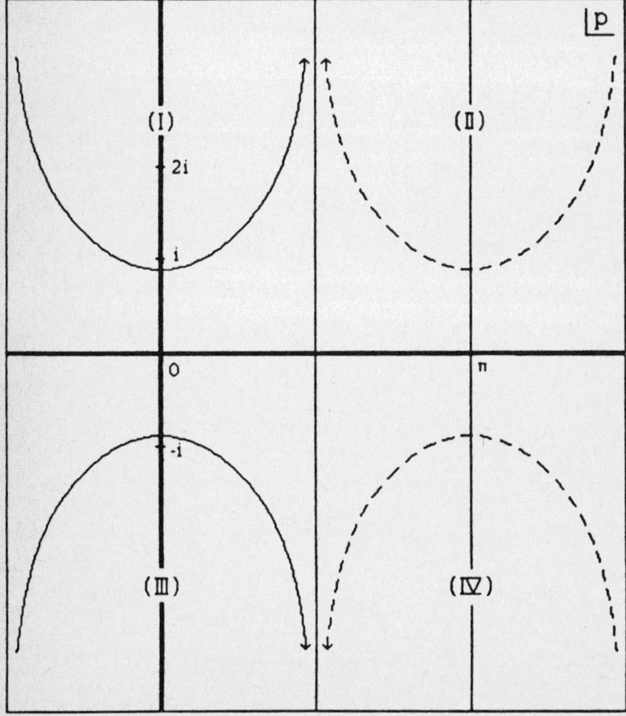

Fig. (8) The complex p-plane with bound and anti-bound state trajectories shown. In
this and the next three figures, we have illustrated the case c = 2.

and complex ω-plane as shown in fig's. (9), (10). These two pictures also show the trajectories of the particles, and of the bound states. The bound states divide into two disjoint families: those with negative energy, and those with positive energy (often called anti-bound). For the particles, such a division is not demanded since the two branches are not disjoint. Again, this structure reminds one of Lorentz invariance. Fig. (11) shows the dispersion curves for particles and bound states.

If $P(\phi)$ is the momentum function for the bound state or particle, we will later need $dP/d\phi$. But $p = \sin^{-1}(\bar{\omega})$, so

$$dP/d\Phi = [1 - \bar{\omega}^2]^{-1/2} = - \omega^{-1}$$

Then

$$dP/d\phi = -\omega^{-1} \text{ , particles ;}$$

$$= -2\mathrm{Re}\,\omega^{-1}\text{, bound states .}$$

I) Three-Body Scattering

The most interesting case of three-body scattering is a single spin incident on a bound state. The diagram for this scattering is shown in fig. (12). The incoming state is $[\uparrow][\uparrow\downarrow - \downarrow\uparrow] = [\uparrow\uparrow\downarrow] - [\uparrow\downarrow\uparrow]$. Thus after the first collision $T_1 \equiv T_{12}(\phi - \phi'-ic/2)$,

$$T_1[\uparrow][\uparrow\downarrow - \downarrow\uparrow] = T_1[\uparrow\uparrow\downarrow] - T_1[\uparrow\downarrow\uparrow] = -[\uparrow\uparrow\downarrow] - r_1[\uparrow\downarrow\uparrow] - t_1[\downarrow\uparrow\uparrow] \quad .$$

After the second collision $T_2 \equiv T_{23}(\phi - \phi' + ic/2)$,

$$T_2T_1[\uparrow][\uparrow\downarrow - \downarrow\uparrow] = - T_2[\uparrow\uparrow\downarrow] - r_1T_2[\uparrow\downarrow\uparrow] - t_1T_2[\downarrow\uparrow\uparrow]$$

$$= - r_2[\uparrow\uparrow\downarrow] - t_2 [\uparrow\downarrow\uparrow] - r_1r_2[\uparrow\downarrow\uparrow] - r_1t_2[\uparrow\uparrow\downarrow] + t_1[\downarrow\uparrow\uparrow]$$

$$= -(r_2 + r_1t_2)[\uparrow\uparrow\downarrow] - (t_2 + r_1r_2)[\uparrow\downarrow\uparrow] + t_1 [\downarrow\uparrow\uparrow] \quad .$$

This is the outgoing state.

Let us simplify the coefficients. For the first,

$$-(r_2 + r_1t_2) = c[c - i(\phi - \phi' - ic/2) + i(\phi - \phi' + ic/2)]/D$$

$$= c(c - c/2 - c/2)/D = 0 \quad .$$

The quantity D is the common denominator,

$$D = [c - i(\phi - \phi' - ic/2)][c - i(\phi - \phi' + ic/2)]$$

$$= [c/2 - i(\phi - \phi')][3c/2 - i(\phi - \phi')] \quad .$$

For the second coefficient,

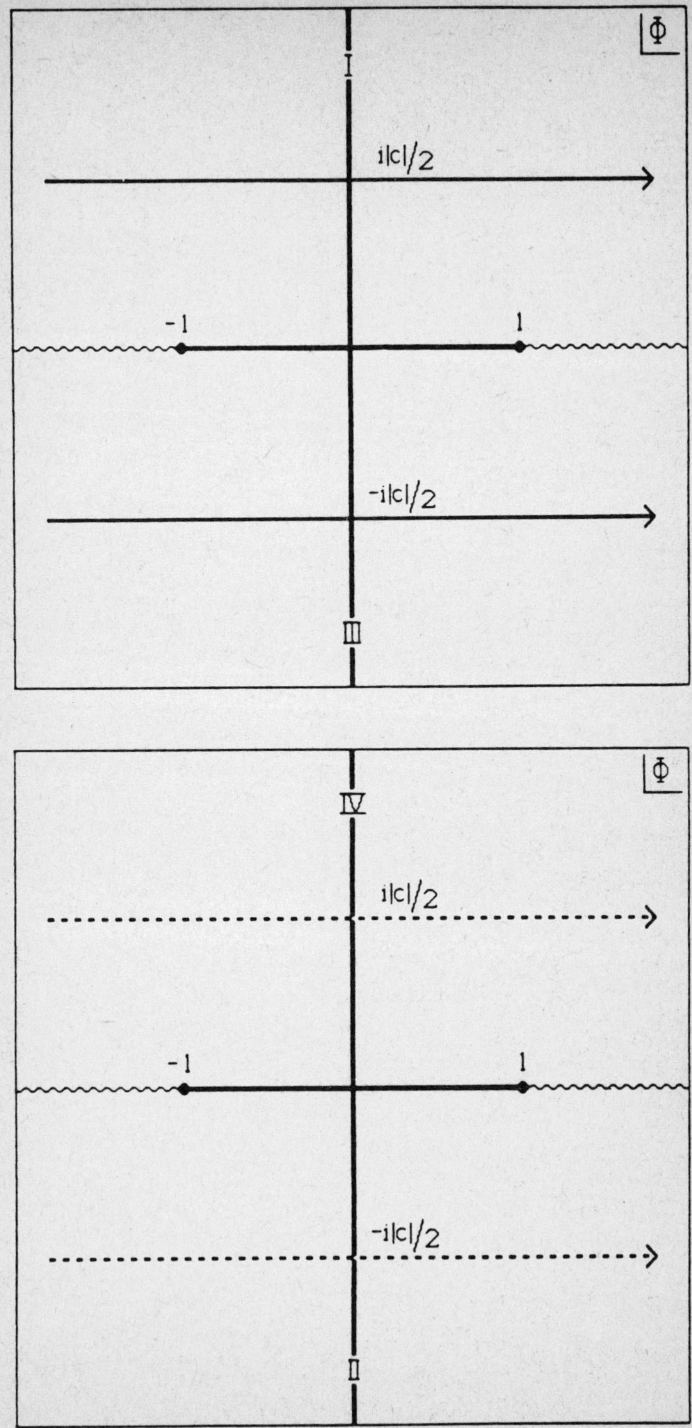

Fig. (9) The complex ϕ -plane.

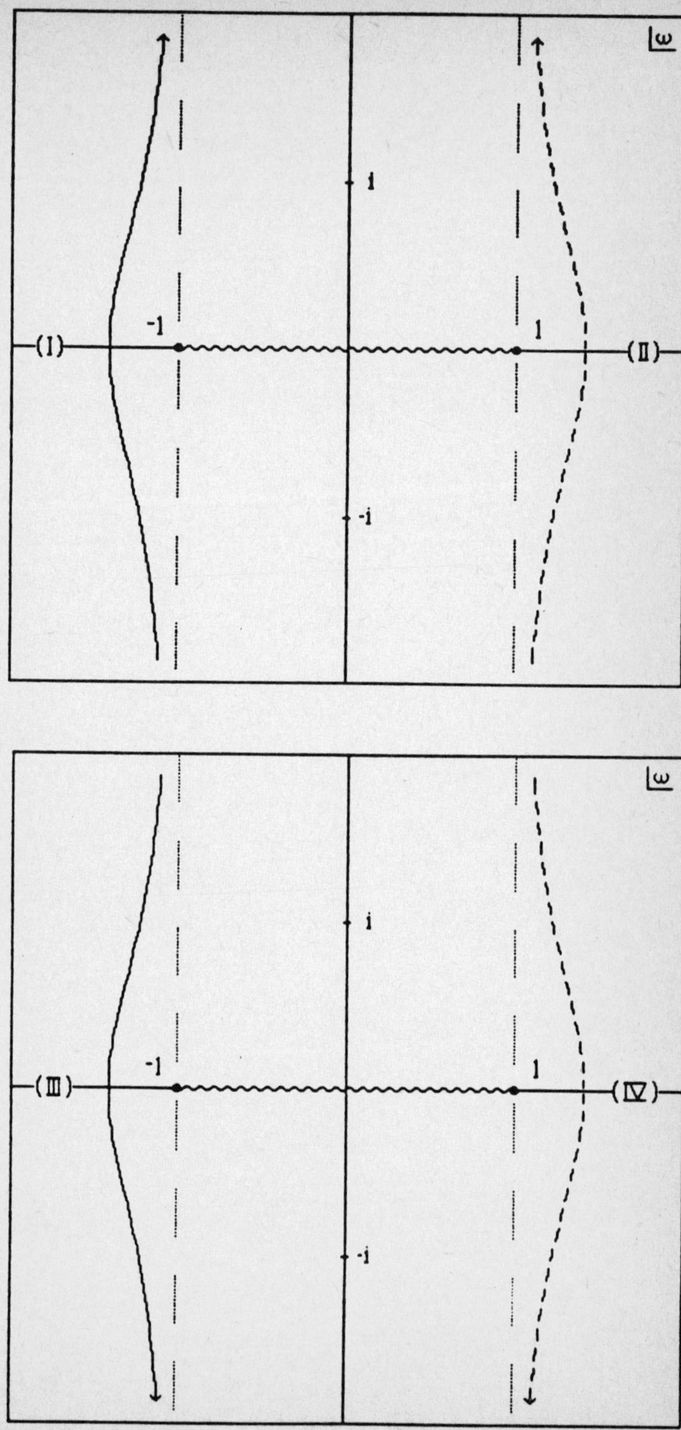

Fig. (10) The complex ω -plane.

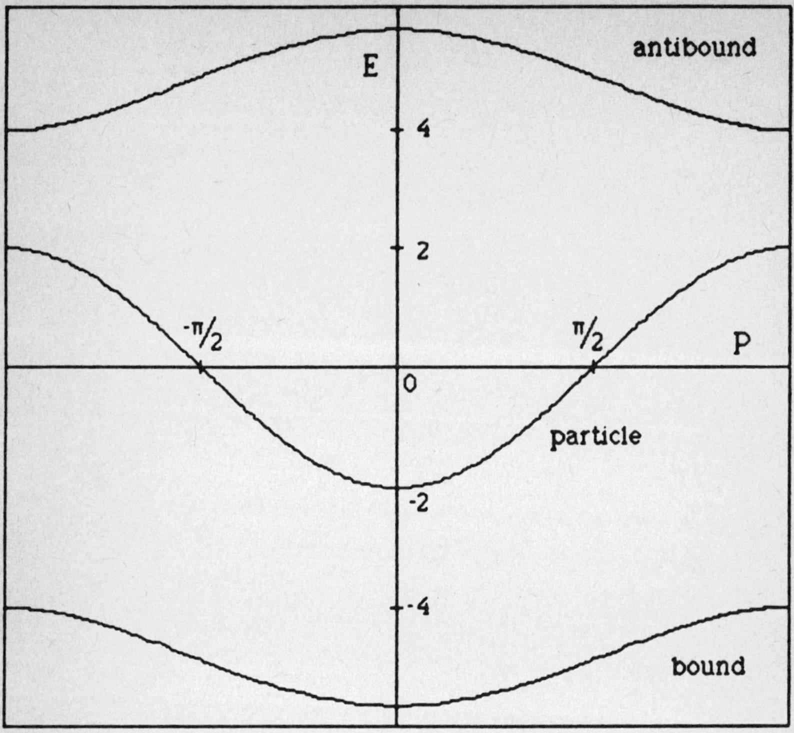

Fig. (11) Dispersion curves for particles, bound states and anti-bound states.

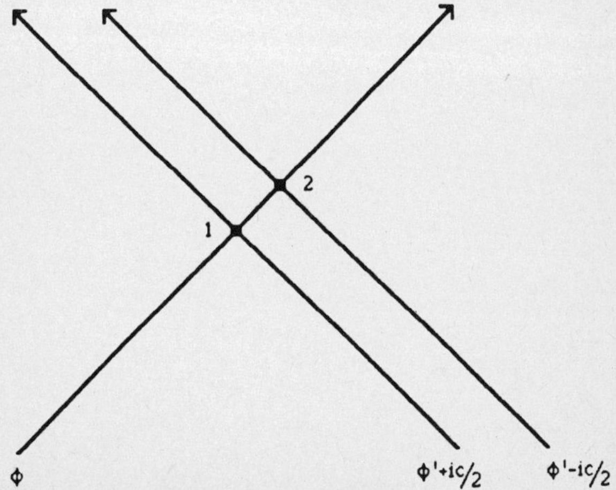

Fig. (12) Three-particle scattering of a particle and a bound state.

$$-(t_2 + r_1 r_2)$$

$$= -\{c^2 + i(\phi - \phi' + ic/2)[c - i(\phi - \phi' - ic/2)]\}/D$$

$$= - \{c^2 - [c/2 - i(\phi - \phi')]^2\}/D$$

$$= - [3c/2 - i(\phi - \phi')][c/2 + i(\phi - \phi')]/D \quad .$$

But using the expression for D,

$$- (t_2 + r_1 r_2) = - \frac{c/2 + i(\phi - \phi')}{c/2 - i(\phi - \phi')}$$

Let us compare this with the expression for t_1, the third coefficient,

$$t_1 = \frac{i(\phi - \phi' - ic/2)}{c - i(\phi - \phi' - ic/2)} = \frac{c/2 + i(\phi - \phi')}{c/2 - i(\phi - \phi')} \quad .$$

We thus see that the second coefficient is the negative of the third, and the out going state is:

$$t_1([\downarrow\uparrow\uparrow] - [\uparrow\downarrow\uparrow]) = -t_1[\uparrow\downarrow - \downarrow\uparrow][\uparrow] \quad .$$

Thus, the particle passes through the bound state with an amplitude

$$- t_1 = - \frac{c + 2i(\phi - \phi')}{c - 2i(\phi - \phi')} = -e^{-i\theta(2(\phi - \phi'))} \quad .$$

The phase shift is given by

$$e^{- i\theta(\phi)} = \frac{c + i\phi}{c - i\phi} \quad .$$

or

$$\theta(\phi) \equiv -2\tan^{-1}(\phi/c) \quad ,$$

There is no disassociation nor reflection in the scattering of a particle from a bound state, consistent with the system being integrable.

J) Four-Body Scattering and a Fluid of Bound States

Let us now derive the phase shift for the scattering of two bound states, as shown in fig. (13). This can be seen as first the scattering of a particle with $\phi - ic/2$ on a bound state ϕ', then the scattering of a particle with $\phi + ic/2$ on the same bound state. Thus, using the results from the previous section, the transmission amplitude is

$$\frac{c + 2i(\phi + ic/2 - \phi')}{c - 2i(\phi + ic/2 - \phi')} \cdot \frac{c + 2i(\phi - \phi')}{c - 2i(\phi - ic/2 - \phi')}$$

$$= - \frac{c + i(\phi - \phi')}{c - i(\phi - \phi')} = -e^{-i\theta(\phi - \phi')} \quad .$$

This is exactly the phase shift for a delta-function boson gas with attractive interaction c. However, in our present case, the pole at $\phi - \phi' = -ic$ does not correspond to a bound state, and thus we can determine the ground state properties by essentially taking the integral equations for the repulsive delta-function problem and continuing c to negative values.

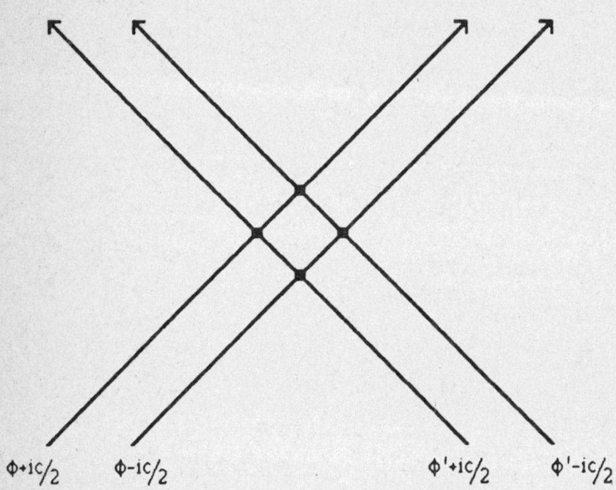

Fig. (13) Four-particle scattering of two bound states.

Of course we are only considering a fluid of bound states, but this should be sufficient to determine the ground state, and some excited states, for the case s = 0 when c is attractive. This is the line AB in fig. (4).

To illustrate, if we carry a bound state aroung the ring of circumference N transmitting it through the other bound states as we go, the periodic boundary condition requires that the net phase shift be an integer multiple of 2π, or

$$NP(\phi) - \sum_{\phi'}{}' \theta(\phi - \phi') = 2\pi I(\phi) \quad .$$

$P(\phi)$ is the momentum of a bound state, $-\theta(\phi)$ is the previous two bound state phase shift, $I(\phi)$ is a set of integers corresponding to the quantum numbers of the state, and the summation is over all the other K-1 bound states. These equations form a set of K coupled equations for the ϕ's. For the ground state we can assume again that the integers are distributed densely about the origin and the ϕ's distribute themselves with a density $\sigma(\phi)$ for $|\phi| < \phi_0$. We normalize $\sigma(\phi)$ so that

$$\int_{-\phi_0}^{\phi_0} \sigma(\phi)d\phi = K/N = n/2 \quad .$$

This is the density of bound states, or half the density of particles. The spin is $s = 0$.

Then the coupled equations reduce to an integral equation, which after differentiation becomes

$$\sigma(\phi) + \int_{-\phi_0}^{\phi_0} L(\phi - \phi') \, \sigma(\phi')d\phi' = \xi(\phi) \quad .$$

The kernal is

$$L(\phi) \equiv \frac{1}{2\pi} \frac{d\theta(\phi)}{d\phi} = \frac{1}{2\pi} \frac{2|c|}{c^2 + \phi^2} \quad ,$$

and the inhomogeneous term is

$$\xi(\phi) \equiv \frac{1}{2\pi} \frac{dP(\phi)}{d\phi} \quad .$$

The energy is given as

$$E/N = \varepsilon_0 = \int_{-\phi_0}^{\phi_0} E(\phi) \, \sigma(\phi)d\phi \quad .$$

The energy $E(\phi)$ and momentum $P(\phi)$ were given in section (G).

The integral equation is well-behaved for a finite limit ϕ_0, and thus the ground state energy $\varepsilon_0(n,0,c)$ is an analytic function of n. However at $\phi_0 = \infty$, the integral equation is singular. In this limit, we may integrate the integral equation from $-\infty$ to $+\infty$, and we find

$$n/2 + n/2 \int_{-\infty}^{\infty} L(\phi)d\phi = \int_{-\infty}^{\infty} \xi(\phi)d\phi = P(+\infty)/\pi \quad .$$

The integral over L is easily seen to be

$$\theta(+\infty)/\pi = -2\tan^{-1}(-\infty)/\pi = 1 \quad .$$

From sect. (G) and fig. (8), we see $P(+\infty) = \pi$. Thus $\phi_0 = +\infty$ corresponds to $n = 1$, or a half-filled band.

For $\phi_0 = +\infty$, corresponding to the point B in the phase diagram fig. (4), the integral equations can be solved by Fourier transforms. The Fourier transform of a function $\sigma(\phi)$ is defined as

$$\tilde{\sigma}(\alpha) = \int_{-\infty}^{\infty} d\phi \; e^{-i\alpha\phi} \; \sigma(\phi) \quad .$$

Then the integral equation becomes

$$\tilde{\sigma}(\alpha) \, [1 + \tilde{L}(\alpha)] = \tilde{\xi}(\alpha)$$

or

$$\tilde{\sigma}(\alpha) = \tilde{\xi}(\alpha)/[1 + \tilde{L}(\alpha)] \quad .$$

The ground state energy per site is

$$\varepsilon_0 = \frac{1}{2\pi} \int_{-\infty}^{\infty} d\alpha \; \tilde{E}(\alpha)\tilde{\sigma}(\alpha) \qquad = \frac{1}{2\pi} \int_{-\infty}^{\infty} d\alpha \; \frac{\tilde{E}(\alpha)\tilde{\xi}(\alpha)}{1 + \tilde{L}(\alpha)} \quad .$$

In the appendix, we evaluate the following Fourier transforms:

$$\tilde{L}(\alpha) = e^{-|\alpha c|} \quad ,$$
$$\tilde{\xi}(\alpha) = e^{-|\alpha c|/2} \, J_0(\alpha) \quad ,$$
$$\tilde{E}(\alpha) = -4\pi e^{-|\alpha c|/2} J_1(\alpha)/\alpha \quad .$$

$J_n(\alpha)$ is the Bessel function of order n. This gives us the final expression

$$\varepsilon_0(1,0,c) = -2 \int_{-\infty}^{\infty} \frac{d\alpha J_0(\alpha)J_1(\alpha)/\alpha}{1 + e^{|\alpha c|}} \quad .$$

This formula is good for all c, and agrees with Lieb and Wu, who worked from the

repulsive case.

There is no reason why we can't push this reasoning further. Suppose in addition

to the K pairs of particles in bound states, we also have M-2K particles all of spin

up, and thus unable to bind. Since a particle with ϕ scatters from a bound state

with ϕ' only by a phase shift $-\theta(2(\phi - \phi'))$, as derived in sect. (I), then if we

transport a particle with $\phi = \psi$ around the ring, it suffers a total phase shift

$$Np(\psi) - \sum \theta(2(\psi - \phi)) = 2\pi J(\psi) \quad .$$

The integer multiple of 2π on the right is required by periodic boundary conditions.

Likewise when we transport a bound state around the ring, the total phase shift will

now be:

$$NP(\phi) - \sum_{\phi'}{}' \theta(\phi - \phi') - \sum_{\psi} \theta(2(\phi - \psi)) = 2\pi I(\phi) \quad .$$

Again we assume the ϕ's and ψ's distribute with densities $\sigma(\phi)$, $\rho(\psi)$ for $|\phi| < \phi_0$,

$|\phi| < \phi_0$, so we have the integral equations

$$\sigma(\phi) + \int_{-\phi_0}^{\phi_0} L(\phi - \phi')\sigma(\phi')d\phi' = \xi(\phi) - 2\int_{-\psi_0}^{\psi_0} L(2(\phi - \psi))\rho(\psi)d\psi \quad,$$

$$\rho(\psi) = \eta(\psi) - 2\int_{-\phi_0}^{\phi_0} L(2(\psi - \phi))\sigma(\phi)d\phi \quad.$$

The quantity η is defined to be

$$\eta(\psi) \equiv \frac{1}{2\pi}\,\frac{dp}{d\psi} = \frac{1}{2\pi\sqrt{1-\psi^2}} = -\frac{1}{2\pi\omega(\psi)} \quad.$$

The normalizations for the densities are

$$\int_{-\phi_0}^{\phi_0} \sigma(\phi)d\phi = K/N = (n-s)/2,$$

$$\int_{-\psi_0}^{\psi_0} \rho(\psi)d\psi = (M-2K)/N = s \quad.$$

Finally, the ground state energy per site is

$$E/N = \varepsilon_0(n,s,c) = \int_{-\phi_0}^{\phi_0} E(\phi)\sigma(\phi)d\phi + 2\int_{-\psi_0}^{\psi_0} \omega(\psi)d\psi \quad.$$

We return to these formulas in sect. (Q).

Again we can easily verify that $\phi_0 = +\infty$ corresponds to n = 1, or the line BC of fig. (4), while $\psi_0 = 0$ corresponds to s = o, or the line AB of fig. (4). On the other hand $\phi_0 = 0$, of course, is the simple example of sect. (C). The integral equation for $\phi_0 = +\infty$ is a singular integral equation, and some care must be exercised in treating it.

K) Interlude

After the first couple of lectures, the participants at the conference became very confused, and asked several good questions about what had been done, and where we were going. An entire lecture was spent answering these questions, and it would be a shame if the reader did not have the same opportunity to benefit from such close questioning. Thus I have put into these published notes this section of questions and answers, at approximately the same point at which it occured during the lectures.

What are really doing?

We are doing just what we said we would do in the introduction. In the Bethe ansatz for the wave function are a number of terms, each with their own amplitude $A(P|Q)$. These amplitudes are connected by real collisions. For instance consider three particles, and let P be the identity permutation 123, with the momenta ordered $p_1 > p_2 > p_3$. This is a state in which the particles are going to collide in the future: an IN state. But which particles are going to collide? For that we have to specify the "channels", i.e., which particle is to the left, which to the right and which in the middle? The label Q tells us that. Let's say particle 1 is left, 2 is middle, 3 to the right, so we have an incoming amplitude $A(123|123)$, and incoming wave:

$$A(123|123)\ e^{i(p_1 x_1 + p_2 x_2 + p_3 x_3)}.$$

Now suppose the first two particles, 1 and 2, were to collide long before 2 and 3 hit. A space-time plot showing the classical trajectories would look like fig. (14). What can happen? Well, only two things: The particles could be reflected with a reflection amplitude $r(p_1, p_2)$, so that the ordering of the particles would be unchanged and remain $Q = 123$; or the first particle could be transmitted through the second with a transmission amplitude $t(p_1, p_2)$, reversing the ordering of the first two particles so we would then have $Q' = 213$.

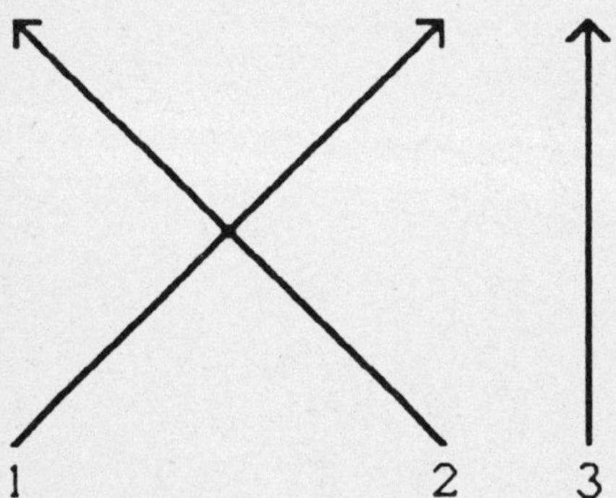

Fig. (14) A two-body collision on the space-time plot.

At the same time --for both possibilities or channels-- the first two momenta are reversed in order, indicating that the two-body collision has taken place. The new ordering of momenta is $p_2 p_1 p_3$ or $P' = 213$.

To summarize: Because of the possibility of this two body collision, the wave function must also have terms

$$r(p_1,p_2)A(123|123)e^{i(p_2x_2 + p_1x_2 + p_3x_3)}$$

$$+ t(p_1,p_2)A(123|123)e^{i(p_2x_1 + p_1x_2 + p_3x_3)}$$

$$\equiv A(213|123)e^{i(p_2x_1 + p_1x_2 + p_3x_3)}$$

$$+ A(213|213)e^{i(p_2x_1 + p_1x_2 + p_3x_3)}$$

We thus make the identifications,

$$A(P'|Q) = r(p_1,p_2)A(P|Q) \quad ,$$

$$A(P'|Q') = t(p_1,p_2)A(P|Q) \quad .$$

This is one example of the connection formulae.

However, there are many more possibilities. We could have started with any other ordering of particles or channels than $Q = I$. Also, the second and third particles in line could have collided first. Or the collision between the first and second could have been followed by a collision between the second and the third, taking us from P' = 213 to P'' = 231. There are many, many possibilities for collisions leading to many, many equations for many amplitudes. All of the possible amplitudes are contained in the Bethe ansatz of sect. (D) and all of the possible equations can be generated by our basic connection formulae of sect. (E).

But don't forget the physics of collisions behind these mathematical equations; the mathematics is essentially an efficient bookkeeping procedure to keep track of all that must be happening to the particles. The matrix notation is especially nice, because we don't have to keep track of all the channels. And although we have many amplitudes --M! M!-- this is far fewer than we might have had, for in general we will have all momenta consistent with the conservation of energy and momentum.

<u>What was that business about mirrors?</u>

Well, quantum mechanics is really a <u>wave</u> mechanics, and to draw classical trajectories as in fig. (14) is clearly inadequate. There is no provision for drawing transmission and reflection, except by multiple figures with different labels for the different channels. And if we were to include the possibility that 2 struck 3 first, yet more figures are needed.

However, if we look in the space where the wavefunction lives--three dimensional cartesian space labeled by coordinates $y_1y_2y_3$--than all possibilities can be drawn on the same diagram. Since the total momentum is conserved, let us separate out the center of mass coordinate $(y_1+y_2+y_3)/\sqrt{3}$, and place it perpendicular to the plane of the paper. That is, we look along the diagonal of the $y_1y_2y_3$ coordinate system and see projected onto the plane of the paper the y_1 axis, identical to the $y_2 = y_3$ plane or mirror, etc. This projection is shown in fig. (15). The different channels now correspond to different wedge-shaped sectors bounded by the planes. It is the same set of coordinates we used in the introduction.

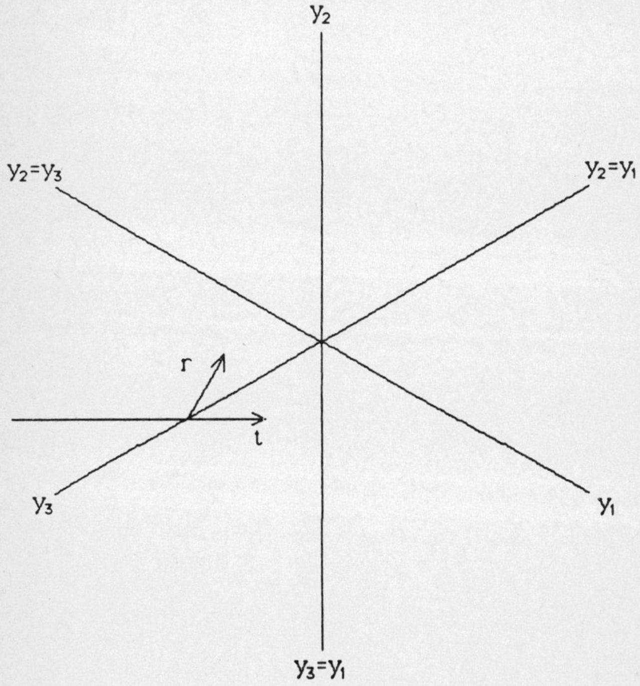

Fig. (15) The same collision as fig. (14) shown as ray tracing in the mirror picture.

Now, the incoming state is just a plane wave. In our example, it is in the left-hand sector, moving to the right. The diagram of fig. (14) then corresponds to ray-tracing where the ray we follow strikes the bottom wall of the wedge. A portion is reflected and the rest is transmitted.

<u>What is this consistency business?</u>

Just this: If we start with incoming momenta P = 123, then there are different ways we could arrive at the outgoing momenta P'' = 321. Fig. (7) shows the two possible ways the particles could scatter.   These correspond in the mirror picture to which of the two mirrors bounding a sector we strike first. However, fig. (7) doesn't show the channels we scatter into at each two-body scattering.

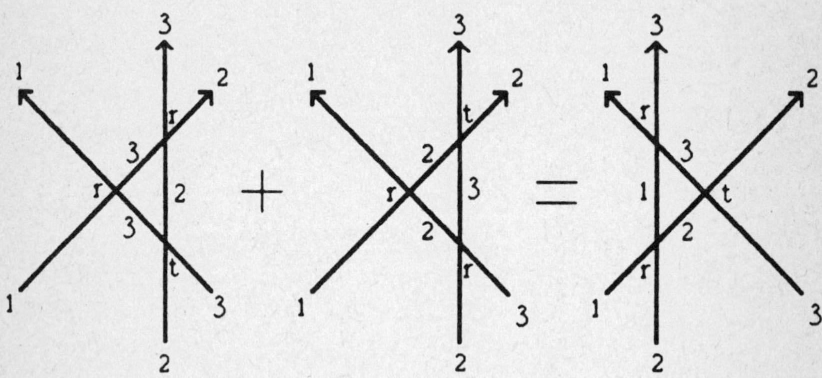

Fig. (16) The particular matrix elements or amplitudes of consistency relation II of fig. (7) that must be related.

Let us take a particular incoming channel Q = 123 and ask how can we get to Q'' = 132. In fig. (7), part (II), from the left diagram, there are two contributions, while from the right there is only one. We must get the same answer in both cases. The requirement for consistency is shown in fig. (16). It is equivalent to the particular equation for the matrix elements of a matrix equation,

$$r(y) \; r \; (x+y) \; t(x) + t(y) \; r(x+y) \; r(x) = r(y) \; t(x+y) \; r(y) \quad .$$

The arguments are $x = \phi_1 - \phi_2$, $y = \phi_2 - \phi_3$. This equation was verified in sect. (F).

As a problem of ray tracing in a kaleidoscope, the consistency requires that the amplitudes in all sectors be the same on the two sides of the shadow of the vertex. For instance, in the sector above, the mirror scattering looks like fig. (17). If the amplitudes are not the same on the two sides of the dashed line, there will be shadow scattering of the vertex, and hence diffraction. If diffraction, Bethe ansatz can not be a solution, for it is based on the assumption that there is no diffraction.

Why do I like the mirrors so much? On one picture you can see both the wave aspects and the ray tracing. You see the possibility of diffraction with the mirrors, while the billard ball diagrams allow you to forget that it is really a wave problem that is being discussed. Also the mirrors show the rigid "shape" of the permutation group which underlies all this integrability business, whereas the billiard ball diagrams are rather flexible and floppy and one can't see how really essential the permutation group is. For instance, why should allowing one particle with a slightly different coupling constant or mass destroy the integrability of the whole system?

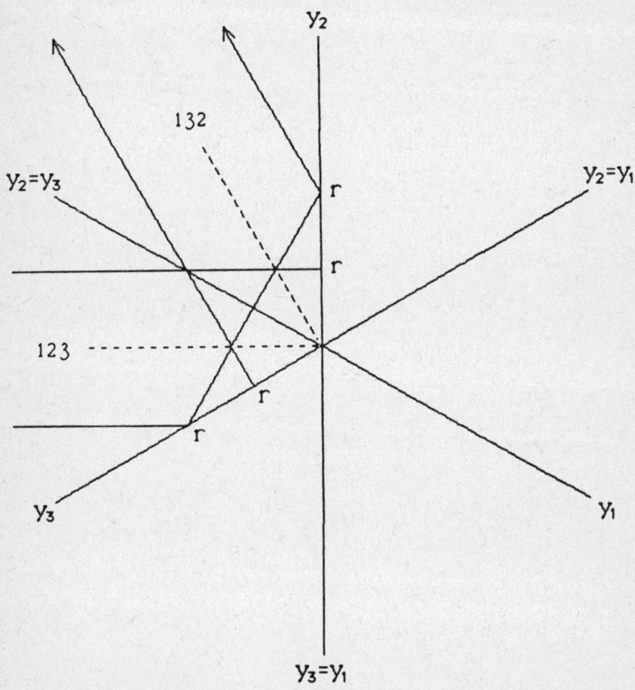

Fig. (17) The same relationship as in fig. (16) shown by ray tracing in a kaleidoscope of mirrors.

<u>Where do we get r and t that go into the connection formulae?</u>

That's easy. Just solve the two-body problem, for r and t are the two-body reflection and transmission amplitudes. That's all we did in sect. (E).

<u>What about the other particles? Why don't we have to worry about $y_1 = y_2 = y_3$?</u>

Just as for the scattering, we can treat the third and other particles as spectators, and so solve for two particles at a time, matching amplitudes pairwise.

The exclusion principle causes the wave function to vanish if two particles of the same spin or type are on the same site. We don't put this in at the beginning, but put it in when we choose a particular representation of the permutation group in the scattering operators T. The connection formulae may be consistent for other representations, but they will not refer to generalizations of the Hubbard model for particles obeying other statistics.

So, due to the exclusion principle, the wave function vanishes when the coordinates of two particles of the same type occupy the same site, and thus we never have to worry about the third particle.

<u>What is that funny notation you started using in sect. (G)?</u>

Remember the A(P) are M! dimensional column vectors with elements A(P|Q). However K of these particles are spin down, while the other M-K particles are spin up. Then [↑↓↑↑...] represents a column vector which has all zero elements <u>except</u> when one of the M-K up spins is first, one of the K down spins is second, one of the M-K up spins is third, etc. All the non-zero elements are equal.

It is only this representation, or one equivalent for it, which can be used for the Hubbard model. It is a complete basis for the Hubbard model amplitudes. And it is physically appealing, for it is just what you would see if you looked at the possible channels.

<u>So what is the bottom line?</u>

The connection equations

$$A(P') = T_{jj+1}\,(\phi - \phi')A(P) \quad ,$$

where P, P' are the same permutations <u>except</u> P has p, p' in positions j, j+1 while P' has p', p in positions j,j+1, are all consistent. These are M! M! (M-1) equations for M!M! unknowns. Actually, the incoming amplitudes in any of the M! channels should be arbitrary, so we should be able to determine M!(M!-1) unknowns given the M! incoming amplitudes A(I). Thus we have a solution to the Hubbard model, at least for the scattering problem.

These were the questions that were asked of me. In addition, I asked myself a question, and tried to answer it:

<u>What do you think you are doing?</u>

In this way I tried to cast the Bethe ansatz method into a more general framework, as follows:

All the permutations $P \epsilon S_M$ are generated by the elementary nearest-neighbor permutations or "mirrors" $P_{jj+1} \equiv \alpha_j$, $j = 1,\ldots,M-1$. Since they are mirrors $\alpha_j^2 = I$. The other properties they have define the group S_M: $(\alpha_j \alpha_\ell)^2 = I$, $|j-\ell|>1$, i.e., they commute unless they are neighbors, when $(\alpha_j \alpha_{j+1})^3 = I$. This is the kaleidoscope relation. To summarize, the only defining relations are $(\alpha_j \alpha_\ell)^{n_{j\ell}} = I$, and these define a Coxeter group.

In terms of the geometry of the mirrors in M-1 dimensional space, the angle between mirrors α_j and α_ℓ is $\pi/n_{j\ell}$. For S_M the angles are either π (a mirror with itself), $\pi/2$ if they commute, or $\pi/3$ for neighbors. We then build up the group by multiplying α's in strings. There is a natural geometry called the graph of the group. The points of the graph represent elements of the group, while the bonds are of M-1 types, one type for each generator α_j. If $P' = \alpha_j P$, then we draw a j-type bond between points P and P'.

For instance, for S_3 , $\alpha_1 = \alpha = P_{12}$, $\alpha_2 = \beta = P_{23}$, and the graph is show in fig. (18). The inner automorphisms of the group are transformations of the graph, and we can choose the graph to be symmetricial under these transformations. A similar example for S_4 is shown in fig. (19). At each point P we can apply any of the M-1 generators α; the graph is homogeneous, in that each point is like any other.

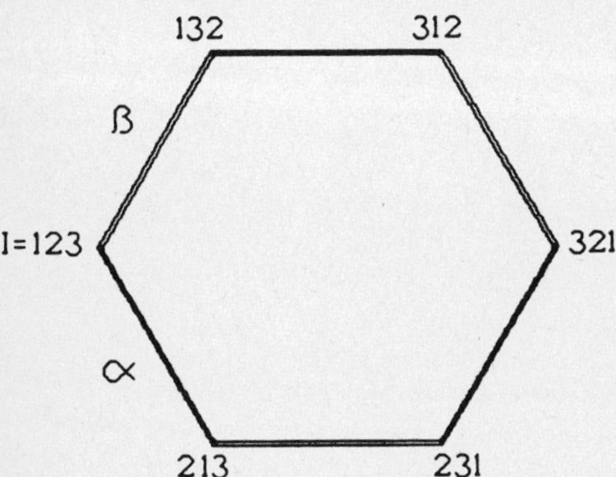

Fig. (18) The graph of the group S_3; the two generators α,β are labeled by solid and doubled lines respectively.

We get from a point P to another point P' by applying a string of generators α. This is represented by a path on the graph from P to P'. There is also a concept of distance between points; this is simply the smallest number of bonds linking two points. The ingoing and outgoing states are always the furthest apart: M(M-1)/2 .

Now, let us look at our connection formulae $A(P') = T_{jj+1}(P) A(P)$. Thus T_{jj+1} is labeled by the generator α_j which we apply to P to take us to $P' = \alpha_j P$. Also A(P) is a function on the graph of the group, and thus $T\alpha(P)$ is a vector function on the graph of the group, "like" a vector field.

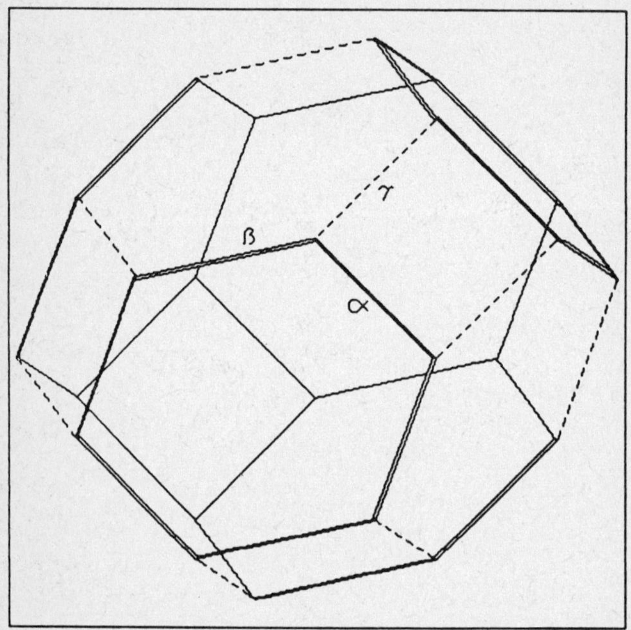

Fig. (19) The graph of the group S_4; the three generators α,β,γ are shown solid, doubled and dashed respectively.

In fact we could have started with any one of M! independent A(I), because we have M! independent incoming channels. Thus we could use the connection formulae to connect a complete, orthonormal, M! dimensional basis. Then, A(P) is a function from the graph of the group into the space of M! × M! unitary matrices.

The connection formulae then are easily inverted to give $T_\alpha(P) = A(\alpha P)A^{-1}(P)$. If we remember that Abelian addition is "like" group multiplication through exponentiation, then the vector field $T\alpha(P)$ is "like" the gradient of the single valued scalar field A(P):

$$\vec{T} = \vec{\nabla}A \quad \text{"like"} \quad T\alpha(P) = A(\alpha P)A^{-1}(P) \quad .$$

Then applying a string of T's to go from A(P) to A(P') is "like" a line integral:

$$A(P') = \int_P^{P'} \vec{T} \cdot d\vec{\ell} + A(P) \quad \text{"like"} \quad A(P') = T_{\alpha N}(\alpha_N P')\ldots T_{\alpha 2}(\alpha_1 P)T_{\alpha 1}(P)A(P) \quad .$$

Consistency says we should get the same A(P') no matter which path we take, or the line integral around closed loops is zero:

$$\oint_{\substack{\text{closed} \\ \text{loop}}} \vec{T} \cdot d\vec{\ell} = 0 \quad \text{"like"} \quad \Pi\, T_{\alpha_N}(\alpha_n P)\ldots T_{\alpha_2}(\alpha_1 P)T_{\alpha_1}(P) = I \quad .$$

If the space were continuous, we could reduce this to $\vec{\nabla} \times \vec{T} = 0$, but the best we can do is to shrink the loops into the smallest possible elementary loops, or co-cycles, corresponding to the defining relations of the group. These are either bonds, squares or hexagons depending on whether $n_{j\ell}$ = 2,4,6. Section (F) showed consistency for n = 2,6. We didn't even feel it necessary to do a proof for n = 4. (Why?) Thus,

$$\vec{\nabla} \times \vec{T} = 0 \quad \text{"like"} \quad \Pi_{\text{cocycle}}\, T(\alpha_N P)\ldots T(P) = I \quad .$$

Thus, to summarize, we are given a vector field "$\vec{T}$" from the 2-body scattering problem. We must show it is exact, or "$\vec{\nabla} \times \vec{T} = 0$". These are the non-diffraction, consistency, Yang-Baxter or factorizable S-matrix equations. Then we can find a solution for the amplitudes by a line integral "$A(P) = \int_I^P \vec{T} \cdot d\vec{\ell} + A(I)$." The result will not depend upon the path.

The remainder of this part (II) is spent deriving the Bethe-Yang, or nested Bethe ansatz; these techniques are necessary for a direct attack on the repulsive Hubbard model. The discussion is long and detailed. The algebraic Bethe ansatz of Faddeev, Takhtadjan and Korepin is much more efficient in arriving at the resulting algebraic equations. However the results so obtained by this method for the Hubbard model, we have already derived very simply in sect. (J), so we will here only establish the equivalence of the two approaches for the ground state energy. Whether one can com-pletely bypass the Bethe-Yang ansatz for all excited states is an open question. Probably we can. We emphasize that this streamlined derivation of sect. (J) is original, and is published for the first time in these lectures.

L) <u>Another Representation</u>

We have made much use of the vectors A(P) connected by relations:

$$A(P) = T_N \ldots T_2 T_1 A(I) \quad .$$

The string of T's are suitably chosen to take us from the identity I to the permutation P, i.e.,

$$P = P_N \ldots P_2 P_1 \quad .$$

Here P_j is a permutation operator acting on the permutations Q of the particle identities.

This is what might be called the reflection representation, since the diagonal elements are the reflection amplitudes. It is also useful to define another representation--the transmission representation--whose diagonal elements are the transmission amplitudes. In this representation we use vectors $A'(P)$,

$$A'(P) = P^{-1} A(P) \quad , \quad A(I) = A'(I) \quad .$$

Then a true and consistent relation for A's becomes a true and consistent relation for the primed A's. For instance,

$$A(P_2 P_1) = T_2 A(P_1) = T_2 T_1 A(I)$$

becomes

$$A'(P_2 P_1) = P_1 P_2 A(P_2 P_1) = P_1 P_2 T_2 P_1 P_1 A(P_1)$$
$$= P_1 P_2 T_2 P_1 A'(P_1) = T_2' A'(P_1) = T_2' P_1 T_1 A(I) = T_2' T_1' A'(I) \quad .$$

We have here defined the new connection operators:

$$T_2' \equiv P_1 P_2 T_2 P_1 \quad ,$$
$$T_1' \equiv P_1 T_1 \quad .$$

In this way, the relation

$$A(P) = T_N \ldots T_1 A(I)$$

becomes

$$A'(P) = T_N' \ldots T_1' A'(I) \quad ,$$

with

$$T_j' \equiv P_1 \ldots P_{j-1} P_j T_j P_{j-1} \ldots P_1$$

Then true operator relations for T's become true operator relations when the T's are primed.

Let us look more closely at a connection operator T'. We have gone from $A(I)$ to $A(P)$, and we now choose to go from $A(P)$ to $A(P')$ by

$$T = T_{\ell\ell+1}(\phi_{P\ell+1} - \phi_{P\ell+1})$$

$$= T_{\ell\ell+1}(\phi_j - \phi_k) = r(\phi_j - \phi_k) + t(\phi_j - \phi_k)P_{\ell\ell+1} \quad .$$

Here $j = P\ell$, $k = P\ell+1$. This then becomes

$$T' = P^{-1}P_{\ell\ell+1} \, T \, P = t(\phi_j - \phi_k) + r(\phi_j - \phi_k)(P')^{-1} P \quad .$$

We examine the permutation,

$$P'' \equiv (P')^{-1} P = P^{-1} P_{\ell\ell+1} P$$

more closely. The permutation P puts $P1$ into 1, $P2$ into 2,..., j into ℓ , k into ℓ + 1,..., PM into M. Then $P_{\ell,\ell+1}$ exchanges j and k. The inverse of P puts everything back where it was, except that j and k have now been exchanged. Thus we conclude that

$$P'' = P_{jk} \quad ,$$

and $T' = t(\phi_j - \phi_k) + r(\phi_j - \phi_k)P_{jk} \quad .$

$$\equiv T'(\phi_j - \phi_k) \equiv T'_{jk}$$

There is now no need to index both the ϕ's and P when we specify T', since they have the same indicies. This is the representation that will be used by Professors Faddeev, Takhtadjan, and Korepin.

M) <u>Periodic Boundary Conditions</u>

We now wish to impose periodic boundary conditions so that the point x and the point $x + N$ are equivalent. Then let $x_M = N + x_0$,

$$x_0 < x_1 < \ldots < x_{M-1} < x_M \quad .$$

For the wavefunction we have

$$\Psi(x_1,\ldots, x_M|Q1,\ldots, QM) = \Psi(x_0, x_1,\ldots, x_{M-1}|QM,Q1,\ldots, QM-1)$$

$$= \Psi(x_0, x_1,\ldots, x_{M-1}|Q'1,\ldots, Q'M) \quad .$$

The permutation Q' is a cyclic permutation of Q:

$$Q' = (QM, Q1,\ldots, QM-1) \quad .$$

We hope that the wavefunction may be made periodic by identifying pairs of terms in the two regions Q,Q'. Thus, for a permutation P we seek another permutation P' such that

$$A(P|Q) \, e^{i(p_{P1}x_1+\ldots+p_{PM}x_M)} = A(P|Q) \, e^{i(p_{P1}x_1+\ldots+p_{PM}x_0+p_{PM}N)}$$

$$= A(P'|Q') \, e^{i(p_{P'1}x_0+\ldots+p_{P'M}x_{M-1})}$$

Thus, P' must be chosen to be a cyclic permutation of P : $P' = (PM,P1,\ldots,PM-1)$.

Further, we must choose the amplitudes so that

$$e^{ip_{p}p_M^N} A(P|Q) = A(P'|Q')$$

This set of M! equations may be written as

$$e^{ip_{p}p_M^N} A(P) = P_{M,M-1}\cdots P_{21} A(P') =$$

$$P_{M,M-1}\cdots P_{21} T_{12}(\phi_{P1} - \phi_{PM})\cdots T_{M-1M}(\phi_{PM-1} - \phi_{PM})A(P)$$

Let $A(P) = T\,A(I)$, so that

$$e^{ip_{p}p_M^N} P^{-1} T\,A(I) = e^{ip_{p}p_M^N} A'(P) =$$

$$P^{-1}P_{M,M-1}\cdots P_{21}T_{12}(\phi_{P1} - \phi_{PM})\cdots T_{M-1,N}(\phi_{PM-1} - \phi_{PM})\,P\,A'(P)$$

$$= T'(\phi_{P1} - \phi_{PM})\cdots T'(\phi_{PM-1} - \phi_{PM})\,A'(P) \quad .$$

For the final simplification, we take

$$A'(P) = T'(\phi_j - \phi_M)\cdots T'(\phi_j - \phi_{j+1})A'(I) \quad ,$$

so that a sufficient condition for us to be able to construct a periodic wavefunction from our previous scattering states is for us to be able to choose p_j's and $A'(I)$ so that

$$e^{iNp_j}A'(I) = T'(\phi_{j+1} - \phi_j)\cdots T'(\phi_M - \phi_j)T'(\phi_1 - \phi_j)\cdots T'(\phi_{j-1} - \phi)A'(I)$$

$$\equiv X_j A'(I) \quad .$$

The process is shown graphically in fig. (20).

This is all very grand, but what is required is that $A'(I)$ be a simultaneous eigenvector of j different matrix operators. This leads us to the question: Are all the matrix equations obtained by imposing periodic boundary conditions consistent?

N) <u>Consistency II</u>

The consistency of the equations

$$\Lambda_j A'(I) \equiv e^{iNp_j} A'(I) = X_j A'(I) \quad ,$$

where X_j is the appropriate string of T' operators, is similar to our previously determined consistency of the T operators. In fact the same diagrams may be used. For instance starting from $A'(I)$, the first two T' operators are shown as the first two scatterings in fig. (20).

This continues until we arrive at $T'(\phi_M - \phi_j)$. However, this particular

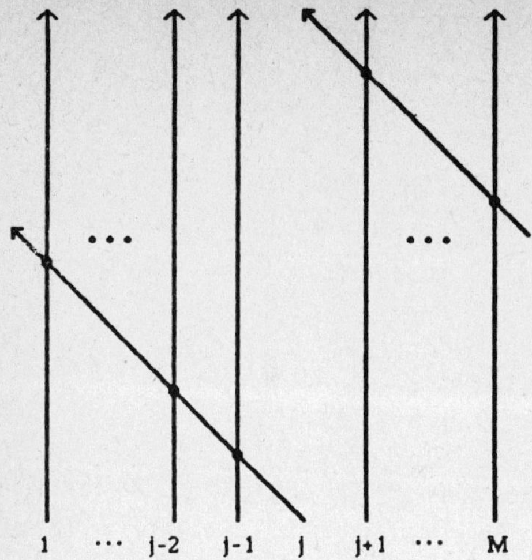

Fig. (20) The graph corresponding to the transport of a particle with ϕ_j around the ring. This is the graphical representation of the algebraic operator X_j, and hence of the requirement of periodic boundary conditions.

operator is not a valid scattering of the first and last particles around the

loop. Thus we have adjoined one more permutation operator, P_{1M}, to the previous

$M - 1$ generators, P_{jj+1} ($j = 1,\ldots,M - 1$), of the permutation group S_M to form

the generators of a new group.   The defining relations for the generators

$$\alpha_j = P_{jj+1}, \; j = 1,\ldots,M ; \quad M + 1 = 1 \; ,$$

are
$$\alpha_j{}^2 = I \; ,$$
$$(\alpha_j\alpha_{j+1})^3 = I \; ,$$
$$(\alpha_j\alpha_k)^2 = I \quad \text{all other pairs } j,k.$$

In contrast to the permutation group S_N which is a finite group, this new group is

infinite, although still discrete. A portion of the graph of this Coxeter group for

the case of three particles is shown in fig. (21).

The consistency of the eigenvalue equations are then consistency conditions on

strings of T' operators through

$$X_j \, X_k = X_k \, X_j \; .$$

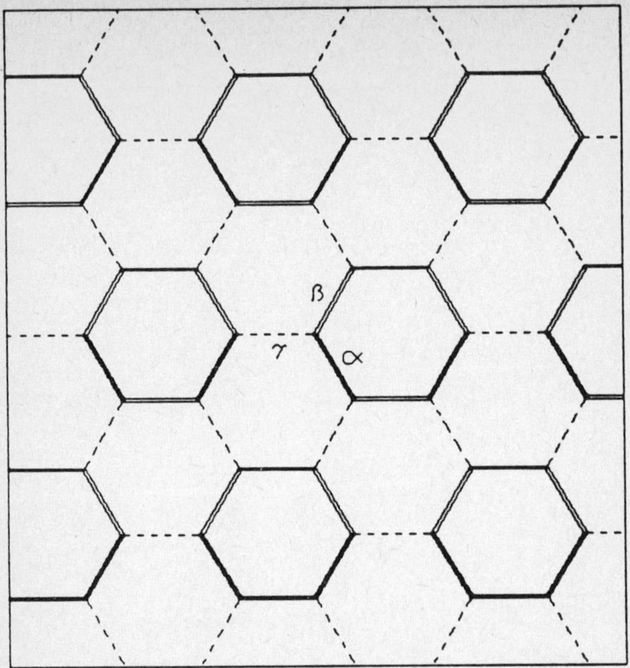

Fig. (21) The graph of the infinite Coxeter group corresponding to S_3 with periodic boundary conditions. The three generators α, β, γ are shown solid, doubled and dashed respectively.

But these consistency conditions are equivalent to the consistency of the elementary tangles or defining relations of the enlarged group. However, <u>locally</u> the defining relations for the new group are the same as the defining relations of the permutation group, so they are in fact already verified.

Note that the operators X_j appear to be simply "cyclic permutations" which take particled j around the ring. However, they can not really be cyclic permutations in S_M, because if we try to write P_{1M} as

$$P_{2M} \, P_{3M} \cdots P_{M-1M} \, P_{1M} \, P_{1M-1} \cdots P_{13} \, P_{12} \quad ,$$

without introducing a new generator and thus enlarging the group, then we simply have an identity $X_j = I$ in the group S_M. Yet the eigenvalue equation, to be non-trivial, states that the different X_j's must have different eigenvalues Λ_j.

In fact in the larger group, the operators

$$P_{j,j+1} \, P_{j+1,\, j+2} \cdots P_{M-1,M} \, P_{M1} \, P_{12} \cdots P_{j-2,j-1} P_{j-1,j} \quad ,$$

on which X_j is based, play the role of translations by primitive lattice vectors on the graph of the infinite group. They commute one with another, and thus the X_j's commute. Then the set of eigenvalue equations,

$$X_j A'(I) = \Lambda_j A'(I) \quad ,$$

is simply an example of Bloch's theorem.

O) Solution of the Auxiliary Problem: One Spin

In order to proceed to the determination of the eigenvalues and eigenvectors of our original Hubbard Hamiltonian, we must first solve explicitly the M (consistent) eigenvalue equations:

$$\Lambda_j \zeta = X_j \zeta; \qquad j = 1,\ldots,M$$

where

$$\Lambda_j \equiv e^{iNp_j} = e^{iN\sin^{-1}(\phi_j)}$$

$$\zeta \equiv A'(I) \quad ,$$

$$X_j = T'(\phi_{j+1} - \phi_j)\ldots T'(\phi_M - \phi_j)T'(\phi_1 - \phi_j)\ldots T'(\phi_{j-1} - \phi_j) \quad .$$

This we call the auxiliary problem. Note that these equations have a common eigenvector ζ. It was the contribution of Yang to show us how to solve these equations. We remind you (see sect. (13)) that:

$$T'(\phi_j - \phi_k) = t(\phi_j - \phi_k) + r(\phi_j - \phi_k)P_{jk} \equiv t_{jk} + r_{jk}P_{jk} \quad ,$$

$$r(\phi) = -\frac{c}{c - i\phi} \quad , \qquad t(\phi) = \frac{i\phi}{c - i\phi} \quad ,$$

thus $t + r = -1$. Whenever T' operates on a vector which is symmetric under permutation of the two particles, it is equivalent to -1. In all that follows, we shall assume a repulsive potential, $c > 0$, so there are no bound states.

As an example, suppose all particles are spin up except one, which is spin down. The eigenvector ζ is then symmetric under permutation of any two spin up particles. The eigenvector ζ has amplitude $\zeta(j) \equiv \zeta_j$, where the coordinate $j = 1,2,\ldots,M$ indicates the single overturned spin. The amplitude is otherwise the same no matter what the arrangement of the up spins. Consider now $T'_{j-1\ j}$, the first factor in X_j, acting on ζ:

$$T'_{j-1\ j}\{\ldots + \zeta_j[j] + \zeta_{j-1}[j-1] + \ldots\}$$
$$= \{\ldots + [t_{j-1\ j}\zeta_j + r_{j-1\ j}\zeta_{j-1}][j]$$

$$+ [t_{j-1\ j}\zeta_{j-1} + r_{j-1}\zeta_j][j-1] +\ldots\}$$

(The notation $[j]$ represents the overturned spin at j.) We thus write this symbolically as

$$\zeta_j \xrightarrow{T'_{j-1\ j}} t_{j-1\ j}\zeta_j + r_{j-1\ j}\zeta_{j-1} \equiv \zeta_j^{(1)},$$

$$\zeta_{j-1} \xrightarrow{T'_{j-1\ j}} t_{j-1\ j}\zeta_{j-1} + r_{j-1\ j}\zeta_j \ .$$

All other ζ_ℓ's are unchanged.

We now note that as no other T' in the string X_j will permute the down spin out of $j-1$ again, the amplitude of $[j-1]$ will remain unchanged, and we may thus write:

$$\Lambda_j\zeta_{j-1} = t_{j-1\ j}\ \zeta_{j-1} + r_{j-1j}\zeta_j \ .$$

Likewise, the new amplitude $\zeta_j^{(1)}$ may be written as

$$\zeta_j^{(1)} = t_{j-1\ j}\ \zeta_j + r_{j-1\ j}\ \zeta_{j-1} \ .$$

We proceed to apply the second T' in the string, $T'_{j-2\ j}$, and find

$$\zeta_{j-2} \xrightarrow{T'_{j-2\ j}} t_{j-2\ j}\zeta_{j-2} + r_{j-2\ j}\zeta_j^{(1)}$$

$$\zeta_j^{(1)} \xrightarrow{T'_{j-2j}} t_{j-2,j}\zeta_j^{(1)} + r_{j-2,j}\zeta_{j-2} \equiv \zeta_j^{(2)}$$

The amplitude of $[j-2]$ will now be unchanged by any further T' in X_j. Thus,

$$\Lambda_j = t_{j-2\ j} + r_{j-2\ j}\ \frac{\zeta_j^{(1)}}{\zeta_{j-2}}$$

and

$$\zeta_j^{(2)} = t_{j-2j}\ \zeta_j^{(1)} + r_{j-2\ j}\zeta_{j-2} \ .$$

The recursive nature of this scheme becomes apparent:

$$\Lambda_j = t_{j-k\ j} + r_{j-k\ j}\ \frac{\zeta_j^{(k-1)}}{\zeta_{j-k}} \qquad\qquad k = 1,2,\ldots,M-1;\ j + M = j$$

and

$$\zeta_j^{(k)} = t_{j-k\ j}\ \zeta_j^{(k-1)} + r_{j-k\ j}\ \zeta_{j-k}$$

Since Λ_j is an eigenvalue, the first equation implies that Λ_j is independent of the variable k or $j-k$ on the right hand side.

Let us evaluate $\zeta_j^{(k-1)}$ by the first equation, using the expressions for t,r:

$$\zeta_j^{(k-1)} = [(\Lambda_j - t_{j-k,j})/r_{j-k,j}]\zeta_{j-k} =$$

$$[-\Lambda_j + i(1 + \Lambda_j)(\phi_{j-k} - \phi_j)/c]\zeta_{j-k}$$

Advancing k to k+1, we also express $\zeta_j^{(k)}$ in terms of ζ_{j-k-1}. We may then substitute these expressions into the second recursion relation to obtain a closed equation for ζ_j's:

$$[-\Lambda_j + i(1 + \Lambda_j)(\phi_{j-k-1} - \phi_j)/c]\zeta_{j-k-1}$$

$$= \frac{-\zeta_{j-k}}{c - i(\phi_{j-k} - \phi_j)} \{c - i(\phi_{j-k} - \phi_j)$$

$$\cdot [-\Lambda_j + i(1 + \Lambda_j)(\phi_{j-k} - \phi_j)/c]\} \quad .$$

Therefore the ratio of the two successive terms of the amplitude are

$$\frac{\zeta_{j-k-1}}{\zeta_{j-k}} = -\frac{c^2 - i(\phi_{j-k}-\phi_j)[-c\Lambda_j+i(1+\Lambda_j)(\phi_{j-k}-\phi_j)]}{[c - i(\phi_{j-k}-\phi_j)][-c\Lambda_j+i(1+\Lambda_j)(\phi_{j-k-1}-\phi_j)]} \quad .$$

The numerator factors into

$$[c - i(\phi_{j-k} - \phi_j)][c + i(1 + \Lambda_j)(\phi_{j-k} - \phi_{j-k} - \phi_j)] \quad ,$$

so that we end up with the ratios as:

$$\frac{\zeta_{j-k-1}}{\zeta_{j-k}} = - \frac{c - c\Lambda_j/(1 + \Lambda_j) + i(\phi_{j-k} - \phi_j)}{c\Lambda_j/(1 + \Lambda_j) + i(\phi_{j-k-1} - \phi_j)} \quad .$$

Now the index j indicates which eigenvalue equation we are discussing, but since the eigenvectors are all the same, this ratio must be independent of j. The only way this can happen is if the combination:

$$- c\Lambda_j/(1 + \Lambda_j) - i\phi_j$$

is constant. We denote this constant as

$$- c\Lambda_j/(1 + \Lambda_j) - i\phi_j \equiv - c/2 - i\lambda \quad ;$$

so that we finally have

$$\frac{\zeta_{j-k-1}}{\zeta_{j-k}} = - \frac{i\lambda - i\phi_{j-k} - c/2}{i\lambda - i\phi_{j-k-1} + c/2} \quad .$$

Upon iterating,

$$\zeta_j = \prod_{\ell=1}^{j-1} - \frac{i\phi_\ell - i\lambda - c/2}{i\phi_{\ell+1} - i\lambda + c/2} \cdot \zeta_1 \quad .$$

When we invert our equation defining the constant λ , we find for Λ_j ,

$$\Lambda_j = -\frac{i\phi_j - i\lambda - c/2}{i\phi_j - i\lambda + c/2} \ .$$

And lastly, the quantity

$$\zeta_j^{(k)} = \frac{i\phi_{j-k-1} - i\lambda - c/2}{i\phi_j - i\lambda + c/2} \ \zeta_{j-k-1} \ .$$

Thus, the two recursion relations are satisfied by explicit construction. The resulting functions giving $\zeta_j, \zeta_j^{(k)}, \Lambda_j$ for this one-overturned spin case will be important in constructing the many overturned spin case. Thus, following Shastry, we define the following functions:

The wave function ζ_j normalized so that $\zeta_1 = 1$, with parameter λ, we denote by $f_j(\lambda)$, and find:

$$f_j(\lambda) \equiv \prod_{\ell=1}^{j-1} - \frac{i\phi_\ell - i\lambda - c/2}{i\phi_{\ell+1} - i\lambda + c/2} \qquad j > 1 \ .$$

Often we will suppress the λ dependence.

Likewise the eigenfunction Λ_j for one-overturned spin we write as $\sigma_j(\lambda)$, with

$$\sigma_j(\lambda) \equiv - \frac{i\phi_j - i\lambda - c/2}{i\phi_j - i\lambda + c/2} = e^{i\theta(2(\phi_j - \lambda))}$$

The following relationship holds:

$$f_j(\lambda) = \frac{i\phi_1 - i\lambda + c/2}{i\phi_j - i\lambda + c/2} \ \sigma_1 \sigma_2 \cdots \sigma_{j-1} \ .$$

The iterated f is :

$$f_j^{(k)}(\lambda) = \frac{f_j}{\sigma_{j-1} \cdots \sigma_{j-k}} = \frac{i\phi_{j-k-1} - i\lambda - c/2}{i\phi_j - i\lambda + c/2} \ f_{j-k-1}$$

Then the recursion relations used to generate these functions become identities satisfied by the functions:

$$t_{j-k,j} \frac{f_j}{\sigma_{j-1} \cdots \sigma_{j-k+1}} + r_{j-k,j} \ f_{j-k} = \frac{f_j}{\sigma_{j-1} \cdots \sigma_{j-k}} \qquad , \quad (I)$$

$$t_{j-k,j} \ f_{j-k} + r_{j-k,j} \frac{f_j}{\sigma_{j-1} \cdots \sigma_{j-k+1}} = \sigma_j f_{j-k} \qquad , \quad (II)$$

The second identity (II) we call the eigenvalue relation, while the first (I) prepares the amplitude of the spin at j for the next interation. Thus we call (I) the preparation identity.

Finally the identities only hold for $j - k > 1$, decreasing the index from j to 1. There are corresponding identities for increasing the index from j to M, but we won't need those for what follows.

These identities may be independently verified, if one wishes.

Let us suppose we concentrate on the eigenvalue when $j = M$. Then we may use the identities for all operators $T'_{\ell j}$. When we are through, we find that all amplitudes satisfy the eigenvalue equation, except for $\zeta_j = \zeta_M$. The amplitude has been iterated so that it finally is equal to

$$f_M^{(M-1)} = \frac{f_M}{\sigma_{M-1} \cdots \sigma_1}$$

But to satisfy the eigenvalue equation, this must be equal to $\sigma_M f_M$. Thus, we have the consistency relation

$$\prod_{j=1}^{M} \sigma_j = 1 \quad ,$$

or

$$\prod_{j=1}^{M} - \frac{i\phi_j - i\lambda - c/2}{i\phi_j - i\lambda + c/2} = 1.$$

Now, each of the M eigenvalue equations

$$\Lambda_j \zeta = X_j \zeta$$

are unitarily equivalent to cyclic permutations of each other, so each eigenvalue Λ_j is simply given by

$$\Lambda_j = \sigma_j = - \frac{i\phi_j - i\lambda - c/2}{i\phi_j - i\lambda + c/2} \equiv e^{iNp(\phi_j)}$$

Thus, we have M + 1 equations for M variables ϕ_j and one variable λ. (Remember $\phi_j = \sin p_j$.)

Before we proceed onward to overturn more spins, let us review the way in which the eigenvalue equation for one overturned spin was satisfied. We will simply give the process in words, and at the same time introduce a rather picturesque terminology. By the k iteration, we mean the result of a product of operators $T'_{j-k,j} \, T'_{j-k+1,j} \cdots T'_{j-1,j}$ acting on the eigenvector ζ.

After the $k - 1$ iteration, the amplitudes of all the down spins $j-k+1, \ldots, j-1$

have been multiplied each by σ_j, by repeated use of the eigenvalue relation (II), and the amplitude of the down spin at j is "prepared" for the k iteration, by the **preparation** identity (I). The amplitudes of all down spins to the left of, and including, the j - k one, are unchanged.

Then upon multiplication by $T'_{j-k,j}$, the process is repeated once again. All amplitudes are unchanged <u>except</u> (I) prepares the j amplitude for the next iteration, while (II) now multiplies f_{j-k} by σ_j. We say the iteration has "passed over" the down spin at j - k, leaving the eigenvalue σ_j. Assuming j = M, then after the M - 1 interation, we are through with the entire X_j operator, and we need only match the j amplitude by the eigenvalue equation.

P) <u>Two-Overturned Spins</u>

For two overturned spins, let us try to write the eigenvector by a Bethe ansatz-type wave function. The role of the plane wave for a single particle, however, is now played by the wavefunction for a single overturned spin $f_j(\lambda)$. That is, if all spins are up except for two overturned spins at j and ℓ , we try for the amplitude of the state $[j,\ell]$, in the form

$$\zeta(j,\ell) = Bf_j(\lambda)f_\ell(\lambda') + B'f_j(\lambda')f_\ell(\lambda) \quad .$$

As we shall see, the two terms are necessary for two overturned spins, since the two down spins are not independent. However, we hope for the next best thing--that they scatter non-diffractively. (Once again, the amplitude will be the same no matter which down spins are where, and no matter which up spins are at the M-2 remaining sites.)

One further notation reduces writing out formulas. We define a conjugation operator which changes unprimed quantities to primed ones and vice versa. Then, with $f_j(\lambda')$ defined as f_j' and $\sigma_j(\lambda')$ defined as σ_j', we have as the Bethe-Yang ansatz for the amplitudes:

$$\zeta(j,\ell) = Bf_jf_\ell' + conj.$$

In what follows, we assume that we are dealing with the equation labeled j = M, so there are no spins to the right of j.

Now after the first iteration, when we multiply ζ by $T_{j'-1\ j}$, all amplitudes

are the same as before unless there is a spin down at j or j-1. If both, the sign of
the amplitude is changed. If only a down spin at j-1, then by (II) f_j' is multiplied
by σ_j' , while if there is only a down spin at j, f_j' is prepared for the next
iteration by (I), and becomes f_j'/σ'_{j-1}. So far, so good.

When we now try to perform the next iteration, if the left-hand down spin is far
to the left, we are in good shape to continue. However, if the next down spin is at
j - 2, while the first was at j-1, then we are in trouble. The reason for our difficulty is seen by closely examining the amplitudes. After the first iteration, the
iteration has passed over the spin at j-1, so the amplitude of the state [j-2, j-1] is:

$Bf_{j-2}\, f_{j-1}'\ \sigma_j' + conj.$

The amplitude of the state [j-1,j] is:

$-Bf_{j-1}\, f_j + conj.$

If f_j' had been prepared for the next iteration, the amplitude would have been:

$B\, f_{j-1}\, f_j'/\sigma'_{j-1} + conj.$

However, what we would really like is for f_{j-1}' to be passed over, and instead
f_j be prepared for the second iteration. Thus, we wish the amplitude of the state
[j-1,j] had been:

$B(f_j/\sigma_{j-1})(f_{j-1}'\ \sigma_j') + conj.$

Then f_j would be prepared for whatever down spin is to the left of the one at j-1,
and we realize our wish for nearly independent spins.

Let us imagine for the moment that we can in fact choose the values of B, B' so
that our wish coincides with reality. Then B, B' satisfy the equation

$-Bf_{j-1}\, f_j' + conj. = Bf_j f_{j-1}'\ \sigma_j'/\sigma_{j-1} + conj$.

We say that the down spin at j hit and passed over the down spin at j-1, leaving σ_j'.
By construction, f_j is now prepared for the second iteration. After this second
iteration, using the identity (II), the amplitude for the state [j-2, j-1] is:

$\sigma_j\sigma_j'\ [Bf_{j-2}\, f_{j-1}' + conj.]$

This amplitude will remain unchanged for the rest of the iteration, and thus we
identify the eigenvalue as

$\Lambda_j = \sigma_j\sigma_j'$.

But our relief is only temporary, for now we have to worry about the amplitude for [j-2, j] after these two interations. In fact, if we look at the amplitude for [j-k, j], after k iterations, we find f_j' prepared only for the kth iteration, when we wish f_j' had been passed over, and f_j had been prepared for the next $k+1^{th}$ iteration. That is we have:

$$- Bf_{j-k}\, f_j'/\sigma_{j-1}'\cdots\sigma_{j-k+1}' + \text{conj.}$$

while we wish for

$$Bf_jf_{j-k}'\sigma_j'/\sigma_{j-1}'\cdots\sigma_{j-k} + \text{conj.}$$

To make our wish come true, and the down spin at j hit and pass over the down spin at j-k, we must be able to choose B, B' to satisfy the identity (III):

$$- Bf_{j-k}f_j'/\sigma_{j-1}'\cdots\sigma_{j-k+1} + \text{conj.}$$

$$= Bf_jf_{j-k}\,\sigma_j'/\sigma_{j-1}\cdots\sigma_{j-k} + \text{conj.} \quad , \qquad\qquad (III)$$

for all $j-1 > k \geqslant 1$.

This is a big order, so let us see if we can satisfy

$$\frac{B}{B'} = - \frac{f_{j-k}'f_j/\sigma_{j-1}\cdots\sigma_{j-k+1}' + f_j'f_{j-k}\sigma_j/\sigma_{j-1}\cdots\sigma_{j-k}'}{\text{conj.}}$$

independently of k. First, multiply numerator and denominator by

$$\sigma_{j-1}\cdots\sigma_{j-k}\ \sigma_{j-1}'\cdots\sigma_{j-k}' \ .$$

Then the expression becomes

$$\frac{B}{B'} = - \frac{\sigma_{j-k}f_jf_{j-k}'\sigma_j'\sigma_{j-1}'\cdots\sigma_{j-k}' + f_j'f_{j-k}\sigma_j'\cdots\sigma_{j-k}}{\text{conj.}}$$

If we return to our original expressions for the functions $f_j(\lambda)$, we find that the numerator has a very large common factor, which is also symmetric in λ, λ', thus invariant under conjugation, and which therefore, will cancel an identical factor from the denominator. The factor is:

$$\frac{\sigma_1\cdots\sigma_{j-1}\sigma_1'\cdots\sigma_{j-1}'(i\phi_1-i\lambda+c/2)(i\phi_1-i\lambda'+c/2)}{(i\phi_j-i\lambda+c/2)(i\phi_{j-k}-i\lambda+c/2)(i\phi_j-i\lambda'+c/2)(i\phi_{j-k}-i\lambda'+c/2)}$$

This leaves the relatively simple relation

$$\frac{B}{B'} = - \frac{N}{N_{\text{conj.}}} \quad ,$$

with

$$N = (i\phi_{j-k}-i\lambda - c/2)(i\phi_j-i\lambda' + c/2) - (i\phi_j-i\lambda - c/2)(i\phi_{j-k}-i\lambda' + c/2) \quad .$$

$$= [i\phi_{j-k} - (i\lambda + c/2)][i\phi_j - (i\lambda' - c/2)]$$

$$- [i\phi_{j-k} - (i\lambda' - c/2)][i\phi_j - (i\lambda + c/2)]$$

$$= \text{cross terms}$$

$$= i(\phi_{j-k} - \phi_j)[i(\lambda - \lambda') + c] \quad .$$

Therefore, much to our surprise, we can make our wishes conform to reality, and we have proven the "hit and pass over" identity (III). As a bonus, we have the explicit formula for the coefficients B, B':

$$\frac{B}{B'} = \frac{i(\lambda - \lambda') + c}{i(\lambda - \lambda') - c} = -e^{-i\theta(\lambda - \lambda')}$$

We review in words: Consider the amplitude of $[j - \ell, j - k]$. It is untouched up until the k iteration. Meanwhile, the amplitude of $[j - \ell, j]$ is gradually being prepared for its big moment with j-k during the kth interation, while at the same time doing its job for other amplitudes along the way. Then on the kth iteration, we pass over j-k leaving a σ behind. At the same time, in the amplitude for $[f-k-j]$, j hits and passes over j-k, also leaving a σ behind. From then until the ℓ iteration, the amplitude for $[j-k,j]$ is being prepared for its big moment with j-ℓ during the ℓ iteration. After the ℓ iteration, the amplitude of $[j-\ell,j-k]$ is the original amplitude times the eigenvalue $\sigma_j\sigma_j'$. It will not be changed further.

Consider the eigenvalue equation with j = M, so we can carry the iterations right out to the end, The only loose ends are the amplitudes of $[j-k,j]$, prepared for a meeting which never comes. We must make them satisfy the eigenvalue equation by hand. Thus,

$$\frac{Bf_M f'_{M-k}\sigma'_M}{\sigma_{M-1}\cdots\sigma_1} + \text{conj.} = \sigma_M\sigma_M'[Bf_{M-k}f_M' + \text{conj.}] \quad .$$

Let's try to satisfy this by

$$\frac{Bf_M f'_{M-k}\sigma'_M}{\sigma_{M-1}\cdots\sigma_1} = B'\sigma_M\sigma_M' f_{M-k}' f_M \quad .$$

The other equation is the conjugate of this one. Thus we have

$$\frac{B}{B'} = \prod_{j=1}^{M} \sigma_j = \frac{1}{\prod\sigma_j'}$$

Let $\lambda = \lambda_1$, $\lambda' = \lambda_2$. Then we have the coupled equations

$$\Lambda_j = e^{ip_j N} = \prod_{\alpha=1}^{2} - \frac{i\phi_j - i\lambda_\alpha - c/2}{i\phi_j - i\lambda_\alpha + c/2} = \sigma_j \sigma'_j \qquad (j = 1,\ldots,M)$$

$$\frac{B}{B'} = \prod_{\beta(\neq\alpha)=1}^{2} \frac{i(\lambda_\alpha - \lambda_\beta) + c}{i(\lambda_\alpha - \lambda_\beta) - c} = \prod_{j=1}^{M} - \frac{i\phi_j - i\lambda_\alpha - c/2}{i\phi_j - i\lambda_\alpha + c/2} \quad . \qquad (\alpha = 1,2)$$

The first equation comes from cyclic permutation of the ϕ's. If ϕ and λ are real, all factors are on the unit circle.

Q) K Overturned Spins

Is there any portion of this scheme which doesn't immediately generalize to more than two overturned spins? No. For K overturned spins, we now try for a solution in the form of the Bethe-Yang or nested Bethe ansatz: The amplitude for K overturned spins out of M total spins, located at positions $[x_1, x_2, \ldots, x_K]$, is given by

$$\zeta(x_1 \ldots x_K) = \sum_{P \in S_K} B(P) \prod_{\alpha=1}^{K} f_{x_\alpha}(\lambda_{P_\alpha}) \quad .$$

The preparation identity (I) and the eigenvalue identity (II) remain the same, while the hit and pass over identity (III) becomes:

$$\frac{B(\ldots\lambda,\lambda'\ldots)}{B(\ldots\lambda',\lambda\ldots)} = \frac{i(\lambda-\lambda')+c}{i(\lambda-\lambda')-c} \quad .$$

All of these relations are obviously consistent, since exchanging λ, λ' gives the inverse relation. Otherwise these are simply scalar relations, in contrast to the previous scattering operators.

The eigenvalue is the product of the single particle eigenvalues, and so is given by

$$\Lambda_j = \prod_{\alpha=1}^{K} \sigma_j(\lambda_\alpha) = e^{iNp(\phi_j)} \quad .$$

The periodic boundary condition for the auxiliary problem gives the consistency relation:

$$\frac{B(\lambda_\alpha \ldots)}{B(\ldots\lambda_\alpha)} = \prod_{j=1}^{M} \sigma_j(\lambda_\alpha) = \prod_{\substack{\beta=1 \\ (\neq\alpha)}}^{K} \frac{i(\lambda_\alpha - \lambda_\beta)+c}{i(\lambda_\alpha - \lambda_\beta)-c} \quad .$$

That's all there is to it!

Let us take the logarithm of these equations. Once again we introduce the phase shift $\theta(x)$ by

$$\frac{ix+c}{ix-c} = -e^{-i\theta(x)} \quad .$$

Then the equations are

$$2\pi I(\lambda_\alpha) - \sum_{\substack{\beta=1 \\ (\neq\alpha)}}^{K} \theta(\lambda_\alpha-\lambda_\beta) = -\sum_{j=1}^{M} \theta(2(\lambda_\alpha-\phi j)) \quad ,$$

$$2\pi J(\phi_j) + \sum_{\alpha=1}^{K} \theta(2(\phi j-\lambda_\alpha)) = Np(\phi_j) \quad .$$

We now look for the ground state of the energy

$$E = 2\sum_{j=1}^{M} \omega(\phi j) \quad ,$$

and assume that λ and ϕ distribute themselves densely about the origin with densities $\sigma(\lambda)$, $\tau(\phi)$ respectively. The normalizations are:

$$\int_{-\lambda_0}^{\lambda_0} \sigma(\lambda)d\lambda = K/N = (n - s)/2 \quad ,$$

$$\int_{-\phi_0}^{\phi_0} \tau(\phi)d\phi = M/N = n \quad .$$

Differentiating the equations once, and replacing the summations by integrals with appropriate densities, we obtain the following two coupled integral equations:

$$\sigma(\lambda) + \int_{-\lambda_0}^{\lambda_0} L(\lambda-\lambda')\sigma(\lambda')d\lambda' = 2\int_{-\phi_0}^{\phi_0} L(2(\lambda-\phi))\tau(\phi)d\phi$$

$$\tau(\phi) - 2\int_{-\lambda_0}^{\lambda_0} L(2(\phi-\lambda))\sigma(\lambda)d\lambda = \eta(\phi)$$

The kernal $L(\phi)$ and $\eta(\phi)$ are defined exactly the same as in section (J).

$$L(\phi) = -\frac{\text{sign}(c)}{2\pi} \frac{d\theta}{d\phi} = \frac{1}{2\pi} \frac{2|c|}{c^2+\phi^2} \quad ,$$

$$\eta(\phi) = \frac{1}{2\pi} \frac{dp}{d\phi} = \frac{1}{2\pi} \frac{1}{\sqrt{1-\phi^2}} = -\frac{1}{2\pi\omega(\phi)} \quad .$$

The energy per site is

$$\varepsilon_0(n,s,c) = 2\int_{-\phi_0}^{\phi_0} \omega(\phi)\tau(\phi)d\phi \quad .$$

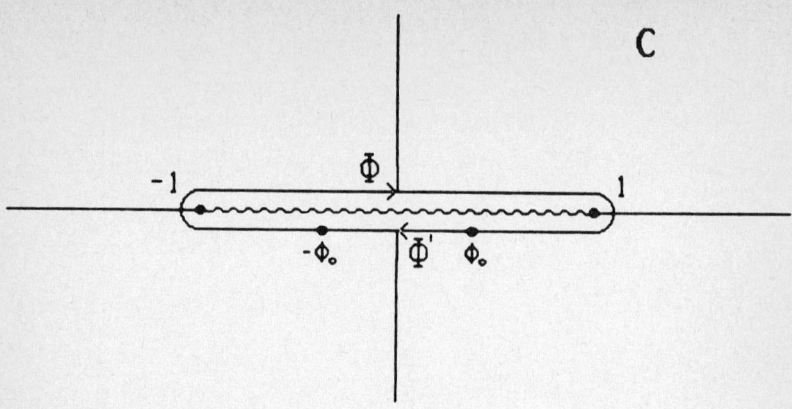

Fig. (22) The contour C in the complex ϕ plane, as explained in the text.

However, in the section (B) on symmetries, we obtained from symmetry (C) the relation

$$\epsilon_0(n,s,c) = \epsilon_0(1-s,1-n,-c) + c(n-s) \ .$$

It was this relation that allowed us to obtain the ground state energy for the repulsive case from the solution for the attractive case in section (J). We now wish to show that although the two sets of integral equations --those of section (J) and the present set-- appear quite different, in fact they are equivalent by the relationship above.

To this end, we introduce a more powerful notation. Let L_n be the integral operator with kernal $nL(n(\phi))$; let Λ and Φ be projection operators which give zero if $|\lambda| > \lambda_0$ or $|\phi| > \phi_0$ respectively, and return the value of a function otherwise. Thus, we rewrite the integral equations as:

$$\sigma + L_1\Lambda\sigma = L_2\Phi\tau \ , \quad \tau = \eta + L_2\Lambda\sigma \ .$$

The normalizations are:

$$\alpha^+ \Phi\tau = n \ , \quad \alpha^+ \Lambda\sigma = (n-s)/2 \ .$$

The function α is simply 1, and the energy is $\epsilon_0 = 2\omega^+ \Phi\tau$.

Now the projection operators Λ, ϕ project out the integral onto finite pieces of the maximum possible ranges of λ and ϕ. What are these maximum ranges? We claim for λ it is $\lambda_0 = +\infty$, and thus $\Lambda=I$ represents an integration of λ over the entire real axis. For ϕ, we claim that the maximum range is for p from $-\pi$ to $+\pi$, or $\phi = \sin p$

along a closed contour C circling a branch cut from -1 to +1 in the clockwise direction. This is the integration path for ϕ when we write $\Phi = C$; it is shown in fig. (22).

The second equation can be substituted into the first, to give us a closed equation for σ. If we write $\Phi = C - \Phi'$ --where by Φ' we mean an integration of the analytically continued functions on the missing piece of the contour C --then the first integral equation becomes

$$\sigma + L_1\Lambda\sigma = L_2(C - \Phi')(\eta + L_2\Lambda\sigma)$$

$$= L_2 C\eta + L_2 C L_2 \Lambda\sigma - L_2\Phi'\eta - L_2\Phi'L_2\Lambda\sigma \ .$$

Since L_η has only poles, well away from the real axis, $L_2 C L_2 = 0$. On the other hand,

$$L_2 C\eta = \frac{1}{(2\pi)^2} \oint \frac{d\phi}{\sqrt{1-\phi^2}} \ \frac{4|c|}{c^2+4(\phi-\lambda)^2}$$

$$= \frac{1}{2\pi} \frac{1}{\sqrt{1-(\lambda+\frac{i|c|}{2})^2}} + c.c.$$

$$= -\frac{1}{2\pi} \, 2 \, \mathrm{Re} \, \omega(\lambda+i|c|/2) = \xi(\lambda) \quad .$$

This $\xi(\lambda)$ is the same function as in sect. (J).

We thus arrive at the final form for the integral equations

$$\sigma + L_1\Lambda\sigma = \xi - L_2\Phi'\rho \quad ,$$

where we defined ρ by $\quad \rho = \eta + L_2\Lambda\sigma \quad .$

The expression for the energy becomes

$$\epsilon_0 = 2\omega^+ (C- \Phi')(\eta + L_2\Lambda\sigma)$$

$$= 2\omega^+ C\eta + 2\omega^+ C L_2\Lambda\sigma - 2\omega^+ \Phi'\rho \quad .$$

The first term is

$$2\omega^+ C\eta = -\frac{1}{\pi} \oint_C d\phi \ \frac{\sqrt{1-\phi^2}}{\sqrt{1-\phi^2}} \ = 0 \quad ;$$

while for the second, we first evaluate

$$L_2 C\omega = -\frac{1}{2\pi} \oint_C d\phi\sqrt{1-\phi^2} \ \frac{4c}{c^2+4(\phi-\lambda)^2}$$

$$= c - \left[\frac{1}{\sqrt{1-(\lambda + i|c|/2)^2}} + c.c. \right]$$

$$= c + 2 \, \mathrm{Re} \, \omega(\lambda + i|c|/2) \quad .$$

Then the final expression for the ground state energy is

$$\varepsilon_0 = c(n-s) + 4\,\text{Re}\;\omega^+(\lambda+i|c|/2)\Lambda\sigma - 2\omega^+\Phi'\rho \quad .$$

Finally, the normalization for τ becomes

$$n = \alpha^+(C-\Phi')(\eta + L_2\Lambda\sigma)$$

$$= \alpha^+C\eta + \alpha^+CL_2\Lambda\sigma - \alpha+\Phi'\rho$$

$$= 1 - \alpha^+\Phi'\rho \quad .$$

These equations are very close to our previous expressions for the attractive case of section (J). However, one further identification is needed. The piece of contour identified by Φ' is <u>not</u> on the same sheet as Φ .

Let us translate the variable to the sheet where say $\omega(\phi)$ has a minimum instead of a maximum. This requires

$$\omega \rightarrow -\omega \quad , \qquad \eta \rightarrow -\eta$$

$$\Phi' \rightarrow -\Phi' \quad , \qquad \rho \rightarrow -\rho \quad .$$

Finally then, the equations are:

$$\sigma + L_1\Lambda\sigma = \xi - L_2\Phi'\rho \;,$$

$$\rho = \eta - L_2\Lambda\sigma \;,$$

$$\varepsilon_0 = c(n-s) + 4\text{Re}\;\omega^+\Lambda\sigma + 2\omega^+\Phi'\rho \;,$$

$$n = 1 - \alpha^+\Phi'\rho \;,$$

$$(n-s)/2 = \alpha^+\Lambda\sigma \quad .$$

Making the identification with corresponding quantities of section (J), we see that the energies are indeed equivalent as required by symmetry.

III.   <u>THE ASYMPTOTIC BETHE ANSATZ</u>

A)   <u>Introduction</u>

In these lectures, as I stated at the outset, I have tried to keep the general structure of the theory to the foreground. In particular I have emphasized the following relationship between concepts:

$$\text{Integrability} \;\rightarrow\; \text{Non-Diffraction} \;\rightarrow\; \text{Bethe Ansatz}$$

Let us very briefly review.

1) Integrability. For a quantum system, integrability implies that we have a complete set of conserved quantities $L_1, L_2, \ldots, L_N$, which commute one with another.

We can take one of these to be the Hamiltonian, and for a translationally invariant system, another will be the momentum. Then it is enough to say they commute one with another; that guarantees they are conserved.

2) Non-Diffraction. If the system admits scattering --that is, if it flies apart --then integrability implies that the system is non-diffractive. Certainly not all systems do scatter, so non-diffraction is not a concept that is always applicable.

However for those systems which are both integrable and scatter, we can say they must scatter non-diffractively. Then by ray tracing and matching amplitudes, we will have certain consistency conditions, or necessary conditions for the scattering to be non-diffractive. These are:

a) conditions on the geometry of the mirrors, so that outgoing momenta align;

b) consistency of the two-body scattering amplitudes, so that there is no shadow scattering from the vertex.

3) Bethe ansatz. If a system is non-diffractive, then asymptotically the wave function must be of the Bethe ansatz form:

$$\Psi \cong \sum A(P) e^{i\sum x_j P_{Pj}} \quad , \quad [A(P') = TA(P)]$$

The Bethe ansatz form is simply a restatement that the scattering is non-diffractive. Note now however, the appearance of the word <u>asymptotically</u>; the Bethe ansatz form will not hold everywhere, as it did for the Hubbard model. This however will be no handicap.

For the Hubbard model, we entered this scheme at the third level. Namely, we were given the local interaction, assumed that the Bethe ansatz held everywhere, and then verified that the wave function satisfied the Schrodinger equation. This also required that we show the consistency of the amplitudes, or the two-body scattering operators.

We could have entered at the second level, and asked: What two-body scattering amplitudes are consistent with the scattering being diffractionless? This demands that we seek solutions to the equations

$$T'_{13} T'_{12} T'_{23} = T'_{23} T'_{12} T'_{13} \quad ,$$

symbolized by fig. (7), or fig. (16), or fig. (17). This is an approach I associate with Zamolodchikov.[35] Then if we find such a solution we can take the two-body scattering operator T'_{ij} and by inverse scattering methods, find the corresponding two-body potential $V(x)$.

Suppose we had done this for the Hubbard model. The solutions to the consistency relations would say that the boson or three-component Hubbard models might be solved by Bethe ansatz. This just emphasizes that the consistency relations are only _necessary_ conditions, not _sufficient_ conditions.

We can also enter the scheme at the first level of integrability. If we could prove quantum integrability, then non-diffraction and the asymptotic Bethe ansatz would follow. We could then use all the machinery of Bethe ansatz to calculate ground state energies, excitations, thermodynamics and scattering of excitations--everything except correlations. The machinery works just as well with the asymptotic wave function as with the wave function, since all of these quantities depend on the spectral properties alone. This is the aproach we will take in this section.

Once again, let me collect the references before I start. Since this section and the following one are interrelated, references for both are included here. Intregrability will be proven by using the Lax[36] technique. This was first used by Moser for the classical versions of some of these N-particle systems. Calogero, Ragnisco and Marchioro[38] later showed that the method could also be applied to the quantum system. These systems had been solved previously by Calogero[39-43] and Sutherland[34,44-49].

Sutherland also related these systems to mathematics which had previously appeared in the theory of random matricies. The book by Mehta[50] is a good review; the original papers are due to Mehta[50,51], Mehta and Gaudin[52], Dyson[53], and Mehta and Dyson[54]. Review articles on these and related topics are Sutherland[34], Calogero[55], and Olshanestsky and Perelomov[56,57].

B) <u>Classical Integrability - The Lax Pair</u>

We wish to find integrable systems of N classical, one-dimensional discrete particles. We shall prove integrability by constructing a Lax pair of Hermitean matrices L, A, such that

$$dL/dt = i[A,L].$$

This ensures that the eigenvalues of L are constants of motions. Now for N particles in one dimension which scatter and are integrable, we know that the N independent constants of motion can simply be chosen as the asymptotic momenta. Therefore it is reasonable to seek L as an NxN Hermitean matrix with diagonal elements p_j (j=1,...N).

Then as the particles scatter and $x_i - x_j \to \pm \infty$, we expect L to diagonalize itself in time, so we will try for L in the form:

$$L_{jj} = p_j \quad ,$$
$$L_{j\ell} = i\alpha(x_j - x_\ell) \quad , \quad (j \neq \ell) \quad .$$

The function $\alpha(x)$ we take to be real and odd, so that L is Hermitean.

We want the Lax equation for the time evolution of L to be equivalent to the equations of motion of the particles:

$$\dot{x}_j = p_j,$$
$$\dot{p}_j = - \sum{}' V'(x_j - x_\ell) \quad .$$

(Time derivatives are denoted by dots, space derivatives by primes, and a prime on a summation denotes that the term with the index equal to the free index is to be omitted.)

Then the diagonal terms of the Lax equation are

$$dp_j/dt = \sum{}' \alpha(x_j - x_\ell)[A_{\ell j} + A_{j\ell}] \quad .$$

Now the following form for the off-diagonal elements of A is reasonable:

$$A_{j\ell} = \beta(x_j - x_\ell) \quad , \quad (j \neq \ell) \quad .$$

We assume $\beta(x)$ is real and even, so A is Hermitean. Then the diagonal terms of the Lax equation become

$$dp_j/dt = 2 \sum{}' \alpha(x_j - x_\ell) \, \beta(x_j - x_\ell) \quad .$$

Comparing with the equations of motion, we make the identification

$$2\alpha(x)\beta(x) = - V'(x) \quad .$$

The off-diagonal terms of the Lax equation lead to

$$i \, d\alpha(x_j - x_\ell)/dt = i(\dot{x}_j - \dot{x}_\ell)\alpha'(x_j - x_\ell)$$
$$= i(p_j - p_\ell)\alpha'(x_j - x_\ell)$$
$$= i(p_\ell - p_j)\beta(x_j - x_\ell) - \alpha(x_j - x_\ell)(A_{jj} - A_{\ell\ell})$$
$$- \sum_k{}'' [\beta(x_j - x_k)\alpha(x_k - x_\ell) - \alpha(x_j - x_k)\beta(x_k - x_\ell)], \quad (j \neq \ell) \quad .$$

A perfectly reasonable way for this equality to occur is for

$$\beta(x) = - \alpha'(x) \quad .$$

Combining this with our previous equation, we find

$$2 \, \alpha'(x)\alpha(x) = V'(x) , \quad \text{or}$$

$$V(x) = \alpha^2(x) + \text{constant}.$$

This leaves us with a final equation, expressing the desire that the unwanted terms would disappear:

$$(A_{jj} - A_{\ell\ell})\alpha(x_j - x_\ell)$$

$$= \sum_k{}'' [\alpha'(x_j - x_k)\alpha(x_k - x\ell) - \alpha(x_j - x_k)\alpha'(x_k - x_\ell)], \quad (j \neq \ell) \quad .$$

The right hand side of this equation is a sum over contributions from all groups of three particles, so we are led to postulate the following form for the diagonal elements of A:

$$A_{jj} = \sum_k{}' \gamma(x_j - x_k) \quad .$$

We assume $\gamma(x)$ to be real and even.

Now it is enough that the previous equation hold for all triples of particles. If we let $x \equiv x_j - x_k$, $y \equiv x_k - x_\ell$, so $x + y = x_j - x_\ell$, then the equation for a typical triple is

$$[\gamma(x) - \gamma(y)]\alpha(x + y) = \alpha'(x)\alpha(y) - \alpha(x)\alpha'(y) \quad .$$

This equation illustrates a basic principle or rule-of-thumb for integrable systems which I call the Rule of Three;

<u>Rule of Three: If it works for three particles, it probably works for N particles.</u>

I know of no natural violations. The Hubbard model is consistent with this rule, for the Bethe ansatz wouldn't work for three bosons, for instance. The origin of this rule is rather deep, and rests on our previous considerations of the permutation group S_N. All scattering in one dimension is basicaly permutations of the particles. The permutation group is defined by binary relations of the form $(\alpha_j\alpha_{j+1})^3 = I$ or $(P_{12}P_{23})^3 = I$. Thus, the defining relations refer to only three particles.

To return to our previous equation for $\gamma(x)$ and $\alpha(x)$, it is a functional equation and must hold for all x, y. Thus clearly it is a necessary condition that it hold infinitesimally. Let us change variables so that $x \to x + dx$, $y \to -x$, so $x + y \to dx$.

The equation now becomes

$$[\gamma(x + dx) - \gamma(x)]\alpha(dx) + \alpha(x)\alpha'(x + dx) + \alpha(x + dx)\alpha'(x) = 0 \quad .$$

We expand in powers of dx, and then the equation must hold at each order. Ignoring the solution $\alpha'(x) = 0$ or $V = $ constant, to lowest order we have $\gamma'(x)\alpha(dx)dx = -2\alpha(x)\alpha'(x)$. Thus we conclude that $\alpha(x) \approx -\alpha/x + \alpha_1 x + \ldots$ as $x \to 0$. Then $\alpha_0\gamma'(x) = 2\alpha(x)\alpha'(x)$ or $\gamma(x) = \alpha^2(x)/\alpha_0 + \gamma_1$. The functional equation is now a closed equation for $\alpha(x)$ alone.

Let us expand to the next order in dx --again only a necessary condition-- to find the equation

$$\alpha'''(x)\ \alpha(x) - 3\alpha''(x)\alpha'(x) + 12\ \alpha_1\ \alpha(x)\alpha'(x)/\alpha_0 = 0.$$

We integrate this once, and with the integrating factor $2\ \alpha'(x)\alpha^{-5}(x)$ a second time, to find

$$[\alpha'(x)]^2 = 6\alpha^2(x)\alpha_1/\alpha_0 + A/2 + A_1\ \alpha^4(x) \quad .$$

We redefine the three arbitrary constants α_1/α_0 , A, A_1 putting $\alpha(x)$ in the standard form

$$\alpha(x) = \frac{g}{sn(x|m)} \quad ,$$

and

$$V(x) = \frac{g^2}{sn^2(x|m)} + \text{constant} \quad .$$

Here $sn(x|m)$ is the Jacobi elliptic function with parameter m.

But is this form sufficient? To answer this, we must return to our original functional equation and see if it is satisfied by this expression. We write the equation as an addition formula

$$\alpha(x + y) = \alpha_0\ \frac{\alpha'(x)\alpha(y) - \alpha(x)\alpha'(y)}{\alpha^2(x) - \alpha^2(y)}$$

Substituting our proposed expression for $\alpha(x)$ into this equation, after some manipulation, the equation is seen to be equivalent to the addition formula for Jacobi functions as given in the standard references.

So we have a family of integrable classical systems--the integrals of motion are the eigenvalues of L, and they are seen to be complete by examining the limits $t \to \pm\ \infty$. We leave a detailed discussion of the physics of these systems until sect. (D).

C) <u>The Quantum Case</u>

Although we have produced N independent integrals of motion (in involution) for the classical case --which for instance we could choose as the eigenvalues of L, or as the trace of powers of L, etc.-- it is difficult to take these over into quantum mechanics directly because of ordering problems. In the expressions for the integrals, the various factors do not necessarily commute, so the corresponding quantum expressions will be ambiguous. The delicate cancellations implied by the Lax equation probably will not occur, and cannot be made to occur.

However, for the problems at hand, Calogero, Ragnisco and Marchioro have shown us how to construct integrals of motion for which the ordering in the quantum case is unambiguous. We then can use many of the cancellations in the classical case without change for the quantum case. Let us consider a standard expression for the determinant of the matrix L:

$$\det L = \sum_P \sum_Q (-1)^{P+Q} \prod_{j=1}^{N} L_{P_j Q_j} / N! \quad .$$

Once again, P and Q are permutations of the integers 1 to N. For the quantum case, the matrix elements of L are now operators. For instance in the standard configuration space basis, the diagonal elements of L are

$$L_{jj} = p_j = -i\hbar \partial/\partial x_j$$

while the off-diagonal elements of L are

$$L_{j\ell} = i\alpha(x_j - x_\ell) \, , \, (j \neq \ell) \quad .$$

The function $\alpha(x)$ is the function derived in sect. (B).

Thus all elements of L commute except L_{jj} with $L_{j\ell}$ or $L_{\ell j}$, all ℓ. But in the expression for the determinant of L, in each term a given row index P_j occurs only once, and a given column index Q_j occurs only once. This is just the familiar "one from row j, one from column ℓ " rule for determinants. Thus if in a given term the diagonal element L_{jj} occurs as a factor, then we will have no other factors with an index j, so L_{jj} commutes with all other factors. Therefore <u>there are no ordering ambiguities in the expression for det L!</u>

Likewise, the determinant

$$D(\lambda) \equiv \det (L - \lambda I) = \sum_{k=0}^{N} \lambda^{N-k} L_k$$

is unambiguously ordered, and thus the N quantities L_k, k = 1 to N are unambiguously ordered. Certainly the quantities L_k are classically N complete integrals of motion; can they also be integrals for the quantum system?

What is involved in answering this question? For any operator Q, the time evolution is governed by the Hamiltonian operator

$$H = \frac{1}{2m} \sum p_j^2 + \sum_{1=j<\ell}^{N} \alpha^2(x_j - x_\ell) \quad ,$$

through the Heisenberg equations of motion dQ/dt = i[[H,Q]] . Here the double commutator indicates a quantum mechanical commutator between operators, as distinguished from the matrix commutators occuring in the Lax equation. For instance the canonical commutation relations are $[[x_j,p_\ell]] = i\hbar\delta_{j\ell}$. Now for the classical case, the invariance of $D(\lambda)$ follows directly from the existence of the Lax pair obeying the Lax equation, and the invariance of the determinant under unitary transformations. For the quantum case, however, we must directly prove that $[[H, D(\lambda)]] = 0$.

Actually we must prove more. The integrals L_k must also commute one with another to be simultaneously diagonalizable. Presumably H is included in the integrals--actually it is a combination of L_1 and L_2 --so it is enough to verify that $[[D(\lambda), D(\lambda')]] = 0$ for all λ, λ'.

The proof is due to Calogero, Ragnisco and Marchioro although as far as I am aware, a complete proof has not yet been published. Such a proof is too long for these lectures, and will have to wait for an up-coming review article. So let us accept on faith for now

$$[[H,D(\lambda)]] = 0 \quad .$$

D) The Variety of These Integrable Systems

After this heavy dose of theory, let us now examine the most general potential which leads to an integrable system. We have found

$$V(x) = g^2 \, sn^{-2} \, (x|m) \quad .$$

We will discover that within this form are included many physically different and interesting potentials, each illustrating a different facet of behavior.

First, we note that the potential has a singularity at the origin of the form

$V(x) \to (g/x)^2$, $x \to 0$. Although for a quantum system of two particles, we can make sense of a weakly attractive x^{-2} singularity, this is not possible classically. And even in the quantum case, we make sense by still requiring the wave function to vanish as two particles approach each other. In fact, one can interpolate between free fermions and free bosons, so that what appears as a weakly attractive system of fermions can equally well by viewed as a weakly repulsive system of bosons. This continuation comes about mathematically, simply by replacing the coupling constant g by λ through $g^2 = \lambda(\lambda-1)$, and allowing $\lambda \geq 0$. The boundary condition on the wave function is $\psi \sim (x_i - x_j)^\lambda$, $(x_i - x_j) \to 0$. Note that both $\lambda = 0$ and $\lambda = 1$ turn off the interaction through $g^2 = 0$, but the first is as free bosons while the second is as free fermions. Continuation to negative λ does not lead to normalizable wave functions in the thermodynamic limit $N \to \infty$.

Although we have used the terms bosons and fermions, we use them here only in referring to the boundary condition on the wave function. Since in all cases $\lambda > 0$, the wave function vanishes a two particles approach, the configuration space breaks up into N! disjoint subspaces, and we may impose any statistics that we desire. The only place that the statistics will show itself its in the off-diagonal elements of the density matricies. The spectrum, for instance, will be unaffected. Because of this boundary condition on the wavefunction, we say the potential is essentially repulsive.

We could have introduced a scale parameter into the general potential, but such a scale could be absorbed into the other parameters of the problem. The potential then has two periods: 2K(m), 2iK'(m). K(m) is the elliptic integral of parameter m, while K'(m) is the elliptic integral of complementary parameter $m' = 1 - m$. These serve as length scales in the problem. The first length is the real period of the problem and thus is the box size L of the periodic problem. The second period is imaginary, and determines the range of the interaction. The final length scale in the problem is the interparticle spacing a = L/N = 2K/N.

Let us first of all consider the limit as the box size $L = 2K(m) \to \infty$, for a fixed N. This then corresponds to $m \to 1$, and $V(x) \to g^2 \sinh^{-2}(x)$ $L \to \infty$. This is a scattering problem, since the box walls have been pushed to infinity, and the system is open.

This system in turn has two limits. The first is the case when the energies are large, so the distance of closest approach is small compared with the range of exponential terms(≈ 1). Then we expand for small x to find $V(x) \to (g/x)^2$, $x \to 0$, $L \to \infty$. This might properly be called the dense limit, since in scattering, it leads to dense configurations.

At the other extreme is the very low energy limit where the closest approach of the particles is very large compared with the range of the potential. Let $a \gg 1$ be the measure of the distance of closest approach, or the least average interparticle spacing during collision. This situation may properly be called the dilute limit. Then if we order the particles $x_1 < x_2 < ... < x_N$, we find $x_j = y_j + ja$ where y_j is of order 1. The potential becomes

$$V(x_j - x_\ell) = g^2 \sinh^{-2} (y_j - y_\ell + a(j - \ell)), \quad j > \ell$$

$$\to 4g^2 e^{-2a(j-\ell)} e^{-2(y_j - y_\ell)}, \quad a \to \infty \quad .$$

We thus rescale g so that $2ge^{-a} \equiv g_1$, and find the limiting form for the potential to be:

$$V \to \delta_{|j-\ell|, 1} \, g_1^2 \, e^{-2|y_j - y_\ell|} \quad .$$

This we recognize as the Toda problem[58].

We reemphasize that all of these problems are scattering problems, where the particles are unconfined by boundaries. We will return to this point in the next section.

At the other extreme, we may take the imaginary period $2iK'(m)$, and thus the range, to be large, forcing the limit $m \to 0$, or $V(x) \to g^2 \sin^{-2}(x)$, $m \to 0$.

This is a periodic potential of period π, which may be viewed as a periodic version of the x^{-2} potential. This can be seen from a partial fraction decomposition

$$g^2 \sin^{-2} x = g^2 \sum_{n=\infty}^{\infty} (x - n\pi)^{-2} \quad ,$$

provided we rescale g so that the potential is finite for distances of order $a = L/N = \pi/N$. That is, $g_1 = g/a$ is finite as $a \to 0$. This forces the potential at $L/2 = \pi/2$ to be of strength $g^2 = g_1^2 a^2 = (g_1 \pi/N)^2$. Thus, there is no effect on a particle from particles half a world away.

Apart from these two limits, we can always take the dense limit where the inter-particle spacing a = L/N is small compared with the other two lengths in the problem --the box size and the range. This is simply a power series expansion in x, and we have $V(x) \rightarrow g^2/x^2$, $a \rightarrow 0$.

We then summarize the various limits and expressions in fig. (23). We emphasize that: (1) All the particles and their pairwise potentials are identical; (2) the potentials are essentially repulsive, so that the wave function is forced to vanish whenever two particles approach each other, and thus the statistics of the particles is <u>almost</u> irrelevant; and finally (3) the potentials and the associated systems are in most cases periodic, although in a suitable limit they may be seen as simply periodic boundary conditions imposed on a local scattering potential.

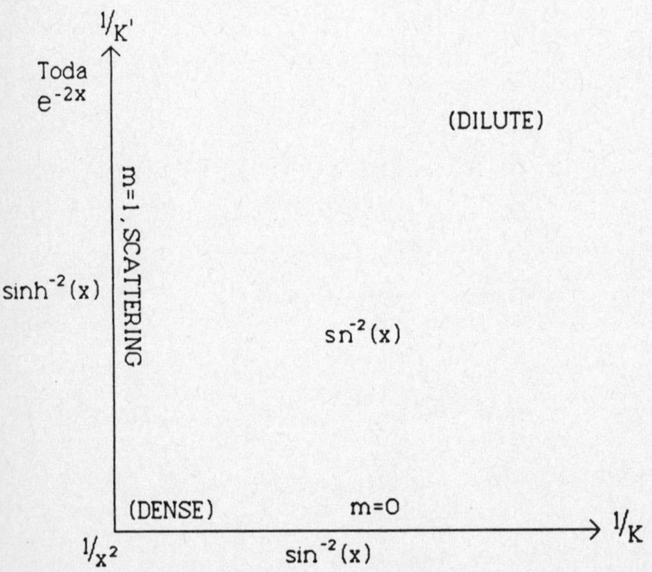

Fig. (23) Various models included in the general integrable systems. These are the repulsive interactions between particles of the same species.

In the expression for the potential, for a physical interpretation we need only that we get real energies for real particle coordinates. There may be other ways to do this. The most interesting case is to consider two types of particles. For the first type, we have coordinates x, while for the second we have coordinates y. In the

Lax pair formalism, and hence in the expression for the potential, we take variables x which are chosen to be the same as the coordinates for the first type particle, but for the second type particle are chosen to be $x = y + iK$.

Then the interactions between like particles are unchanged, while between unlike particles, we have a potential

$$g^2 sn^{-2}(y - x + iK') = g^2 m sn^2(y-x|m) = g^2 m - g^2 m cn^2(y-x|m) \equiv g^2 m + V_{+-}(y-x)$$

The first line follows from a standard relation for elliptic functions, the second shows the attractive nature of the potential, while the last line introduces a notation where $\sigma = \pm$ corresponds to the first or second type of particle respectively, and $V_{\sigma\sigma'}(x)$ is the potential between particles of type σ and σ'. Thus, the previous potential is now called $V_{++}(x) = V_{--}(x)$.

We rather quickly repeat the previous considerations for the potential $V_{+-}(x) = - g^2 m cn^2(x|m)$. First, there is no singularity at the origin and the potential is attractive. Second, as the period increases with $m \to 1$, we approach the scattering problem with $V_{+-}(x) \to - g^2 cosh^{-2}(x)$, $L \to \infty$. This is a familiar potential which is unusual in that it includes reflectionless potentials. Third, as we take $m \to 0$--that is the range going to zero--we find the potential vanishes, and the two systems decouple. Fourth, for the dense system, if we scale the coupling so that the repulsive interaction is finite, then the attractive interaction is scaled away. Fifth, for the dense limit we obtain a harmonic attraction between unlike particles. Finally in the dilute limit of the $m = 1$ case--a scattering problem which previously gave us the Toda lattice--there are two scales. On the first, the unlike particles bind in pairs and are otherwise free, while on the second scale, the bound pairs shrink to points which then interact by a Toda-type potential. Again we illustrate the various limits by fig. (24).

E) The Asymptotic Bethe Ansatz -- Calculations

In this section, we now focus our attention entirely on the scattering problems. For now, let us also assume that we have only particles of a particular type, so that there are no bound states. For these systems, first of all, the previous algebraic analysis simplifies in many respects.

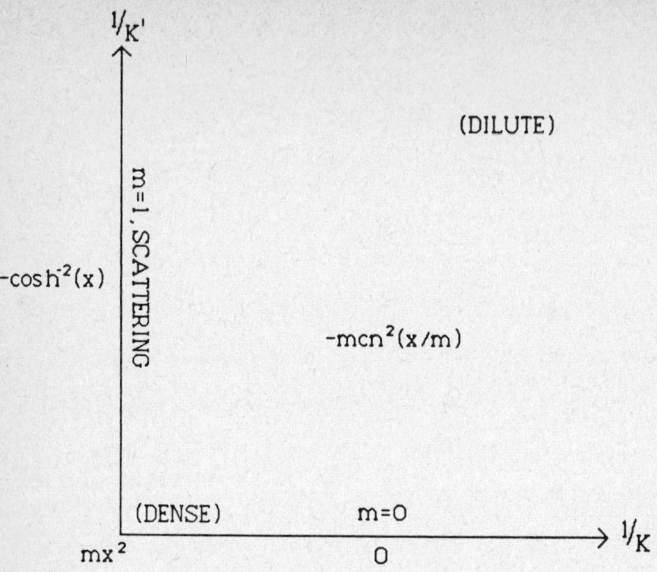

Fig. (24) Attractive interactions between particles of different species, which are also integrable.

We have shown (or accepted on faith) that the conserved quantities L_k are constants of motion, and hence by the Jacobi identity, the commutators are constants of motion. But if we evaluate the commutators at large times, the L_k are diagonal, and hence the L_k mutually commute one with another, and are in involution. Further, L_k is a symmetric homogeneous function of degree k in the p_j, in the same limit $t \to \infty$, and thus the L_k for k = 1 to N are complete.

If we consider any asymptotic region $x_1 \ll x_2 \ll \ldots \ll x_N$, then we may take a state which is an eigenstate of the individual momenta. However, if we fix the N invariants L_k, then we constrain the p_j to be only rearrangements of the incoming p's: $p_{P1} \ldots p_{PN}$, where P is a permutation of the integers 1 to N. The outgoing state obviously has momenta $p_1 < p_2 < \ldots < p_N$, and the incoming state has $p_1' > \ldots > p_N'$, so $p_j' = p_{N-j}$.

To summarize the previous considerations, in the asymptotic region $x_1 \ll x_2 \ll \ldots \ldots \ll x_N$, the wavefunction is of the form

$$\Psi \cong \sum_{P} A(P)\, e^{i\sum x_j P_{Pj}} \ .$$

P is one of the $N!$ permutations of the integers $1,\ldots,N$ denoted $P_1,\ldots,PN$. Thus, we say the wavefunction is non-diffractive, or asymptotically of Bethe ansatz form. We find immediately the total momentum to be $P = \sum p_j$. It is convenient to redefine the Hamiltonian slightly, so that $H = - \sum \partial^2/\partial x^2_j + V$. Thus the energy is $E = \sum p_j^2$. In fact, the previous operator $D(\lambda) \equiv \det(L - \lambda I)$ has eigenvalues $\prod_{j=1}^{N} (p_j - \lambda)$.

Although all the above is exact, all the spectra are continuous, since the systems are unbounded, while the questions we really wish to answer are for dense or thermodynamic systems, not scattering systems. To answer such questions, we must count the states and a convenient trick is to put the system in a very large periodic box. Before we can do this however we must first determine the amplitudes $A(P)$.

Imagine that we do a scattering experiment with wave packets arranged so that more than two particles are never near at the same time. Then the p's will be rearranged in pairs, for this is one dimension, and every pair must collide at some time. Thus, the phase difference between corresponding coefficients will be given simply by the two body phase shift $\theta(p)$:

$$\frac{A(\ldots p', p\ldots)}{A(\ldots p, p'\ldots)} = -e^{-i\theta(p -\ p')} \ .$$

We emphasize that the two body phase shift $\theta(k)$ is determined (by definition) only from the scattering of two particles.

But if we were to obtain a different result if one allowed the packets to overlap in threes, or if they collided in a different order in time, then this difference would be diffraction and would produce p's different from the original p's or permutations there of, and by the previous reasoning does not occur.

Now, to count the states, as we said, we place the particles on a large ring of length L. Let us follow a particle with p_j around the ring. On the way, it collides with every other particle, and the order of the collisions does not matter. At each collision--say with p_ℓ--it suffers a change of phase give by the two-body phase shift $- \theta(p_j - p_\ell)$. After circling the ring once, it has suffered a net phase shift equal to Lp_j from the plane wave, plus the sum of two-body phase shifts $- \sum_{\ell} \theta(p_j - p_\ell)$.

This must be an integer multiple of 2π, and thus--considering each particle in turn--we have the N coupled equations which quantize the p_j's:

$$Lp_j - \sum_{\ell=1}^{N'} \theta(p_j - p_\ell) = 2\pi I_j \quad .$$

The integers I_j are the quantum numbers for the state. Our previous expressions for momentum and energy of course still hold if we only use p_j's which are solutions to the above algebraic equations.

Before we get deeply involved in the specialized techniques for calculating with the Bethe ansatz wavefunction let us first answer the question:
<u>But isn't this just an approximation valid only for a dilute system?</u> <u>You haven't solved it exactly, have you?</u>

Yes, this is an exact solution, <u>in the thermodynamic limit</u>, meaning the limit as $N,L \to \infty$, $N/L = d$ fixed. For instance we will get all results correctly that do not depend on N: ground state energy density, finite temperature thermodynamics, dispersion curves of elementary excitations, scattering of elementary excitations--everything but boundary effects. Of course we cannot calculate correlation functions by these techniques, for we know our wave function only asymptotically; this has nothing to do with the fact that we have only a finite number of particles in an infinite volume.

The argument goes as follows: An expansion of the thermodynamics, (including $T = 0$), in terms of the density is known as a virial expansion. Some of the most beautiful and profound results in statistical mechanics show us how to do such an expansion--the Mayer expansion--and prove in all rigor, that there is a finite radius of convergence, as in the work of Groeneveld and Lebowitz. When there is no region of convergence-- which is the situation for the important case of the Coulomb inter-action--there is a good reason; we first must allow the interaction to screen itself before we can physically expect convergence. And when we reach a singularity on the real axis, and the convergence breaks down, there is also a good reason--a phase transition. A liquid phase is not a slightly more dense gas phase. The gas phase, which is the region of convergence of the virial series, is usually separated by a curve of essential singularities from the liquid phase as in the work of Lee and Yang.

The term of order N in the Mayer series for the virial expansion involves only N or fewer particles, and requires only calculations in an infinite volume. These results have been extended to the quantum case, as for instance in the famous formula of Uhlenbeck and Beth for the second virial coefficient in terms of the two-body phase shift. The general question has been treated by Dashen and Ma. Can one determine the thermodynamics of a system using only the on-mass-shell scattering amplitudes? They answer yes, and show how. An affirmative answer of course is expected by physical intuition.

Thus, we expect that in principle we may calculate the thermodynamics in the low density phase of our system using only the N-body scattering data in an infinite volume. This is not just a low-density limit, since probably in most one dimensional cases, all densities are in the low density phase. The techniques of Yang and Yang show us how to carry out this procedure explicitly for the asymptotic Bethe ansatz. After we have the results in hand, we may examine them for convergence. If they have a singularity for a real value of the density, then there we must stop; the equations will signal their own limitations. This returns us to the discussion at the beginning of section (II) where we emphasized the importance of mapping out the fundamental region of parameter space.

To summarize: We claim that the results based on the asymptotic Bethe ansatz are exact in the thermodynamic limit, i.e. ignoring boundary effects. Additional support comes from explicit calculations by other methods. But these are unnecessary, since the general arguments presented above are convincing.

We return to the Bethe ansatz machinery. As emphasized, the only ingredient needed in a calculation is the two-body phase shift $\theta(p)$, so let us calculate it. We choose as an example the dense limit when the potential becomes proportional to $(x_j - x_\ell)^{-2}$. Then in the center of mass of the two particle system, the Schrodinger equation becomes

$$-\psi'' + \lambda(\lambda-1)\psi/r^2 = p^2\psi \quad .$$

Now since there is no length scale in the problem, while $1/p$ is a length, and since $\theta(p)$ is dimensionless, it can only be that $\theta(p)$ is a constant. Yet $\theta(p)$ must be an odd function of p, so $\theta(p) = $ constant $p/|p|$. What is the constant? For this we must solve the problem, or at least consult Landau and Lifshitz.

The equation of course is very similar to the radial equation for a free particle, and so it probably comes as no surprise to find the solution is expressed in terms of a Bessel function $J_n(x)$ as

$$\psi = \sqrt{r}\ J_{\lambda\ -1/2}(pr) \to \cos(pr - \pi\lambda/2),\ r \to \infty.$$

Then $\theta(p) = \pi\lambda\,\mathrm{sign}(p)$,

Looking back to the equations which determine the p's we find

$$Lp_j - \pi\lambda \sum_\ell{}' \mathrm{sign}(p_j - p_\ell) = 2\pi I_j$$

Now it is reasonable to expect that as we increase the interaction strength λ, the p's do not cross one another. Let us assume this for now and check it later. Then the ordering of the p's will be the ordering of the quantum numbers I_j, and so the equations are solved by

$$p_j = 2\pi I_j/L + \pi\lambda/L \sum_\ell{}' \mathrm{sign}(I_j - I_\ell)$$

$$= 2\pi I_j/L + \pi\lambda/L\ \{\text{number of I's less than } I_j - \text{number of I's greater than } I_j\}\ .$$

If we number the particles so that $I_1 < \ldots < I_N$, then

$$p_j = 2\pi/L[I_j + \lambda(j - (N + 1)/2)]\quad .$$

Therefore the spacing between successive p's is $p_{j+1} - p_j > 2\pi\lambda/L$. This confirms that the ordering of the p's is preserved.

We could just as well have defined new quantum numbers

$$J_j \equiv I_j + (j - (N+1)/2)$$

These new quantum numbers are again integers (N odd) ordered as before, but now they must be unequal, whereas the previous I's could be equal. Thus the J's are free fermion quantum numbers, while the I's are free boson numbers. The p's can as well be written in terms of the J's as

$$p_j = 2\pi J_j/L + \pi(\lambda-1) \sum_\ell{}'\ \mathrm{sign}(J_j - J_\ell)\quad .$$

In this equation, $\lambda=1$ (free fermions) turns off the interaction, while for the previous equation, $\lambda=0$ (free bosons) turns off the interaction.

In fig. (25) we show the evolution of the p's for a given state as a function of the interaction strength λ. We have chosen a state which has one hole and one particle in the free fermion description.

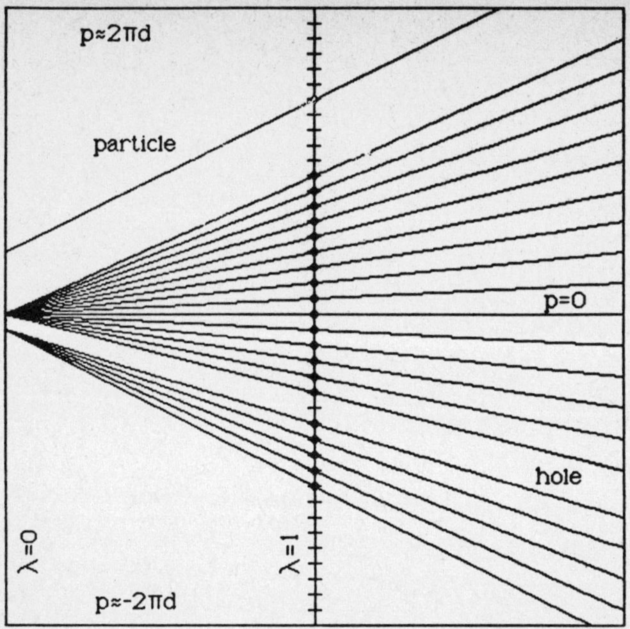

Fig. (25) The evolution of the p's as a function of λ in the dense limit, from free bosons at λ =0 through free fermions at λ =1. The state contains one hole and one particle.

We now illustrate the utility of the asymptotic Bethe ansatz by performing three classic calculations for this example. The calculations for this system are greatly simplified compared with corresponding calculations for a system like the delta-function bosons, because the kernal of the integral operator becomes $\theta'(p)/2\pi =$ $= (\lambda - 1)\delta(p)$. Thus the integral equations become algebraic equations. The derivation of the equations will be presented in the lectures of Prof. Korepin, and are originally due to Lieb and Liniger, Lieb, and Yang and Yang.

1) <u>The ground state energy</u>:

$$1/2\pi = \rho(p) + 1/2\pi \int_{-B}^{B} \theta'(p - p')\rho(p')dp'$$

$$= \rho(p) + (\lambda - 1)\rho(p) \quad , \quad |p| < B.$$

$$\rho(p) = 1/2\pi\lambda \quad ,$$

so
$$d \equiv N/L = \int_{-B}^{B} \rho(p)dp = B/\pi\lambda \quad ,$$

$$\varepsilon_0 = E_0/L = \int_{-B}^{B} \rho(p)p^2 dp = B^3/3\pi\lambda$$

or

$$\varepsilon_0(d) = \pi^2\lambda^2 d^3/3 \quad .$$

2) <u>The excitations near the ground state ($T = 0$):</u>

We take a particle from p_1, $|p_1| < B$, to p_2, $|p_2| > B$. Thus we have a hole labeled by p_1 and a particle labeled by p_2 with $|p_1| < B < |p_2|$. The momentum and energy change are

$$\Delta P = 2\pi \int_{p_1}^{p_2} \rho(p)dp,$$

and

$$\Delta E = \varepsilon(p_2) - \varepsilon(p_1).$$

The $\rho(p)$ is given from the previous equation as

$$\rho(p) = 1/2\pi \begin{cases} 1/\lambda, & |p| < B \quad ; \\ 1, & |p| > B \quad . \end{cases}$$

The function $\varepsilon(p)$ is the solution to the equation

$$- \mu + p^2 = \varepsilon(p) + 1/2\pi \int_{-B}^{B} \theta'(p - p')\varepsilon(p')dp'$$

$$= \varepsilon(p)\begin{cases} \lambda, & |p| < B \quad ; \\ 1, & |p| > B \quad . \end{cases}$$

The chemical potential μ is given by the usual relation $\mu = \partial\varepsilon_0/\partial d = B^2 = (\pi\lambda d)^2$, so

$$\varepsilon(p) = (p^2 - B^2)\begin{cases} 1/\lambda, & |p| < B \quad ; \\ 1, & |p| > B \quad . \end{cases}$$

Thus the momentum and energy become

$$\Delta P = p_2 + \pi d(1-\lambda)\mathrm{sign}(p_2) - p_1/\lambda \quad ,$$

$$\Delta E = p^2 - (\pi\lambda d)^2 - [p_1^2 - (\pi\lambda d)^2]/\lambda \quad .$$

We write these as a single dispersion curve $E(P)$, representing particles for p_2 and holes for p_1. Fig. (26) shows the dispersion relations for both. Note that $|dE/dP|$ at $E = 0$ is the same for both branches and equal to $2\pi\lambda d$.

3) <u>The thermodynamics:</u>

The thermodynamics is given as in Yang and Yang by an expression for the pressure,

$$P(\mu,T) = T/2\pi \int_{-\infty}^{\infty} dp \ \ln(1 + e^{-\varepsilon/T}) \quad .$$

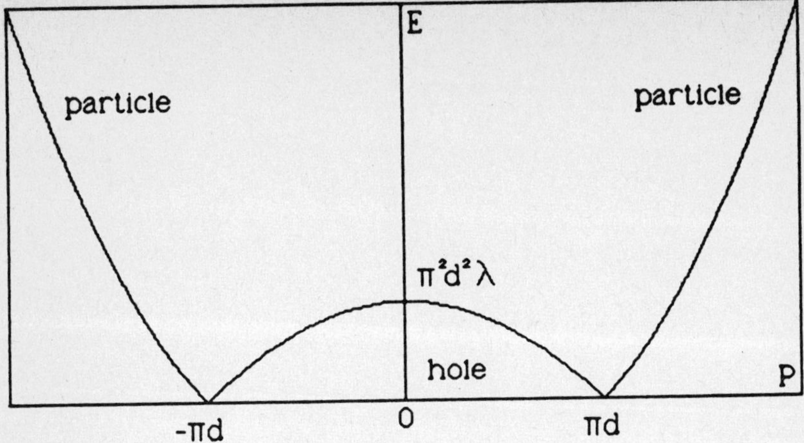

Fig. (26) The particle and hole dispersion curves for the dense limit.

The function $\varepsilon(p)$ satisfies the integral equation

$$\varepsilon(p) = -\mu + p^2 + T/2\pi \int_{-\infty}^{\infty} \theta'(p-p') \ln(1 + e^{-\varepsilon(p')/T})dp'$$

$$= -\mu + p^2 + T(\lambda - 1)\ln(1 + e^{-\varepsilon(p)/T}) \quad .$$

This is just the finite temperature version of the previous equation for ε.

Let us define $\zeta \equiv 1 + e^{-\varepsilon/T}$, so $2\pi P/T = \int_{-\infty}^{\infty} dp \ln \zeta(p)$, while $\zeta^\lambda - \zeta^{\lambda-1} = e^{(\mu-p^2)/T}$. This form is explicit enough to allow us to determine $P(\mu,T)$, or better P as a function of $z \equiv e^{\mu/T}$ and T. The dependence on T is trivial. We find that as a function of z, P has the following singularities:

a) $\lambda > 1$, a branch cut along the negative real axis to $-\infty$ beginning at the branch point

$$z_0 = (\lambda-1)^{\lambda-1}/\lambda^\lambda \quad ;$$

b) $1 > \lambda > 0$, two branch cuts extending to infinity from the branch points

$$z_0 = e^{\mp i\pi\lambda}/\lambda^\lambda(1 - \lambda)^{1-\lambda} \quad .$$

In both cases a power series expansion good for $|z| < z_0$ is

$$2\sqrt{\pi} \, P/T^{3/2} = \sum_{n=1}^{\infty} z^n B_n \quad ,$$

with

$$B_n = (-1)^{n+1}(n\lambda)!/n^{3/2}\lambda n![n(\lambda - 1)]! \quad .$$

Here we of course interpret $x! = \Gamma(1 + x)$.

Certainly the dynamics for this example of a many-body system is simple compared with the usual situation. Looking at fig. (26) one might even be tempted to say it is trivial. But the above expression for the thermodynamics does contain the classical limit as $\lambda \to + \infty$, so if anyone really thinks this is a "trivial" problem, I challenge them to evaluate the <u>classical</u> partition function

$$Z = \frac{1}{N!} \prod_{j=1}^{N} \int_{0}^{L} dx_j \exp[- \frac{\lambda}{T} \sum_{j<\ell} (x_j-x_\ell)^{-2}]$$

and check our result!

All of the other scattering problems, as discussed in sect. (D), can be solved by similar use of the asymptotic Bethe ansatz, although we will not expect such simple results. The two species problem with an attractive potential and bound states would of course be interesting, taking us once again into the Bethe-Yang or nested Bethe ansatz, as with the Hubbard model.

The Toda system is also quite interesting, for one can take the classical limit and recover the beautiful results of Toda on the classical solitons. One finds that the hole excitation--in many ways analogous to a quantized density wave or phonon-- becomes in the classical limit Toda's small amplitude cnoidal wave; while the particle excitation becomes the soliton. As far as I am aware, this was the first concrete connection to be made between the quantum Bethe ansatz and the classical, integrable soliton supporting systems.

In closing this part (III) on the asymptotic Bethe ansatz, as we look back over the calculations, we see something very striking: Although we spent much effort in establishing at the beginning the integrability of our systems by the Lax machinery-- indeed, we didn't even have time and space to show quantum integrability in these lectures--subsequently when we calculated physical properties of the system, we made no use at all of any of the details! It was enough only to know that the systems were integrable, and the proofs were simply to keep us honest. But if we had said, "Maybe the systems are integrable, and if they are we can calculate the properties like so by the asymptotic Bethe ansatz," the results would have been the same.

This leads us to a question: Let us assume that a particular (repulsive) potential is integrable, and then calculate the various spectral properties--ground state energy, excitations, thermodynamics--using only the two-body phase shift as input. In what sense is this an approximation to the real system? It is not simply a question of the range of the potential, for the method is exact for certain long range potentials. I have no answer, but would just emphasize that the calculation is quite easy and practical.

IV. CORRELATION FUNCTIONS

A) Factorization of the Ground State

This section - the final one in this introduction to the Bethe ansatz - will not appear to be related to the previous discussion of integrable systems until much later. And even then the connection will be rather mysterious.

So let us begin with a game. The object of this game is to find interesting questions for which we have the answer. That is, we start with a candidate for a ground state wave function, and then ask if it is the solution to any interesting Hamiltonian. This form for the wave function should not be so simple as to be trivial, yet should also be simple enough that we have some hope of calculating correlations. Therefore, _Game_: For which systems might the ground state wavefunction be of _product_ or _factorized_ or _Jastrow_ form,

$$\Psi = \prod_{1=j<\ell}^{N} \phi(x_j - x_\ell) \ ?$$

We can take the case of bosons so $\phi(-x) = \phi(x)$. Also for the ground state, we need that $\phi(x) > 0$ for $x \neq 0$. We have also restricted ourself once again to one dimension.

Looking at the Schrodinger equation

$$-\sum_{j=1}^{N} \partial^2\Psi/\partial x_j^2 + V\Psi = E_0\Psi \ ,$$

we can immediately divide by Ψ, since $\Psi > 0$, to write the solution of this inverse problem as

$$V - E_0 = \Psi^{-1} \sum_{j=1}^{N} \partial^2\Psi/\partial x_j^2 .$$

This was too easy; the result cannot be very interesting. So we are led to propose an additional rule for our game,

Rule: The **potential** is to be a sum of pair potentials,

$$V - E_0 = \sum_{1=j<\ell}^{N} U(x_j - x_\ell) \quad .$$

Now the game looks more interesting! Let us differentiate the wave function as indicated to arrive at the expression

$$\Psi^{-1} \sum_j \partial^2 \Psi / \partial x_j^2 = 2 \sum_{j<\ell} \psi''/\psi(x_j - x_\ell)$$

$$+ 2 \sum_{\text{triples}} [(\psi'/\psi(x_1 - x_2) \, \psi'/\psi(x_1 - x_3) + \text{cyclic permutations})]$$

The first term on the right hand side of the equation is of the correct form. And for the second term, we once again see an example of the Rule of Three: If it works for 3 particles, it works for N particles.

Let us introduce the logarithmic derivative,

$$\phi(x) = \psi'(x)/\psi(x) \quad ,$$

an odd function. Then with $x \equiv x_3 - x_2$, $y \equiv x_2 - x_1$, we have the following functional equation to solve:

$$\phi(x)\phi(x + y) + \phi(y)\phi(x + y) - \phi(x)\phi(y) = [f(x) + f(y) + f(x + y)]/2 \quad .$$

Then $f(x)$ will be even, and we find a pair potential

$$U(x) = 2\psi''/\psi + (N - 2)f \quad .$$

How do we proceed to find the solutions to this functional equation? We use the same old trick: If the equation holds for all x and y, then it must hold for small y, so we expand in y. To lowest order we find

$$\phi^2(x) + \phi(x)\phi'(x)y + \phi'(x)\phi(y)y + \dots$$

$$= [2f(x) + f(y) + f'(x)y + \dots]/2 \quad .$$

If we write $f(y) \simeq a + by^2 + \dots$, then we must allow $\phi(y) \simeq \alpha/y + \beta y + \dots$.

Thus, the zero order equation is

$$\phi^2(x) + \alpha\phi'(x) = f(x) + a/2 \quad .$$

This result will allow us to take the next highest order equation and close the equations for ϕ.

The first order equation turns out to be the same as the zero order one, so we go to second order and find

$$2\phi(x)\phi''(x) + 4\beta\phi'(x) + 2\alpha\phi'''(x)/3 = f''(x) + 2b \quad .$$

Combining this with the previous equation, we find that $\phi(x)$ must be a solution of the following third order non-linear differential equation:

$$\alpha\phi'''(x) + 6[(\phi'(x))^2 - 2\beta\phi'(x) + b] = 0 \quad .$$

If we define the combination $\alpha\eta(x) \equiv c - \phi'(x)$, then the equation is a standard form

$$\eta''(x) = 6\eta^2(x) - g2/2$$

for the Weierstrauss elliptic P function. This then leads to $\phi' = \beta - \alpha P(x)$ and $\phi = \alpha\zeta(x) + \beta x$, so $f = \phi^2 + \alpha\phi' - \alpha\beta$. The function $\zeta(x)$ is the Weierstrauss zeta function defined in part by $\zeta'(x) = -P$.

As are all elliptic functions, the Weierstrauss P function is doubly periodic with periods 2K and 2iK'. However,

$$\zeta(x + 2K) = \zeta(x) + 2\zeta(K) \quad .$$

Thus, the potential will not be periodic in general--a distinct disadvantage for the physical interpretation. Thus, we are led to choose

$$2\alpha\zeta(k) + 2K\beta = 0 \quad ,$$

making the potential periodic with box size 2K, and 2iK' the range.

Then,

$$\phi(x) = \alpha Z(x) = \alpha \frac{H'(x|m)}{H(x|m)}$$

$$= \frac{d}{dx} \ell n H^\alpha(x|m) \quad ,$$

or

$$\psi(x) = H^\alpha(x|m) \quad .$$

We have introduced two more functions from the battery of elliptic-related functions-- the Jacobi zeta function Z(x) and the Jacobi theta function H(x).

Again, the differential equation is only a necessary and not a sufficient one, so we still must show that the functional equation is equivalent to known addition formulas for elliptic functions. It is.

And once again this form contains a wealth of different physical models. If we consider dense systems, we expand for small x to find $\phi(x) = \alpha/x$. One easily verifies that this form satisfies the functional equation, for

$$\frac{1}{x}\,\frac{1}{x+y} + \frac{1}{y}\,\frac{1}{x+y} - \frac{1}{x}\,\frac{1}{y} = \frac{y + x - (x+y)}{xy(x+y)} = 0 \quad .$$

Then $f(x) = 0$ and $\psi = x^{\alpha}$. The pair potential is

$$U = 2\psi''/\psi = 2\alpha(\alpha-1)/x^2 \quad .$$

The ground state wave function for the N-body system interacting by this pair potential is

$$\Psi = \prod_{j<\ell} |x_j - x_\ell|^{\alpha} \quad .$$

Clearly this is not normalizable; it corresponds to a scattering state with lowest energy.

We can, however, attempt to confine it by a harmonic well, so that the ground state becomes

$$\psi = \prod_{j<\ell} |x_j - x_\ell|^{\alpha} e^{-\Omega \sum x_j^2/2} \quad .$$

Does this work? We need only try it to see.

$$U - E_0 = \sum_{j<\ell} \frac{2\alpha(\alpha-1)}{(x_j - x_\ell)^2} + \sum_j (\Omega^2 x_j^2 - \Omega)$$

$$+ 2\alpha \sum_{j<\ell} \left(\frac{-\Omega x_j}{x_j - x_\ell} + \frac{-\Omega x_\ell}{x_\ell - x_j} \right)$$

$$= \Omega^2 \sum x_j^2 + 2\alpha(\alpha-1) \sum_{j<\ell} (x_j - x_\ell)^{-2} - E_0 \quad .$$

The ground state energy Eo is given by

$$E_0 = N\Omega + N(N-1)\Omega\alpha \quad .$$

This wave function, with the x_j's treated as complex variables is currently in vogue as a trial wave function for the quantum Hall effect, where it is known as Laughlin's form.

B) <u>Calculation of Ground State Correlations</u>

Another limit of interest is the case when the range becomes large, or $K' \to \infty$. so that

$$\phi(x) \to \lambda\cot x$$

We verify directly that the functional equation is satisfied with

$$\cot(x+y)[\cot x + \cot y] - \cot x \cot y = 1 \quad ,$$

for this is simply the standard addition formula for the cotangent:

$$\cot(x+y) = \frac{1 + \cot x \cot y}{\cot x + \cot y} \quad .$$

Integration gives $\psi = \sin^\lambda x \quad ,$

and
$$U(x) = \frac{2\lambda(\lambda - 1)}{\sin^2 x} - \frac{2\lambda^2}{3}(N + 1) \quad .$$

The final form for the wavefunction is

$$\Psi = \prod_{j<\ell} |\sin(x_j - x_\ell)|^\lambda \quad .$$

Remember that $\sin^{-2}(x)$ is a periodic version of the $1/x^2$ potential, a system that we proved integrable in sect. (III). We now have discovered an exact expression for the ground state wave function.

Suprisingly, this is a form that has appeared previously in a totally different context. In the study of random matrices, it is the (unnormalized) distribution function for the eigenvalues of a matrix drawn from Dyson's circular ensemble. However, in this context the values of λ are fixed at the special values 1/2, 1, 2. We can however exploit directly a very elegant machinery for evaluating correlations in these ground state wavefunctions.

The techniques however are involved. In these lectures we will be content simply to illustrate the method by calculating the normalization of the wave function using a method due to Good[59].

We first rewrite the wave function as

$$\Psi = \sqrt{C} \prod_{j<\ell} |e^{i\theta_j} - e^{i\theta_\ell}|^\lambda \quad ,$$

so that

$$C^{-1} = \prod_j \int_0^{2\pi} d\theta_j \prod_{k<\ell} |e^{i\theta_k} - e^{i\theta_\ell}|^{2\lambda} \equiv (2\pi)^N D \quad .$$

Let us restrict ourselves for now to $\lambda = k$, an integer. Then introducing $z_j \equiv e^{i\theta_j}$, D is identified as the constant term in $\prod_{\ell \neq j} |1 - z_j/z_\ell|^k \quad .$

Let us ask a harder question: What is the constant term $D(\vec{k})$ in

$$F(\vec{z}|\vec{k}) \equiv \prod_{\ell \neq j} (1 - z_j/z_\ell)^{k_j} \quad ?$$

Here $\vec{k}$ is a vector with components k_j, k_j a positive integer. The vector $\vec{z}$, of course, has components z_j.

We now claim that the following expression is simply a fancy way of writing the number one:

$$\sum_j \prod_\ell{}' (1 - z_j/z_\ell)^{-1} = 1.$$

We can see this by applying the Lagrange interpolation formula to one. Or we can verify it directly by noting that the function on the left hand side of the equation is symmetric and homogeneous of degree zero. On the other hand the denominator is of degree $N(N-1)/2$ with zeros at $z_j = z_\ell$. Thus the function can only be a constant. If we set $z_1 = 0$, then the function is

$$\prod_\ell z_\ell/(z_\ell - 0) = 1 \quad .$$

Thus the result.

Then if we multiply F by one, we find

$$F(\vec{z}|\vec{k}) = \sum_{j=1}^{N} F(\vec{z}|...k_j -1...) \quad ,$$

so

$$D(\vec{k}) = \sum_{j=1}^{N} D(...k_j -1...) \quad .$$

Now if $k_j = 0$, z_j occurs only with negative powers, so in D itself we can forget that variable entirely. It is easily verified that $D(0) = 1$. Thus, starting from $D(0)$ we can generate all $D(\vec{k})$.

It is easiest to simply verify the expression

$$D(\vec{k}) = \frac{(\sum k_j)!}{\prod k_j!} \quad .$$

Then $D(... k_j -1...) = D(\vec{k})k_j/\sum k_j$,

so

$$D(\vec{k}) = \sum_j D(...k_j-1...) \quad .$$

We also check the initial conditions,

$$D(...0...)'= D(... ...) \quad , \quad D(0) = 1.$$

Thus, we have shown for λ integer,

$$C^{-1} = (2\pi)^N \frac{\Gamma(1 + \lambda N)}{[\Gamma(1 + \lambda)]^N} \quad .$$

By Carlson's theorem, the formula must be good for all $\lambda > 0$.

C) How to Triangulate the Hamiltonian

Having shown how factorizing the ground state into a product form allows calculation of at least some of the ground state correlations--calculations very difficult
to perform for other exactly soluble models such as the delta function boson gas--let
us now demonstrate how the factorization allows us to calculate exactly the excited
energies as well. As an example, let us take the same system as last section--the
periodic inverse square potential. In terms of the angular variables $\theta_j = 2\pi x_j/L$,
the Hamiltonian is

$$H = - \sum_j \partial^2/\partial\theta_j^2 + \lambda(\lambda - 1)/2 \sum_{j < \ell} \sin^{-2} ((\theta_j - \theta_j)/2) \quad .$$

Remember that we found the (unnormalized) ground state to be

$$\Psi_0 = \prod_{j < \ell} |\sin((\theta_j - \theta_\ell)/2|^\lambda \quad ,$$

with ground state energy

$$E_0 = \lambda^2 N(N^2 - 1)/12$$

Since the ground state wave function Ψ_0 has the correlations due to the interaction built into, and since we know it exactly, it is reasonable to look for the
excited states Ψ in the form $\Psi = \Phi\Psi_0$. Then the function Φ may be only weakly
correlated, and look much like free particles. Again this is a common variational
form for real systems. This in turn causes the Hamiltonian H to be transformed by a
<u>similarity</u> transformation Ψ_0, to a new Hamiltonian $\Psi_0^{-1} H\Psi_0$ for the wave function Φ.
This is possible, since Ψ_0 is the ground state wave function with no zeros, hence Ψ_0^{-1}
exists. However, we emphasize that since the transformation is a similarity transformation, not a unitary transformation, the transformed Hamiltonian will no longer
be Hermitean! Of course the eigenvalues are still the eigenvalues of H, and thus real.

We break the transformed Hamiltonian up as

$$\Psi_0^{-1} H \Psi_0 = H_1 + H_2 + E_0 \quad .$$

The E_0 is from H acting on Ψ_0, so that the eigenvalues of $H_1 + H_2$ are the energies
of the excited states measured with respect to the ground state energy. The H_1

are those terms where the derivatives act entirely on Φ, and hence is

$$H_1 = - \sum_j \partial^2/\partial\theta_j^2 \quad ,$$

Finally, H_2 is the sum of cross terms of derivatives--the products of a first deriva-

tive of Φ with the corresponding logarithmic derivative of Ψ_0--and is expressed as

$$H_2 = - i\lambda \sum_{j>\ell} \frac{e^{i\theta_j} + e^{i\theta_\ell}}{e^{i\theta_j} - e^{i\theta_\ell}} (\partial/\partial\theta_j - \partial/\partial\theta_\ell) \quad .$$

Since our hope is that Ψ_0 contains most of the correlation between particles,

while Φ is nearly free, it is reasonable to examine the matrix elements of H_1 and H_2

between free boson eigenstates

$$\Phi_B\{n_j\} = N\{n_j\} \sum_P \exp\{i\sum\theta_j n_{P_j}\} \quad .$$

The normalization $N\{n_j\}$ is

$$N\{n_j\} = [N! \prod_{n=-\infty}^{\infty} \nu(n)!]^{-1/2} \quad ,$$

where $\nu(n)$ is the occupation number of the single particle orbital n.

The first term of the transformed Hamiltonian--H_1--of course is diagonal in this

basis with eigenvalue

$$\varepsilon_1 = \sum_j n_j^2 \quad .$$

Let us examine the effect of one term of H_2 on Φ_B. In particular we choose the

pair j, ℓ with $\theta_j = \theta$, $\theta_\ell = \phi$. Then we examine the effect of this term of H_2 on two

terms of Φ_B: one labeled by the permutation P with $P_j = n$, $P_\ell = m$ $n \geqslant m$, and the

second labeled by the permutation P' which is the same as P except $P'_j = P_\ell = m$ and

$P'_\ell = P_j = n$.

Thus, we have

$$-i\lambda N\{n_j\} \exp [\sum{}'' \theta_j n_{P_j}]$$

$$(\frac{e^{i\theta} + e^{i\phi}}{e^{i\theta} - e^{i\phi}})(\partial/\partial\theta - \partial/\partial\phi)[e^{i(n\theta+m\phi)}+e^{i(m\theta+n\phi)}]$$

$$= \lambda(n - m)N\{n_j\}\exp[\sum{}'' \theta_j n_{P_j}]e^{im(\theta+\phi)}$$

$$(e^{i\theta} + e^{i\phi}) (\frac{e^{ik\theta} - e^{ik\phi}}{e^{i\theta} - e^{i\phi}}) \quad , \quad k = n - m > 0 \quad .$$

But the last factor is just the sum of a geometric series, so the last two factors give

$$(e^{i\theta} + e^{i\phi})[e^{i(k-1)\theta} + e^{i(k-2)\theta +i\phi} + \ldots + e^{i(k-1)\phi}]$$

$$= e^{ik\theta} + e^{ik\phi} + 2 \sum_{\ell=1}^{k-1} e^{i[(k-\ell)\theta + \ell\phi]} \ .$$

Examining this form, we see that H_2 has diagonal elements ε_2 given by

$$\varepsilon_2 = \lambda \sum_{j<\ell} |n_j - n_\ell| \ .$$

However, if we look at the off-diagonal elements of H_2, we see that H_2 connects a state with quantum numbers $\{n_j\}$ to all states obtained by "squeezing" a pair n_j, n_ℓ to $n_j' = n_j - k$, $n_\ell' = n_\ell + k$, where $|n_j' - n_\ell'| < |n_j - n_\ell|$. Thus, if H_2 connects $\phi_B\{n_j\}$ to $\phi_B\{n_j'\}$, then it <u>will not</u> connect $\phi_B\{n_j' \}$ to $\Phi_B\{n_j\}$, except for the diagonal element. This is a contraction mapping, or more simply, H_2 is triangular, in the free boson basis.

Since the determinant of a triangular matrix is simply the product of the diagonal elements, the eigenvalues of $H_1 + H_2$ are the diagonal elements

$$\varepsilon_1 + \varepsilon_2 = \sum n_j^2 + \lambda \sum_{j<\ell} |n_j - n_\ell|$$

$$= \sum_j [n_j^2 + \lambda \sum_\ell{}' \ \text{sign} \ (n_j - n_\ell)] \ ,$$

and the energy eigenvalues are

$$E_0 + \varepsilon_1 + \varepsilon_2 = \sum_j [n_j + \frac{\lambda}{2} \sum_\ell{}' \ \text{sign}(n_j - n_\ell)]^2 \ .$$

In the last expression, we have made use of the identity (N odd) ,

$$\sum_{j=1}^{N} \ (\sum_{\ell=1}^{N}{}' \ \text{sign}(n_j - n_\ell))^2$$

$$= \sum_{\ell= -(N-1)/2}^{(N-1)/2} (2\ell)^2 = 8 \sum_{\ell=1}^{(N-1)/2} \ell^2 = N(N^2-1)/3 \ .$$

Most suprisingly, we recognize that this is exactly the same result we obtained for the inverse-square potential by the asymptotic Bethe ansatz in sect. (III E), allowing for the difference $(2\pi/L)^2$ in the energy due to the variable $\theta = 2\pi x/L$ instead of x. This is the independent verification of the asymptotic Bethe ansatz we referred to. But we had no right to expect the answers to be exactly the same, only to order N^3. Such surprises occur rather frequently.

So this takes us full circle back to the integrable systems. We emphasize that there are integrable systems whose ground state cannot be factored, such as $\sinh^{-2}(x)$. Also, there are integrable systems which can be solved exactly, although not by the asymptotic Bethe ansatz, for they do not support scattering; we have just seen such a case in the $\sin^{-2}(x)$ system. On the other hand, there are systems whose ground state can be factored but seem not to be integrable--although non-integrability has not been proven. The general elliptic case is such a system; it corresponds to a one-dimensional plasma and crystallizes in the ground state into a Wigner crystal. And finally, there are integrable (quantum) systems for which one can do nothing except say they are integrable, and that they do not support scattering. The general $\mathrm{sn}^{-2}(x)$ potential is such a system.

Acknowledgements

I would like to thank the Tata Institute of Fundamental Research and in particular Professors V. Singh, S.S. Jha and B. Sriram Shastry for their hospitality and the opportunity to present these lectures; Professor J.R. Schreiffer and the Institute for Theoretical Physics, Santa Barbara where some of this work was done; Chris McDonald for preparing the manuscript and Jason Sutherland for drawing the figures; and finally all the participants at Asia Plateau for much good physics and lively discussion.

Appendix

We need to evaluate three Fourier transforms for sect. (J).

$$(1)\quad \tilde{L}(\alpha) = \frac{1}{2\pi} \int_{-\infty}^{\infty} d\phi\; e^{-i\alpha\phi}\; \frac{2|c|}{c^2 + \phi^2} = \frac{1}{2\pi i} \int_{-\infty}^{\infty} d\phi\; e^{-i\alpha\phi}\; \frac{1}{\phi - i|c|} - \text{c.c.}$$

If $\alpha \gtrless 0$, we complete the contour in the (lower, upper) half-plane and use Cauchy's theorem to find

$$\tilde{L}(\alpha) = e^{-|\alpha c|}\;.$$

$$(2)\quad \tilde{\xi}(\alpha) = \frac{1}{2\pi} \int_{-\infty}^{\infty} d\phi\; e^{-i\alpha\phi}\; \{[1 - (\phi - i|c|/2)^2]^{-1/2} + \text{c.c.}\}$$

The integral has branch cuts as shown in fig. (27).

Then depending on whether $\alpha \gtrless 0$ we complete the contour in the (lower, upper) half plane, so the contour becomes (C_1, C_2), as shown in Fig. (27).

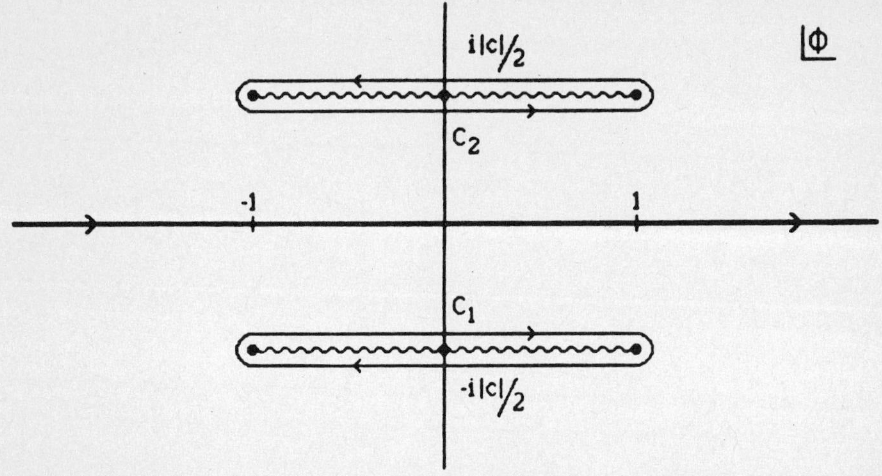

Fig. (27) The contours C_1, C_2 used to evaluate the integrals in the appendix.

Thus

$$\tilde{\xi}(\alpha) = \frac{1}{2\pi} \oint_{C_j} d\phi \; e^{-i\alpha\phi} \{[1-(\phi - i|c|/2)^2]^{-1/2} + c.c.\}$$

$$= e^{-|\alpha c|/2}/\pi \int_{-1}^{1} d\phi \; e^{-i\alpha\phi}(1 - \phi^2)^{-1/2}$$

Let $\phi = \sin\theta$, so

$$\tilde{\xi}(\alpha) = 2e^{-|\alpha c|/2}/\pi \int_{0}^{\pi/2} d\theta \; \cos(\alpha\sin\theta)$$

$$= e^{-|\alpha c|/2} J_0(\alpha) \; ,$$

using a standard form for the Bessel function.

(3)
$$\tilde{E}(\alpha) = -2 \int_{-\infty}^{\infty} d\phi e^{-i\alpha\phi} \{[1 - (\phi - i|c|/2)^2]^{1/2} + c.c.\}$$

The transformations in this case are the same, so

$$\tilde{E}(\alpha) = -2 \oint_{C_j} d\phi \; e^{-i\alpha\phi} \{[1 - (\phi - i|c|/2)^2]^{1/2} + c.c.\}$$

$$= -4 \; e^{-|\alpha c|/2} \int_{-1}^{1} d\phi \; e^{-i\alpha\phi} \sqrt{1 - \phi^2}$$

$$= -8 \, e^{-|\alpha c|/2} \int_{0}^{\pi/2} d\theta \, \cos(\alpha\sin\theta)\cos^2\theta$$

$$= -4\pi \, e^{-|\alpha c|/2} \, J_1(\alpha)/\alpha \quad .$$

Again we have used a standard form for the Bessel function $J_1(\alpha)$.

Bibliography

1. H.A. Bethe, Z. Phys. 71, 205 (1931).
2. L. Hulthén, Arkiv. Mat. Astron. Fys. 26A, No. 11 (1938).
3. R. Orbach, Phys. Rev. 112, 309 (1958).
4. L.R. Walker, Phys. Rev. 116, 1089 (1959).
5. J. des Cloizeaux and J. Pearson, Phys. Rev. 128, 2131 (1962).
6. R. Griffiths, Phys. Rev. 133, A 768 (1964).
7. R. Griffiths, Phys. Rev. 136, A 751 (1964).
8. J. des Cloizeaux and M. Gaudin, J. Math. Phys. 7, 1384 (1966).
9. J. Bonner and M. Fisher, Phys. Rev. 135, A 640 (1964).
10. J. Bonner, thesis, University of London (1968).
11. C.N. Yang and C.P. Yang, Phys. Letters 20, 9 (1966).
12. C.N. Yang and C.P. Yang, Phys. Rev. 147, 303 (1966).
13. C.N. Yang and C.P. Yang, Phys. Rev. 150, 321 (1966).
14. C.N. Yang and C.P. Yang, Phys. Rev. 150, 327 (1966).
15. C.N. Yang and C.P. Yang, Phys. Rev. 151, 258 (1966).
16. E. Lieb and W. Liniger, Phys. Rev. 130, 1605 (1963).
17. E. Lieb, Phys. Rev. 130, 1616 (1963).
18. J. McGuire, J. Math. Phys. 5, 622 (1964).
19. J. McGuire, J. Math. Phys. 6, 432 (1965).
20. J. McGuire, J. Math. Phys. 7, 123 (1966).
21. J. Zinn-Justin and E. Brezin, C.R. 263, 670 B (1966).
22. M. Gaudin, Phys. Letters 24 A (1967).
23. M. Gaudin, thesis, L'Université de Paris (1967).
24. C.N. Yang, Phys. Rev. Lett. 19, 1312 (1967).
25. B. Sutherland, Phys. Rev. Lett. 20, 98 (1968).
26. C.N. Yang and C.P. Yang, J. Math. Phys. 10, 1115 (1969).
27. J. Hubbard, Proc. Roy. Soc. A 276, 238 (1963).
28. E. Lieb and F.Y. Wu, Phys. Rev. Lett. 20, 1445 (1968).
29. E. Lieb and D. Mattis, Mathematical Physics in One Dimension (Acad. Press, N.Y. 1966).
30. M. Gaudin,
31. M. Gaudin, La Fonction d'On de de Bethe pour les Modèles Exacts de la Méchanique Statistique (Commissariat a l'Energie Atomique, Paris 1983).
32. R.J. Baxter, Exactly Solved Models in Statistical Mechanics (Academic Press, London 1982).

33. C.N. Yang, Proceedings of the VII Winter School of Theoretical Physics in Karpacz. Reprinted in C.N. Yang, Selected Papers 1945-1980 (Freeman, San Francisco 1983).

34. B. Sutherland, Rocky Mountain Journal of Mathematics $\underline{8}$, 413 (1978).

35. A.B. Zamolodchikov, Comm. Math. Phys. $\underline{69}$, 165 (1979).

36. P.D. Lax, Comm. Pure Appl. Math. $\underline{21}$, 497 (1968).

37. J. Moser, Adv. in Math. $\underline{16}$, 197 (1975).

38. F. Calogero, O. Ragnisco and C. Marchioro, Lett. Nuovo Cimento, $\underline{13}$, 383 (1975).

39. F. Calogero, J. Math. Phys. $\underline{10}$, 2191 (1969).

40. F. Calogero, J. Math. Phys. $\underline{10}$, 2197 (1969).

41. F. Calogero, J. Math. Phys. $\underline{12}$, 419 (1971).

42. F. Calogero, Lett. Nuovo Cimento $\underline{13}$, 411 (1975).

43. F. Calogero, Lett. Nuovo Cimento $\underline{13}$, 507 (1975).

44. B. Sutherland, J. Math. Phys. $\underline{12}$, 246 (1971).

45. B. Sutherland, J. Math. Phys. $\underline{12}$, 251 (1971).

46. B. Sutherland, Phys. Rev. $\underline{A4}$, 2019 (1971).

47. B. Sutherland, Phys. Rev. $\underline{A5}$, 1372 (1972).

48. B. Sutherland, Phys. Rev. Lett. $\underline{34}$, 1083 (1975).

49. B. Sutherland, Phys. Rev. Lett. $\underline{35}$, 185 (1975).

50. M.L. Mehta, Random Matrices (Academic Press, New York 1967).

51. M.L. Mehta, Nucl. Phys. $\underline{18}$, 395 (1960).

52. M.L. Mehta and M. Gaudin, Nucl. Phys. $\underline{25}$, 447 (1960).

53. F.J. Dyson, J. Math. Phys. $\underline{3}$, 166 (1962).

54. M.L. Mehta and F.J. Dyson, J. Math. Phys. $\underline{4}$, 713 (1963).

55. F. Calogero in Nonlinear Phenomena (Springer, Berlin 1983).

An Introduction to the Theory of
τ Functions

Tetsuji Miwa
RIMS, Kyoto University,
Kyoto 606, JAPAN

1.1   The aim of this note is to give an expository introduction to
the theory of τ functions by focussing our attention on two problems
in mathematical physics, the two dimensional Ising model [1] and the
Kadomtsev-Petviashvili equation [2].

Let us begin with two original results by Wu et al [3] and Hirota
[4] which led us to the notion of τ functions.

1.2   Wu, McCoy, Tracy and Barouch found that the scaling limit of
the two point correlation function of the two dimensional Ising model
is related to a non linear ordinary differential equation known as
a Painlevé equation of the third kind. Let us denote by $\tau_{\pm}(t)$ the
two point correlation function from above (τ_{+}) or from below (τ_{-})
the critical temperature. (The precise definition will be given
later.) The scaled distance between two spins is denoted by a continuous parameter t. Their result reads as

$$\begin{pmatrix} \tau_{+}(t) \\ \tau_{-}(t) \end{pmatrix} = \begin{pmatrix} \cosh \dfrac{\psi(t)}{2} \\ \sinh \dfrac{\psi(t)}{2} \end{pmatrix} e^{\int_{\infty}^{t} \frac{s}{4}[\psi'(s)^2 - \sinh^2 \psi(s)]ds} \tag{1}$$

where $\psi(t)$ solves

$$(t\psi')' = \frac{t}{2} \sinh 2\psi. \tag{2}$$

The correlation functions $\tau_{\pm}(t)$ are our first example of τ functions.

1.3   Hirota introduced the bilinear transform method for solving
soliton equations. For the sine-Gordon equation

$$\partial_{\xi}\partial_{\eta} u = \frac{1}{2} \sinh 2u \tag{3}$$

he obtained the N soliton formula in terms of new dependent variables

$f(\xi,\eta)$ and $g(\xi,\eta)$ which is transformed into $u(\xi,\eta)$ by

$$u = \log \frac{f+g}{f-g} \; . \tag{4}$$

(Here we choose the hyperbolic sine in accordance with (2).) Hirota's variables f and g are our second example of τ functions.

1.4 The equations (2) and (3) are connected by

$$u(\xi, \eta) = \psi(t),$$

$$2\sqrt{\xi\eta} = t.$$

Moreover, from (1) we have

$$\psi = \log \frac{\tau_+ + \tau_-}{\tau_+ - \tau_-} \; .$$

Namely, the correlation functions $\tau_\pm(t)$ and Hirota's variable f, g play the identical role in differential equations (2) and (3), respectively.

What is the mathematical structure which supports this coincidence? In what follows we reveal the answer to this question. The common key is to consider an infinite dimensional Lie group in the free fermion algebra. For the two dimensional Ising model this key lies on the fact that spin operators and transfer matrices belong to the group. For soliton equations this key enables us to identify the solution space with the group orbit of the vacuum vector.

2.1 The two dimensional Ising model is defined on a lattice

$$\Lambda = \{(m,n) \mid m \in \mathbb{Z}/M\mathbb{Z}, \; n \in \mathbb{Z}/N\mathbb{Z}\}$$

with M columns and N rows. We denote by

$$\sigma : \quad \Lambda \quad \longrightarrow \quad \{\pm 1\}$$
$$\qquad\qquad \cup\!\!\!| \qquad\qquad\qquad \cup\!\!\!|$$
$$\qquad (m,n) \qquad\qquad \sigma_{mn}$$

a spin configuration of spins on the lattice Λ. The Hamiltonian is given by

$$E(\sigma) = -E_1 \sum_{m,n} \sigma_{mn}\sigma_{m+1\ n} - E_2 \sum_{m,n} \sigma_{mn}\sigma_{m\ n+1}$$

where $E_1\ E_2 > 0$. The partition function is

$$Z = \sum_{\sigma} e^{-\beta E(\sigma)} \tag{5}$$

where $\beta = 1/kT$. The sum in (5) extends over 2^{MN} spin configurations. The free energy per site f is given by

$$f = -(kT/MN) \log Z.$$

The k point correlation function is

$$\tau((m_1,n_1),\cdots,(m_k,n_k)) = \frac{1}{Z} \sum_{\sigma} \sigma_{m_1 n_1} \cdots \sigma_{m_k n_k}\ e^{-\beta E(\sigma)}.$$

In 2 and 3 we compute the free energy and the correlation functions in the thermodynamic limit $M, N \to \infty$.

<u>2.2</u> The transfer matrix method reduces the problem to the algebra of $2^M \times 2^M$ matrices. Let us rewrite the summand $e^{-\beta E(\sigma)}$ in (5) as

$$e^{-\beta E(\sigma)} = \cdots e^{K_1 \sum_m \sigma_{mn}\sigma_{m+1\ n}}\ e^{K_2 \sum_m \sigma_{mn}\sigma_{m\ n+1}} \cdots, \tag{6}$$

where $K_j = E_j/kT$. In (6) the sum over Λ is splitted into partial sums over rows.

We denote by

$$s : \{0,1,\cdots,M-1\} \longrightarrow \{\pm 1\} \tag{7}$$
$$\Uparrow \qquad\qquad\qquad \Uparrow$$
$$m \qquad\qquad \longmapsto\quad s(m)$$

a spin configuration on a row. The transfer matrix V_1 and V_2 are $2^M \times 2^M$ matrices defined by

$$V_1(s,s') = \delta_{ss'}\ e^{K_1 \sum_m s(m)s'(m+1)},$$

$$V_2(s,s') = e^{K_2 \sum_m s(m)s'(m)}.$$

Then, (5) reads as

$$e^{\beta E(\sigma)} = \cdots V(s_n, s_{n+1}) \cdots,$$

where $V = V_1 V_2$ and $s_n(m) = \sigma_{mn}$. The partition function Z is now written as

$$Z = \text{trace } V^N. \tag{8}$$

We introduce the spin operator

$$\sigma_m^z(s,s') = \delta_{ss'} \, s(m).$$

Then, the correlation function is written as

$$\tau((m_1,n_1), \cdots, (m_k,n_k)) = \frac{1}{Z} \text{ trace } V^{n_1} \sigma_{m_1}^z V^{n_2-n_1} \sigma_{m_2}^z \cdots V^{N-n_k},$$

$$\tag{9}$$

with the restriction $n_1 \leq n_2 \leq \cdots \leq n_k$.

By (8) and (9) our problems reduce to the following.

Problem 1. The diagonalization of V.

Problem 2. The computation of matrix elements of σ_m^z with respect to eigenvectors for V.

2.3 The vector space on which the transfer matrix acts has the structure of tensor product of two dimensional spaces. Namely, a spin configuration s in (7) corresponds to a vector $v(s)$ in

$$\mathcal{F}_{\text{Ising}} = \underbrace{\mathbb{C}^2 \otimes \cdots \otimes \mathbb{C}^2}_{M},$$

given by

$$v(s) = v_{s(0)} \otimes \cdots \otimes v_{s(M-1)}$$

where $v_1 = \begin{pmatrix} 1 \\ 0 \end{pmatrix}$ and $v_{-1} = \begin{pmatrix} 0 \\ 1 \end{pmatrix}$. Let us denote by A_{Ising} the algebra of linear transformations on $\mathcal{F}_{\text{Ising}}$,

$$A_{\text{Ising}} = \text{End}_{\mathbb{C}}(\mathcal{F}_{\text{Ising}}).$$

If we denote by $M(2,\mathbb{C})$ the algebra of 2×2 matrices, we have

$$A_{Ising} \simeq M(2,\mathbb{C}) \otimes \cdots \otimes M(2,\mathbb{C}).$$

The spin operator σ_m^z is written as

$$\sigma_m^z = \begin{pmatrix} 1 & \\ & 1 \end{pmatrix} \otimes \cdots \otimes \begin{pmatrix} 1 & \\ & 1 \end{pmatrix} \otimes \begin{pmatrix} 1 & \\ & -1 \end{pmatrix} \otimes \begin{pmatrix} 1 & \\ & 1 \end{pmatrix} \otimes \cdots \otimes \begin{pmatrix} 1 & \\ & 1 \end{pmatrix}.$$

Here, except for $\begin{pmatrix} 1 & \\ & -1 \end{pmatrix}$ at the m-th position, factors are $\begin{pmatrix} 1 & \\ & 1 \end{pmatrix}$.
We define σ_m^x by

$$\sigma_m^x = \begin{pmatrix} 1 & \\ & 1 \end{pmatrix} \otimes \cdots \otimes \begin{pmatrix} 1 & \\ & 1 \end{pmatrix} \otimes \begin{pmatrix} & 1 \\ 1 & \end{pmatrix} \otimes \begin{pmatrix} 1 & \\ & 1 \end{pmatrix} \otimes \cdots \otimes \begin{pmatrix} 1 & \\ & 1 \end{pmatrix}.$$

In terms of σ_m^z, σ_m^x we can rewrite V_1 and V_2 as follows.

$$V_1 = e^{K_1 \sum_m \sigma_m^z \sigma_{m+1}^z},$$

$$V_2 = (2 \sinh 2K_2)^{M/2} e^{K_2^* \sum_m \sigma_m^x},$$

where K_2^* is defined by the reciprocal relation

$$e^{-2K_2} = \tanh K_2^*.$$

<u>2.4</u> The Jordan-Wigner transformation is to introduce new generators p_m, q_m of the algebra A_{Ising} by

$$p_m = \sigma_0^x \cdots \sigma_{m-1}^x \sigma_m^z , \tag{10}$$

$$q_m = \sigma_0^x \cdots \sigma_{m-1}^x \sigma_m^x \sigma_m^z . \tag{11}$$

If we denote by $[a,b]_+$ the anti-commutator $ab+ba$, they obey the following anti-commutation relations.

$$[p_m, p_{m'}]_+ = 2\delta_{mm'}, \tag{12}$$

$$[p_m, q_{m'}]_+ = 0, \tag{13}$$

$$[q_m, q_{m'}]_+ = -2\delta_{mm'}. \tag{14}$$

Let us rewrite V_1, V_2 and σ_m^z in terms of p_m, q_m. The result is

$$V_1 = e^{K_1 \sum_{m=0}^{M-1} p_m q_{m-1} + K_1 \varepsilon p_{M-1} q_0} ,$$

$$V_2 = (2 \sinh 2K_2)^{M/2} e^{K_2^* \sum_m q_m p_m} , \tag{15}$$

$$\sigma_m^z = q_0 p_0 \cdots q_{m-1} p_{m-1} \, p_m . \tag{16}$$

where $\varepsilon = q_0 p_0 \cdots q_{M-1} p_{M-1}$. In order to maintain translational invariance with respect to p_m, q_m, we ignore the factor ε in the boundary term of V_1. We assume that in the thermodynamic limit $M, N \to \infty$ this modification of V_1 causes no effect. Thus we set

$$V_1 = e^{K_1 \sum_m p_m q_{m-1}} . \tag{17}$$

We denote by W_{Ising} the linear span of p_m, q_m,

$$W_{Ising} = (\oplus_m \mathbb{C} p_m) \oplus (\oplus_m \mathbb{C} q_m).$$

The vector space W_{Ising} is endowed with a symmetric inner product,

$$\langle w, w' \rangle = [w, w']_+ . \tag{18}$$

Note that because of (12) $\sim$ (14) the right hand side of (18) is a c-number. We call a vector in W_{Ising} a free fermion.

2.5   For the moment we forget the Ising model and start again from a vector space W with a non-degenerate symmetric inner product $\langle w, w' \rangle$.

The Clifford algebra, which we denote by A, is an algebra generated by free fermions $w \in W$ on which we impose the following relation,

$$[w, w']_+ = \langle w, w' \rangle . \tag{19}$$

An element in A is called an operator. An operator $a \in A$ is of the form

$$a = \sum_{k=0}^{\dim W} \ \sum_{1 \le i_1 < \cdots < i_k \le \dim W} c_{i_1 \cdots i_k} \, \psi_{i_1} \cdots \psi_{i_k} , \tag{20}$$

where $c_{i_1 \cdots i_k} \in \mathbb{C}$ and $\{\psi_1, \cdots, \psi_{\dim W}\}$ is a basis of W. The sum in (20) can be split into two parts, with even k and odd k. Correspondingly, A is a sum of the even part A_+ and the odd part A_-. Let us denote by $*$ the anti-automorphism of A defined by

$$a^* = \sum_{k=0}^{\dim W} \sum_{1 \le i_1 < \cdots < i_k \le \dim W} c_{i_1 \cdots i_k} \psi_{i_k} \cdots \psi_{i_1} . \tag{21}$$

2.6 The Clifford group $G = G(W)$ is defined by

$$G = \{g \in A \mid \exists g^{-1} \in A, \; gWg^{-1} = W\}.$$

Example 1. Let $X \in A$ be a quadratic form of free fermions,

$$X = \sum_{j,k} a_{jk} \psi_j \psi_k .$$

By using (19) one can show that if w belongs to W

$$adX \cdot w = [X, w]$$

belongs to W. Hence

$$e^X w e^{-X} = e^{adX} w$$

belongs to W. Thus e^X is an element of G. Note that the formulas (15) and (17) tell us that the transfer matrices V_1 and V_2 belong to $G(W_{Ising})$.

Example 2. Let $\psi \in W$ be a free fermion such that $\langle \psi, \psi \rangle \ne 0$. Then $\psi^{-1} = 2\psi/\langle \psi, \psi \rangle$ and

$$-\psi w \psi^{-1} = w - \frac{2\langle w, \psi \rangle}{\langle \psi, \psi \rangle} \psi . \tag{22}$$

This is a reflection in W, and ψ_1 belongs to G. The formula (16) tells us that the spin operator σ_m^z belongs to G.

2.7 The fact that the transfer matrices and the spin operator belong to $G(W_{Ising})$ is the key of the solvability of the two dimensional Ising model. We prepare some basic facts about the Clifford group G.

$\underline{A.}$ An operator g belongs to G if and only if

$$g = w_1 \cdots w_k$$

for some $w_1, \cdots, w_k \in W$ such that

$$\langle w_j, w_j \rangle \neq 0 \qquad (j=1,\cdots,k).$$

$\underline{B.}$ For $g \in G$ we define a c-number $nr(g) \in \mathbb{C}$ by

$$nr(g) = gg^*,$$

where g^* is defined in (21). Then we have

$$nr(g_1 g_2) = nr(g_1)nr(g_2),$$

$$nr(cg) = c^2 nr(g),$$

where $c \in \mathbb{C}$ and $g, g_1, g_2 \in G$.

$\underline{C.}$ G is a disjoint union of $G_+ = G \cap A_+$ and $G_- = G \cap A_-$.

$\underline{D.}$ For $g \in G$ we define $T_g \in O(W)$ (the orthogonal group over W) by

$$T_g(w) = \pm gwg^{-1}, \tag{23}$$

where the sign is $\pm$ if $g \in G_\pm$. Then we have the exact sequence

$$1 \longrightarrow GL(1,\mathbb{C}) \longrightarrow G \longrightarrow O(W) \longrightarrow 1 \tag{24}$$

$\underline{2.8}$ The exact sequence (24) is our guiding principle in solving Problem 1 and 2 stated in 2.2. Namely, instead of dealing with V_1, V_2, σ_m^z in G we consider $T_{V_1}, T_{V_2}, T_{\sigma_m^z}$ in $O(W_{Ising})$. Since T_g determines g up to constant multiple, this strategy will work well.

The proof of A $\sim$ D in 2.7 is not very difficult. We only note that if the inner product is non degenerate any orthogonal transformation is decomposed into a product of reflections. With the formula (22), this fact proves the surjectivity of $G \rightarrow O(W)$ and the assertion $\underline{A.}$

2.9   The trace on A is a $\mathbb{C}$-valued linear function on A characterized by the following.

$$\text{trace } ab = \text{trace } ba, \tag{25}$$

$$\text{trace } a = 0 \quad \text{if} \quad a \in A_-, \tag{26}$$

$$\text{trace } 1 = 2^{\dim W/2}. \tag{27}$$

For A_{Ising} the trace as $2^M \times 2^M$ matrix actually satisfies (25) $\sim$
(27).

The formula for the trace of an element $g \in G_+$ is

$$(\text{trace } g)^2 = nr(g)\det(1+T_g). \tag{28}$$

2.10   We compute the partition function Z exploiting the formula
(28). Since V_1 commutes with σ_m^z the formulas (8) and (9) are
valid with the choice of V,

$$V = V_1^{\frac{1}{2}} V_2 V_1^{\frac{1}{1}} .$$

Using a Fourier transformed basis,

$$p(\theta_\mu) = \sum_{\mu=0}^{M-1} e^{-im\theta_\mu} p_m, \tag{29}$$

$$q(\theta_\mu) = \sum_{\mu=0}^{M-1} e^{-im\theta_\mu} q_m, \tag{30}$$

where $\theta_\mu = 2\pi\mu/M$ we obtain

$$(T_V p(\theta_\mu), T_V q(\theta_\mu))$$

$$= (p(\theta_\mu), q(\theta_\mu)) \begin{pmatrix} \cosh \gamma(\theta_\mu) & a(\theta_\mu) \sinh \gamma(\theta_\mu) \\ a(\theta_\mu)^{-1} \sinh \gamma(\theta_\mu) & \cosh \gamma(\theta_\mu) \end{pmatrix}.$$

Here $\gamma(\theta)$ and $a(\theta)$ are defined by

$$\cosh \gamma(\theta) = \cosh 2K_1 \cosh 2K_2^* - \sinh 2K_1 \sinh 2K_2^* \cos \theta, \tag{31}$$

105

$$a(\theta)^2 = \frac{(1-\alpha_1 e^{i\theta})(1-\alpha_2^{-1} e^{i\theta})}{(1-\alpha_1 e^{-i\theta})(1-\alpha_2^{-1} e^{-i\theta})} \ ,$$

with

$$\alpha_1 = \tanh K_1 \tanh K_2^* ,$$

$$\alpha_2 = \tanh K_2^* / \tanh K_1 .$$

We choose the branch of $\gamma(\theta)$, $a(\theta)$ such that

$$\gamma(0) \geq 0,$$

$$a(0) = \pm 1 \quad \text{for} \quad \alpha_2 \gtrless 1.$$

Noting that $nr(V_1) = 1$ and $nr(V_2) = (2 \sinh 2K_2)^M$, we have

$$Z^2 = (2 \sinh 2K_2)^{MN} \prod_{\mu=0}^{M-1} 2(1 + \cosh N\gamma(\theta_\mu)),$$

In the thermodynamic limit $M, N \to \infty$ we have

$$-\beta f = \frac{1}{2} \log(2 \sinh 2K_2) + \frac{1}{2} \int_0^2 \frac{d\theta}{2\pi} \gamma(\theta)$$

2.11   The Kramers-Wannier duality observes that the free energy computed at (K_1, K_2) coincides with the one at (K_2^*, K_1^*). The line in (K_1, K_2) plane defined by $K_1 = K_2^*$ (or equivalently $K_2 = K_1^*$) is called critical. The critical line corresponds to $\alpha_2 = 1$. We call the region where $\alpha_2 > 1$ (resp. $\alpha_2 < 1$) the disordered (resp. ordered) phase. For a fixed E_1, E_2 the critical line corresponds to the critical temperature T_c and $T \gtrless T_c$ for $\alpha_2 \gtrless 1$.

3.1   For the computation of correlation functions we need to consider a quantity of the form

$$a = \frac{\text{trace } a\ g_0}{\text{trace } g_0} \ , \tag{32}$$

where

$$a = \sigma^z_{m_1 n_1} \cdots \sigma^z_{m_k n_k} ,$$

$$\sigma^z_{mn} = V^n \sigma^z_m V^{-n},$$

$$g_0 = V^N.$$

In general, let us consider a linear function

$$
\begin{array}{ccc}
< \ > \ : \ A & \longrightarrow & \mathbb{C} \\
\Uparrow & & \Uparrow \\
a & \longmapsto & <a>
\end{array}
$$

given by (32) where we choose g_0 from the closure of G_+ such that

$$\text{trace } g_0 \neq 0.$$

We call $<a>$ the expectation value of a with respect to g_0.

<u>3.2</u>   Wick's theorem tells us how to compute $<a>$ when $<ww'>$ for $w, w' \in W$ is known. Note that if $a \in A_-$ then $<a> = 0$. For $w_1, \ldots, w_{2k} \in W$ we have

$$<w_1 \cdots w_{2k}> = \sum_{\substack{i_1 < j_1, \ldots, i_k < j_k \\ i_1 < \cdots < i_k \\ \{i_1,j_1,\ldots,i_k,j_k\}=\{1,\ldots,2k\}}} \text{sgn}\begin{pmatrix} 1 & \cdots & 2k \\ i_1 j_1 & \cdots & i_k j_k \end{pmatrix} <w_{i_1} w_{j_1}> \cdots <w_{i_k} w_{j_k}> .$$

<u>3.3</u>   For a basis $\{\psi_1, \cdots, \psi_{\dim W}\}$ of W let us define matrices J and K of the size $\dim W$ as follows.

$$J_{jk} = <\psi_j, \psi_k>,$$

$$K_{jk} = <\psi_k \ \psi_k>.$$

Assume that $g_0 \in G_+$. Let us denote by T_0 the matrix representation of T_{G_0} with respect to the above basis. The following formula is available in order to compute K:

$$K = J(1 + T_0)^{-1}. \tag{33}$$

Note that if $w, w' \in W$

$$<ww'> + <w'w> = <w,w'>. \tag{34}$$

Conversely, for any bilinear form $< \ >$ on $W \times W$ satisfying (34) there exists $g_0 \in \overline{G_+}$ such that $<ww'>$ is the expectation value of ww' with respect to g_0.

$\underline{3.4}$ The table of expectation values for the basis (29), (30) is computed by using the formula (33). Here we write the result in the thermodynamic limit $M, N \to \infty$. In the limit $M \to \infty$ (29), (30) reduce to

$$p(\theta) = \sum_{m \in \mathbb{Z}} p_m e^{-im\theta},$$

$$q(\theta) = \sum_{m \in \mathbb{Z}} q_m e^{-im\theta},$$

where p_m, q_m ($m \in \mathbb{Z}$) are characterized by (12)$\sim$(14). They satisfy the anticommutation relation:

$$\begin{pmatrix} <p(\theta),p(\theta')> & <p(\theta),q(\theta')> \\ <q(\theta),p(\theta')> & <q(\theta),q(\theta')> \end{pmatrix} = \begin{pmatrix} 2 & 0 \\ 0 & -2 \end{pmatrix} \cdot 2\pi\sigma(\theta+\theta').$$

The expectation values in the limit $N \to \infty$ read as

$$\begin{pmatrix} <p(\theta)p(\theta')> & <p(\theta)q(\theta')> \\ <q(\theta)p(\theta')> & <q(\theta)q(\theta')> \end{pmatrix} = \begin{pmatrix} 1 & -a(\theta)^{-1} \\ a(\theta) & -1 \end{pmatrix} 2\pi\delta(\theta+\theta').$$

Wick's theorem enables us to compute the correlation function $<\sigma_0^z \sigma_m^z>$ as follows. We denote by a_m the Fourier coefficients of $a(\theta)$:

$$a_m = \int \frac{d\theta}{2\pi} \, e^{im\theta} a(\theta).$$

Then we have

$$<p_m p_n> = -<q_m q_n> = \delta_{mn}, \quad <q_m p_n> = -<p_n q_m> = a_{m-n}.$$

From (16) we have

$$\sigma_0^z \sigma_m^z = (-)^m q_0 p_1 q_1 p_2 \cdots q_{m-1} p_m .$$

Hence, using Wick's theorem we have

$$\langle \sigma_0^z \sigma_m^z \rangle = (-)^m \det \begin{pmatrix} a_{-1} & a_{-2} & \cdots & a_{-m} \\ a_0 & a_{-1} & \cdots & a_{-m+1} \\ \vdots & \vdots & & \vdots \\ a_{m-2} & a_{m-3} & \cdots a_{-1} \end{pmatrix} .$$

This result is insufficient because it is limited to the equal row correlations. It is also rather cumbersome in the large m limit which we need to consider for the continuum theory.

<u>3.5</u> Let us denote by $\Lambda(W)$ the Grassmann algebra over W. For a given expectation value $\langle \ \rangle$ we can define an isomorphism

$$: \ : \quad \Lambda(W) \longmapsto A(W)$$
$$\psi \qquad\qquad\qquad \psi$$
$$\lambda \qquad\qquad\qquad :\lambda:$$

called the normal product as follows. First we define the left and the right derivatives in $\Lambda(W)$:

$$R_w(w_n \cdots w_1) = \sum_{j=1}^{n} (-)^{j-1} w_n \cdots \langle w_j w \rangle \cdots w_1 ,$$

$$L_w(w_1 \cdots w_n) = \sum_{j=1}^{n} (-)^{j-1} w_1 \cdots \langle w w_j \rangle \cdots w_n ,$$

where $w_1, \cdots, w_n \in W \subset \Lambda(W)$. We define the normal product recursively by

$$:w\lambda: \ = w:\lambda: \ - \ :L_w(\lambda):$$

or

$$:\lambda w: \ = \ :\lambda:w \ - \ :R_w(\lambda): \ .$$

Especially, we have

$$w : e^\rho : \; = \; :(w+L_W(\rho))e^\rho :, \tag{35}$$

$$: e^\rho : w \; = \; :(w+R_W(\rho))e^\rho : . \tag{36}$$

These formulas with $\rho \in \Lambda^2 W$ will be exploited shortly.

<u>3.6</u> Let g be an element of G_+ such that $\langle g \rangle \neq 0$. Let T be the matrix representation of T_g given in (23) with respect to the basis $\psi_1, \cdots, \psi_{\dim W}$. This is equivalent to say

$$g\psi_k = \sum_j \psi_j g^T{}_{jk}.$$

We assume that g is of the form

$$g = \langle g \rangle : e^\rho :,$$

$$\rho = \frac{1}{2} \sum_{j,k} R_{jk} \psi_j \psi_k,$$

where $R = -{}^t R$. Then, from (35) and (36) we have

$$1 + RK = T - R\,{}^t KT.$$

Namely, we have

$$R = (T-1)({}^t KT+K)^{-1}. \tag{37}$$

Thus one can compute the normal product form of g if one can invert ${}^t KT+K$.

The following simple example is instructive to understand the above formula. We consider

$$W = \mathbb{C}p \oplus \mathbb{C}q$$

with $p^2 = -q^2 = 1$ and $pq = -qp$. We define the expectation value $\langle\ \rangle$ by

$$K = \begin{pmatrix} \langle pp \rangle & \langle pq \rangle \\ \langle qp \rangle & \langle qq \rangle \end{pmatrix} = \begin{pmatrix} 1 & 1 \\ -1 & -1 \end{pmatrix}.$$

Consider $g = pq$, then the formula (37) give

$$g = :e^{pq}: \; .$$

In fact, we can check that

$$:e^{pq}: \; = \; :1+pq: \; = pq.$$

3.7   Let us consider the disordered phase where $\alpha_2 > 1$. Note that the spin operator σ_m^z belongs to the odd part of the Clifford group. We define an even element μ_m^z by

$$\mu_m^z = \sigma_m^z p_m.$$

We compute the quadratic kernel R for μ_m^z by exploiting the formula (37). Then σ_m^z is obtained by the formula (35). We use $p(\theta)$ and $q(\theta)$ as the basis. The orthogonal transformation T corresponding to μ_0^z is characterized by

$$\mu_0^z \psi_m = \begin{cases} \psi_m \mu_0^z & m \geq 0 \\[2ex] -\psi_m \mu_0^z & m < 0 \end{cases}$$

where $\psi_m = p_m$ or q_m. If we set

$$P = (1-T)/2,$$

the kernel P in terms of $p(\theta)$ and $q(\theta)$ reads as

$$P(\theta,\theta') = \frac{e^{-i(\theta-\theta')}}{1-e^{-i(\theta-\theta'-i0)}} \begin{pmatrix} 1 & \\ & 1 \end{pmatrix} . \tag{38}$$

Namely, P is an operator which projects a function on the unit circle onto its negative component of Laurent expansion. The following lemma is useful to invert ${}^t KT+K$.

Lemma. We set

$$E = J^{-1}(K - {}^{t}K).$$

If we find $X_{\pm}$ such that

$$PX_{+}(1-P) = 0, \tag{39}$$

$$(1-P)X_{-}P = 0, \tag{40}$$

$$X_{-} = X_{+}E,$$

then we have

$$R = -2X_{-}^{-1}PX_{+}J^{-1}. \tag{41}$$

For P of (38), (39) and (40) is satisfied if X_{+} (resp. X_{-}) is a multiplication operator of a function of positive (resp. negative) Laurent expansion. In fact, we have

$$E(\theta,\theta') = \begin{bmatrix} & -a(\theta) \\ -a(-\theta) & \end{bmatrix} 2\pi\delta(\theta-\theta')$$

where $a(\theta)$ is splitted as

$$a(\theta) = b(\theta)/b(-\theta),$$

$$b(\theta) = \sqrt{(1-\alpha_{1}e^{i\theta})(1-\alpha_{2}^{-1}e^{i\theta})} \quad .$$

Therefore we have

$$X_{+} = \begin{bmatrix} 1/b(\theta) & \\ & b(\theta) \end{bmatrix} \quad ,$$

$$X_{-} = \begin{bmatrix} & -1/b(-\theta) \\ -b(-\theta) & \end{bmatrix} \quad .$$

Finally, from (41) we have

$$R(\theta,\theta') = \frac{e^{-i(\theta+\theta')}}{1-e^{-i(\theta+\theta')}} \begin{bmatrix} & -\dfrac{b(-\theta')}{b(-\theta)} \\ \dfrac{b(-\theta)}{b(-\theta')} & \end{bmatrix} \quad .$$

By using (35) we have the following formula for σ_0^z.

$$\sigma_0^z = \langle \mu_0^z \rangle : \psi_0 e^{\rho_0} : \ ,$$

$$\rho_0 = \frac{1}{2} \iint \frac{d\theta}{2\pi} \frac{d\theta'}{2\pi} \frac{e^{-i(\theta+\theta')}}{1-e^{-i(\theta+\theta'-i0)}} \frac{b(-\theta)}{b(-\theta')} q(\theta)p(\theta'), \qquad (42)$$

$$\psi_0 = \int \frac{d\theta}{2\pi} \frac{p(\theta)}{b(\theta)} \ . \qquad (43)$$

The m independent constant $\langle \mu_m^z \rangle$ can be computed by the formula

$$\lim_{m\to\infty} \langle \mu_m^z \mu_0^z \rangle = \langle \mu_m^z \rangle^2 \ ,$$

and by using Szego's theorem. The result is

$$\langle \mu_m^z \rangle = (1-\sinh^2 2K_1 \sinh^2 2K_2)^{1/8} (\cosh K_1)^{-1}.$$

<u>3.9</u>   For the computation of correlation functions (9) we need

$$\sigma_{mn}^z = V^n \sigma_m^z V^{-n}.$$

The normal product form of σ_{mn}^z is immediately given once we introduce the creation and the annihilation operators which diagonalize
T_V:

$$p(\theta) = \sqrt{a(\theta)}^{-1}(\psi^\dagger(-\theta) + \psi(\theta)) \ ,$$

$$q(\theta) = \sqrt{a(\theta)} (-\psi^\dagger(-\theta) + \psi(\theta)) \ .$$

Then we have

$$T_H \psi(\theta) = e^{i\theta} \psi(\theta) \ ,$$

$$T_V \psi(\theta) = e^{\gamma(\theta)} \psi(\theta),$$

$$T_H \psi^\dagger(\theta) = e^{-i\theta} \psi^\dagger(\theta),$$

$$T_V \psi^\dagger(\theta) = e^{-\gamma(\theta)} \psi^\dagger(\theta).$$

Here T_H denotes the horizontal translation $p_m \mapsto p_{m+1}$, $q_m \mapsto q_{m+1}$. The formula for σ^z_{mn} is given by rewriting (42), (43) in terms of $\psi(\theta)$ and $\psi^\dagger(\theta)$ and by replacing them by $e^{im\theta + n\gamma(\theta)}\psi(\theta)$ and $e^{-im\theta - n\gamma(\theta)}\psi^\dagger(\theta)$, respectively. The result is neatly written by introducing the elliptic curve of the mass shell. We mean by the mass shell the set of eigenvalues of T_H and T_V:

$$M = M_+ \cup M_- \, ,$$

$$M_\pm = \{(z,\omega)\,|\,z = e^{\pm i\theta}, \ \omega = e^{\pm\gamma(\theta)}, \ \theta \in \mathbb{R}/2\pi\mathbb{Z}\}.$$

In the analogy to the quantum field theory it is natural to consider that $\psi(\theta)$ and $\psi^\dagger(\theta)$ live on M_+ and M_-, respectively. The complexification of M, which we denote by $M^{\mathbb{C}}$, is an elliptic curve defined by (See (31).)

$$\sinh 2K_1(z+z^{-1}) + \sinh 2K_2(\omega+\omega^{-1}) = 2\cosh 2K_1 \cosh 2K_2.$$

The analytic coordinate U on $M^{\mathbb{C}}$ is defined by integrating the formula of the abelian differential,

$$dU = \frac{dz}{\pi i z(\omega - \omega^{-1})} = \frac{d\theta}{2\pi \sinh\gamma(\theta)} \, .$$

We define the free fermion operator on M by

$$\psi(U) = \begin{cases} \sqrt{\sinh\gamma(\theta)} \ \psi(\theta) & U \in M_+ \\[2ex] \sqrt{\sinh\gamma(\theta)} \ \psi^\dagger(\theta) & U \in M_- \, . \end{cases}$$

Then we have

$$\langle\psi(U)\psi(U')\rangle = \begin{cases} \delta(U, \iota(U')) & U \in M_+, \ U' \in M_- \\[2ex] 0 & \text{otherwise} \end{cases} .$$

Here ι denote the involution of $M^{\mathbb{C}}$

$$U = (z,\omega) \longmapsto \iota(U) = (z^{-1}, \omega^{-1}),$$

and the delta function $\delta(U,U')$ is characterized by

$$\int_M dU\,\delta(U,\imath(U'))f(U)g(U') = f(\imath(U'))g(U').$$

With this preparation the spin operator σ_{mn}^z for $T > T_c$ is given by

$$\sigma_{mn}^z = (1-\sinh^2 2K_1\sinh^2 2K_2)^{1/8}: \psi_{mn}e^{\rho_{mn}}: ,$$

$$\psi_{mn} = \frac{1}{\sqrt{\sinh 2K_2}} \int_M dU\, z^m\omega^n\psi(U),$$

$$\rho_{mn} = \frac{1}{2}\iint_{M\times M} dU dU' \frac{\omega-\omega'}{1-z^{-1}z'^{-1}} (zz')^m(\omega\omega')^n\psi(U)\psi(U'). \tag{44}$$

<u>3.10</u>   For $T < T_c$ the correlation function depends on the choice of boundary condition. We consider a finite lattice of $2M+1$ sites in the horizontal direction and define

$$\mu_{mn}^z = V^n\sigma_{-M}^z\sigma_m^z V^{-n}.$$

This means to impose the boundary condition

$$\mu_{-Mn}^z = 1.$$

Then we have

$$\mu_{mn}^z = (1-\sinh^{-2}2K_1\sinh^{-2}2K_2)^{1/8}:e^{\rho_{mn}}: ,$$

with ρ_{mn} given by (44).

<u>3.11</u>   From (44) we have

$$\langle\mu_{00}^z\mu_{mn}^z\rangle = (1-\sinh^{-2}2K_1\sinh^{-2}2K_2)^{1/4} \times$$

$$\times (1 - \frac{1}{2}\iint_{M_+\times M_+} dU_1 dU_2 \left(\frac{\omega_1-\omega_2}{1-z_1^{-1}z_2^{-1}}\right)^2 (z_1z_2)^{-m-1}(\omega_1\omega_2)^{-n-1}+\dots) \tag{45}$$

This expansion is suitable when we consider the large distance limit $m,n \to \infty$. In fact, on M_+ we have $|\omega| < 1$. We can deform the integration contour M_+ so that we have $|z| < 1$ too. Therefore the second term in (45) damps exponentially. In other words, at a fixed temperature $T \neq T_c$, only the first term will contribute in the continuum limit where $m,n \to \infty$, the lattice unit $\varepsilon \to 0$ and $m\varepsilon$, $n\varepsilon$ fixed.

Non trivial limit is obtained by scaling the temperature T together with ε so that $|T-T_c| = \varepsilon$. More precisely, for $T > T_c$ we consider the following limit.

$$\alpha_1 = \alpha + \varepsilon,$$

$$\alpha_2 = 1 + \varepsilon,$$

$$n = x^0/\sqrt{-1}\,\kappa\varepsilon, \tag{46}$$

$$m = x^1/\varepsilon.$$

We put $\sqrt{-1}\,\kappa$ in (46) in order that (x^0, x^1) constitutes the coordinate in the 2 dimensional Minkowski space. We set

$$x^{\pm} = \frac{x^1 \pm x^1}{2}.$$

The coordinate $(z,\omega) = (e^{\pm i\theta}, e^{\pm \gamma(\theta)})$ on M is scaled to $(\pm p, \pm\sqrt{p^2+1})$ by

$$\theta = p\varepsilon.$$

We set

$$u = \pm\sqrt{p^2+1} + p$$

and

$$\psi(u) = \begin{cases} (\sinh \gamma(\theta)/\kappa)^{\frac{1}{2}}\ \psi(\theta)\ , \\ (\sinh \gamma(\theta)/\kappa)^{\frac{1}{2}}\ \psi^{\dagger}(-\theta) \end{cases}$$

for $u \gtrless 0$. Then we have

$$[\psi(u),\ \psi(u')]_+ = 2\pi|u|\delta(u+u').$$

The spin operator is scaled as

$$\sigma^z_{mn} \longrightarrow \left(\frac{2\epsilon(1+\alpha)}{1-\alpha}\right)^{1/8}\sigma(x),$$

where

$$\sigma(x) = {:}\phi(x)e^{\rho(x)}{:} \ , \tag{47}$$

$$\phi(x) = \int \frac{du}{2\pi|u|} \ e^{-i(x^-u+x^+u^{-1})}\psi(u), \tag{48}$$

$$\rho(x) = \frac{1}{2} \iint \frac{du}{2\pi|u|} \frac{du'}{2\pi|u'|} \frac{-i(u-u')}{u+u'-i0} \ e^{-i(x^-(u+u')+x^+(u^{-1}+u'^{-1}))} \times$$

$$\times\psi(u)\psi(u'). \tag{49}$$

Similarly, for $T < T_c$ the spin operator μ^z_{mn} is scaled to

$$\mu(x) = {:}e^{\rho(x)}{:} \ . \tag{50}$$

The free fermions p_m and q_m are scaled to the free Majorana fermion in such a way

$$V^n p_m V^{-n} = \sqrt{\frac{\epsilon}{2}} \ (\psi_+(x)+\psi_-(x)),$$

$$V^n q_m V^{-n} = -i\sqrt{\frac{\epsilon}{2}} \ (\psi_+(x)-\psi_-(x)),$$

where

$$\psi_\pm(x) = \int \frac{du}{2\pi|u|} \ \sqrt{0+iu}^{\pm 1} \ e^{-i(x^-u+x^+u^-)}\psi(u). \tag{51}$$

<u>4.1</u> Our main goal is to characterize the correlation functions $\langle\sigma(a_1)\cdots\sigma(a_n)\rangle$ or $\langle\mu(a_1)\cdots\mu(a_n)\rangle$ in terms of a system of non linear differential equations. The idea is to consider the wave functions

$$w_\pm^{(j)}(x) = \frac{\langle\psi_\pm(x)\mu^{(j)}(a_1)\cdots\mu^{(j)}(a_n)\rangle}{\langle\mu(a_1)\cdots\mu(a_n)\rangle} \ .$$

Here we use the following notation:

$$\mu^{(j_1,\cdots,j_\ell)}(a_k) = \begin{cases} \sigma(a_k) & \text{if } k \in \{j_1,\cdots,j_\ell\} \\ \mu(a_k) & \text{if } k \notin \{j_1,\cdots,j_\ell\} \end{cases} \ .$$

Note that

$$\begin{pmatrix} 1 & \partial_0 - \partial_1 \\ -\partial_0 - \partial_1 & 1 \end{pmatrix} \begin{pmatrix} \psi_+(x) \\ \psi_-(x) \end{pmatrix} = 0.$$

Therefore as far as $w_\pm^{(j)}(x)$ is well-defined we have

$$\begin{pmatrix} 1 & \partial_0 - \partial_1 \\ -\partial_0 - \partial_1 & 1 \end{pmatrix} \begin{pmatrix} w_+^{(j)}(x) \\ w_-^{(j)}(x) \end{pmatrix} = 0. \tag{52}$$

The fact is $w_\pm^{(j)}(x)$ has singularities of square root type at $x^+ = a^+$ and $x^- = a^-$, the $2n$ vector

$$\vec{w}_{op}(x) = \begin{pmatrix} w_{op}^{(1)}(x) \\ \cdot \\ \cdot \\ \cdot \\ w_{op}^{(n)}(x) \end{pmatrix} , \qquad w_{op}^{(j)}(x) = \begin{pmatrix} w_+^{(j)}(x) \\ w_-^{(j)}(x) \end{pmatrix}$$

satisfies a system of linear differential equations containing (52), and the consistency condition for the linear system gives the non linear system.

4.2   In order to derive the linear system a crucial role is played by the following topological commutation relations between the spin fields σ, μ and the free fields $\psi_\pm$:

$$\sigma(x)\psi_\pm(x') = \begin{cases} \psi_\pm(x')\sigma(x) & \text{if } x^+ > x'^+ \text{ and } x^- < x'^- \\[2ex] -\psi_\pm(x')\sigma(x) & \text{if } x^+ < x'^+ \text{ and } x^- > x'^- \end{cases} , \tag{53}$$

$$\mu(x)\psi_\pm(x') = \begin{cases} -\psi_\pm(x')\mu(x) & \text{if } x^+ > x'^+ \text{ and } x^- < x'^- \\[2ex] \psi_\pm(x')\mu(x) & \text{if } x^+ < x'^+ \text{ and } x^- > x'^- \end{cases} . \tag{54}$$

Let us consider the analytic continuation of wave functions in the Euclidean plane, where x^0 is purely imaginary. For simplicity we consider the case $n = 1$:

$$w_{\pm}(x) = \langle \psi_{\pm}(x)\sigma(a)\rangle \ .$$

We have

$$w_{\pm}(x) = \int_0^{\infty} \frac{du}{2\pi|u|} \ \sqrt{0+iu}^{\pm 1} \ e^{-i((x-a)^- u + (x-a)^+ u^{-1})} \ .$$

The integral is convergent if $\mathrm{Im}(x^0 - a^0) < 0$. Similarly,

$$w'_{\pm}(x) = \langle \sigma(a)\psi_{\pm}(x)\rangle,$$

$$= \int_{-\infty}^{0} \frac{du}{2\pi|u|} \ \sqrt{0+iu}^{\pm 1} \ e^{-i((x-a)^- u + (x-a)^+ u^{-1})},$$

is convergent if $\mathrm{Im}(x^0 - a^0) > 0$. In fact, by modifying the integration contour one can show that $w_{\pm}(x)$ and $w'_{\pm}(x)$ are real analytically continuable across the boundary $\mathrm{Im}(x^0 - a^0) = 0$, except at $x=a$. Now, the relation (53) tells us that

$$w_{\pm}(x) = \begin{cases} w'_{\pm}(x) & \mathrm{Im}(x^0 - a^0) = 0, \quad x^1 < a^1 \\[2ex] -w'_{\pm}(x) & \mathrm{Im}(x^0 - a^0) = 0, \quad x^1 > a^1 \end{cases} \ .$$

Therefore we conclude that the Euclidean continuation of $w_{\pm}(x)$ has a branch point at $x=a$ and when x goes around it once $w_{\pm}(x)$ changes its sign. This property of $w_{\pm}(x)$ is called the monodromy property.

<u>4.3</u> Here we prepare several notations and formulas concerning the Euclidean Dirac equation.

We set

$$z = \frac{x^1 - x^0}{2} \ ,$$

$$\bar{z} = \frac{x^1 + x^0}{2} \ .$$

We denote by Γ the 2×2 matrix of differential operator

$$\Gamma = \begin{pmatrix} & \partial \\ -\bar{\partial} & \end{pmatrix} \ ,$$

where $\partial = \dfrac{\partial}{\partial z}$ and $\bar{\partial} = \dfrac{\partial}{\partial \bar{z}}$. We also set

$$M = z\partial - \bar{z}\bar{\partial} + \frac{1}{2}\begin{pmatrix} 1 & \\ & -1 \end{pmatrix} \ . \tag{55}$$

Then we have

$$[M,\Gamma] = 0. \tag{56}$$

Consider the following system of linear equations.

$$Mw = \ell w, \tag{57}$$

$$(\Gamma - 1)w = 0. \tag{58}$$

The first equation (57) implies

$$w_{\pm} = e^{i(\ell \mp \frac{1}{2})\theta} f_{\pm}(r), \tag{59}$$

where $z = re^{i\theta}/2$. If $\ell \in Z$ such $w_{\pm}$ has the same monodromy property as the wave functions considered in 4.2. Substituting (59) into (58) we have

$$[\frac{d^2}{dr^2} + \frac{1}{r}\frac{d}{dr} - (1 + \frac{(\ell \mp \frac{1}{2})^2}{r^2})]f_{\pm}(r) = 0.$$

Solutions are given by the modified Bessel functions.

$$f_{\pm}(r) = cI_{\ell \mp \frac{1}{2}}(r) + c*I_{-(\ell \mp \frac{1}{2})}(r).$$

Therefore we have two independent solutions to the system (57), (58):

$$w_{\ell} = \begin{pmatrix} e^{i(\ell - \frac{1}{2})\theta} I_{\ell - \frac{1}{2}}(r) \\ e^{i(\ell + \frac{1}{2})\theta} I_{\ell + \frac{1}{2}}(r) \end{pmatrix},$$

$$w^{*}_{-\ell} = \begin{pmatrix} e^{i(\ell - \frac{1}{2})\theta} I_{-\ell + \frac{1}{2}}(r) \\ e^{i(\ell + \frac{1}{2})} I_{-\ell - \frac{1}{2}}(r) \end{pmatrix}.$$

For $r \to 0$ we have

$$I_{\ell}(r) \sim \frac{r^{\ell}}{\ell!}.$$

Therefore for $z, \bar{z} \to 0$ we have

$$w_\ell \sim \frac{z^{\ell-\frac{1}{2}}}{(\ell-\frac{1}{2})!} \begin{pmatrix} 1 \\ 0 \end{pmatrix} ,$$

$$w_\ell^* \sim \frac{\bar{z}^{\ell-\frac{1}{2}}}{(\ell-\frac{1}{2})!} \begin{pmatrix} 0 \\ 1 \end{pmatrix} .$$

We also note that

$$\partial w_\ell = w_{\ell-1}, \quad \bar{\partial} w_\ell = w_{\ell+1} , \tag{60}$$

$$\partial w^* = w_{\ell+1}^*, \quad \bar{\partial} w^* = w_{\ell-1}^* . \tag{61}$$

<u>4.4</u>   The topological commutation relations (53), (54) are insuf-
ficient to characterize the singular behavior of wave functions
completely, because the short distance behavior, when $x \to a$, of
$\sigma(x)\psi_\pm(a)$ or $\mu(x)\psi_\pm(a)$ is not specified. A detailed computation
using (35), (36) and (47) $\sim$ (51) shows the following operator product
expansion.

$$\begin{pmatrix} \psi_+(x)\sigma(a) \\ \psi_-(x)\sigma(a) \end{pmatrix} = \tfrac{i}{2}\mu(a)(w_0(x-a)-w_0^*(x-a))$$

$$+ \sum_{j=1}^\infty \mu_j(a)w_j(x-a) - \mu_{-j}(a)w_j^*(x-a), \tag{62}$$

$$\begin{pmatrix} \psi_+(x)\mu(a) \\ \psi_-(x)\mu(a) \end{pmatrix} = \tfrac{1}{2}\sigma(a)(w_0(x-a)+w_0^*(x-a))$$

$$+ \sum_{j=1}^\infty \sigma_j(a)w_j(x-a) + \sigma_{-j}(a)w_j^*(x-a). \tag{63}$$

Here we set

$$\phi_\ell(x) = \int \frac{du}{2\pi|u|}(0+iu)^\ell e^{-i(x^-u+x^+u^-)}\psi(u),$$

$$\sigma_\ell(x) = :\phi_\ell(x)e^{\rho(x)}: ,$$

$$\mu_\ell(x) = \; :\phi_\ell(x)\phi_0(x)e^{\rho(x)}: \; .$$

From $(47)\sim(50)$ we have

$$\partial\mu(x) = -i\mu_1(x), \quad \overline{\partial}\mu(x) = i\mu_{-1}(x), \tag{64}$$

$$\partial\sigma(x) = \sigma_1(x), \quad \overline{\partial}\sigma(x) = \sigma_{-1}(x). \tag{65}$$

Comparing (64) and (65) with (62) and (63) we have

$$w_{op}^{(j)}(x) = \tfrac{1}{2}(w_0(x-a_j)-w_0^*(x-a_j)) - i(\partial_j \log<\mu(a_1)\cdots\mu(a_n)>w_1(x-a_j)$$

$$+ \overline{\partial}_j \log<\mu(a_1)\cdots\mu(a_n)>w_1^*(x-a_j)) + \cdots \quad \text{for} \quad x \to a_j, \tag{66}$$

$$= \tfrac{1}{2}\,\tau_{jk}(w_0(x-a_k)+w_0^*(x-a_k)) + \cdots$$

$$\text{for} \quad x \to a_k \quad (j \neq k), \tag{67}$$

where

$$\tau_{jk} = \frac{<\mu^{(j,k)}(a_1)\cdots\mu^{(j,k)}(a_n)>}{<\mu(a_1)\cdots\mu(a_n)>} \; .$$

Therefore, the behavior of wave functions at branch points contains
all the necessary informations about the correlation functions.

<u>4.5</u> We denote by $W_{a_1,\cdots a_n}$ the space of wave functions satisfying
the following properties.

$$(\Gamma-1)w = 0, \tag{68}$$

$$w = \sum_{j=0}^{\infty} (c_j^{(k)}w_j(x-a_k)+c_j^{*(k)}w_j^*(x-a_k)) \qquad \text{at} \quad x = a_k, \tag{69}$$

$$|w| = O(e^{-2|x|}) \qquad \text{at } x = \infty. \tag{70}$$

For $w,w' \in W_{a_1,\cdots a_n}$ $w_+\overline{w}_+'+w_-\overline{w}_-'$ is a single-valued function in the
Euclidean space. Thus we can define a positive definite inner product
$<w,w'>$ by

$$<w,w'> = \tfrac{1}{2}\int idzd\overline{z}(w_+\overline{w}_-'+w_-\overline{w}_-').$$

Using the Stokes formula we have

$$<w,w'> = -\sum_{j=1}^{n} c_0^{(j)} \overline{c_0^{*',(j)}}.$$

This implies that if $c_0^{(j)} = 0$ $(j = 1,\cdots,n)$ then $w = 0$. In particular, the dimension of $W_{a_1,\cdots,a_n}$ is less than or equal to n. In fact, we can prove that

$$\dim W_{a_1,\cdots,a_n} = n.$$

<u>4.6</u> The canonical basis $(w_{can}^{(1)},\cdots,w_{can}^{(n)})$ for $W_{a_1,\cdots,a_n}$ is such that

$$w_{can}^{(j)} = \delta_{jk}w_0(x-a_k) + \alpha_{jk}w_1(x-a_k) + \cdots$$

$$+ \beta_{jk}w_0^*(x-a_k) + \cdots , \tag{71}$$

at $x = a_k$. One can prove that

$$\alpha = {}^t\alpha,$$

$$\beta = {}^t\overline{\beta} = \overline{\beta}^{-1} : \text{ negative definite.}$$

For a basis $(w^{(1)},\cdots,w^{(n)})$ of $W_{a_1,\cdots,a_n}$ we define $n\times n$ matrices $C = (c_{jk})$ and $C^* = (c_{jk}^*)$ by

$$w^{(j)} = c_{jk}w_0(x-a_k) + \cdots + c_{jk}^*w_0^*(x-a_k) + \cdots , \tag{72}$$

at $x = a_k$. For the canonical basis

$$C = 1, \quad C^* = \beta. \tag{73}$$

For the operator basis $(w_{op}^{(1)},\cdots,w_{op}^{(n)})$

$$C = \tfrac{i}{2}(1-T), \quad C^* = \tfrac{-i}{2}(1+T), \tag{74}$$

where

$$iT = \begin{pmatrix} 0 & \tau^{12} & \cdots & \tau^{1n} \\ -\tau^{12} & 0 & \cdots & \tau^{2n} \\ \vdots & \vdots & & \vdots \\ -\tau^{1n} & -\tau^{2n} & \cdots & 0 \end{pmatrix}.$$

Comparing (73) and (74) we have

$$T = (\beta+1)(\beta-1)^{-1}. \tag{75}$$

Note that Wick's theorem implies that

$$\frac{<\sigma(a_1)\cdots\sigma(a_n)>}{<\mu(a_1)\cdots\mu(a_n)>} = \text{Pfaffian }(iT). \tag{76}$$

<u>4.8</u> Differentiating (72) and using (60), (61) we have

$$\partial w^{(j)} = c_{jk}w_{-1}(x-a_k) + \cdots + c_{jk}^*w_1^*(x-a_k) + \cdots , \tag{77}$$

$$\overline{\partial}w^{(j)} = c_{jk}w_1(x-a_k) + \cdots + c_{jk}^*w_{-1}^*(x-a_k) + \cdots , \tag{78}$$

at $x = a_k$. Now, consider $Mw^{(j)}$, where M is defined in (55). The commutativity (56) implies that $Mw^{(j)}$ satisfies (68), (70), and (69) with the summation from $j = -1$ to ∞. By subtracting a suitable linear combination of (72), (77) and (78) from $Mw^{(j)}$ one can cancel the terms $w_{-1}(x-a_k)$, $w_{-1}^*(x-a_k)$ and $w_0(x-a_k)$ at $x = a_k$. Namely, there exist $n \, n$ matrices $B = (b_{jk})$, $B^* = (b_{jk}^*)$ and $E = (e_{jk})$ such that

$$Mw^{(j)} - \sum_k (b_{jk}\partial w^{(k)}+b_{jk}^*\overline{\partial}w^{(k)}+e_{jk}w^{(k)}) \tag{79}$$

belongs to $W_{a_1,\cdots,a_n}$ and the expansion at $x = a_k$ does not contain $w_0(x-a_k)$. This means that (79) vanishes. Therefore we can conclude the following linear differential equation.

$$M\vec{w} = (B\partial+B^*\overline{\partial}+E)\vec{w}. \tag{80}$$

We use the following notations.

$$a_j = (a_j^1 - a_j^0)/2,$$

$$\overline{a_j} = (a_j^1 + a_j^0)/2,$$

$$df = \sum_{j=1}^{n} \frac{\partial f}{\partial a_j} da_j + \sum_{j=1}^{n} \frac{\partial f}{\partial \overline{a_j}} d\overline{a_j} \ .$$

In the Euclidean plane a_j^1 is real and a_j^0 is purely imaginary. Since $[\frac{\partial}{\partial a_j}, \Gamma] = [\frac{\partial}{\partial \overline{a_j}}, \Gamma] = 0$, a similar argument leads us to the following equation.

$$d\vec{w} = (\Phi\partial + \Phi^*\overline{\partial} + \Psi)\vec{w}, \tag{81}$$

where Φ, Φ^*, Ψ are 1 forms in $a_1, \cdots, a_n, \overline{a_1}, \cdots, \overline{a_n}$.

<u>4.9</u> For $\vec{w} = w_{can}$ (71) the linear system (68) (80) and (81) read as

$$(\Gamma - 1)\vec{w}_{can} = 0 \tag{82}$$

$$M\vec{w}_{can} = (A\partial - \beta\overline{A}\beta^{-1}\overline{\partial} + \gamma)\vec{w}_{can}, \tag{83}$$

$$d\vec{w}_{can} = (-dA\partial - \beta d\overline{A}\beta^{-1}\overline{\partial} + \theta)\vec{w}_{can}, \tag{84}$$

where

$$A = \begin{pmatrix} a_1 & & \\ & \ddots & \\ & & a_n \end{pmatrix}, \qquad \overline{A} = \begin{pmatrix} a_1 & & \\ & \ddots & \\ & & \overline{a_n} \end{pmatrix} ,$$

$$\gamma = [\alpha, A], \qquad \theta = -[\alpha, dA].$$

Computing the consistency condition of (82), (83) and (84) we have the following non linear system

$$d\beta = \theta\beta + \beta\theta^*, \tag{85}$$

$$d\gamma = [\theta, \gamma] + [dA, \beta\overline{A}\beta^{-1}] + [A, \beta d\overline{A}\beta^{-1}], \tag{86}$$

where

$$[\theta, A] + [\gamma, dA] = 0, \tag{87}$$

$$[\theta^*, \overline{A}] + [\beta^{-1}\gamma\beta, d\overline{A}] = 0, \tag{88}$$

$$\text{diag } \theta = \text{diag } \theta^* = 0. \tag{89}$$

By using (66), (67) and (75) we derive the formula

$$d \log \langle \mu(a_1) \cdots \mu(a_n) \rangle$$

$$= \frac{1}{4} \text{ trace}((T-\gamma)\theta + (T-\beta^{-1}\gamma\beta)\theta^*)$$

$$+ \frac{1}{2} \text{ trace}(d(A\overline{A}) - \beta\overline{A}\beta^{-1}dA - \beta^{-1}A\beta d\overline{A}). \tag{90}$$

The formulas $(85)\sim(89)$ and $(76),(90)$ are our result which characterizes the correlation functions.

For $n = 2$ we have

$$\beta = \begin{pmatrix} -\cosh \psi & i \sinh \psi \\ -i \sinh \psi & -\cosh \psi \end{pmatrix},$$

$$\gamma = \begin{pmatrix} & if \\ -if & \end{pmatrix}.$$

If we set

$$a_1 - a_2 = te^{i\theta}/2,$$

the non linear system (85), (86) reduces to

$$\frac{\partial f}{\partial \theta} = \frac{\partial \psi}{\partial \theta} = 0,$$

$$f = \frac{t}{2} \frac{\partial \psi}{\partial t},$$

$$\frac{\partial}{\partial t}\left(t\frac{\partial \psi}{\partial t}\right) = \frac{t}{2} \sinh 2\psi.$$

<u>5.1</u>   Let us go back to Hirota's transformation discussed in 1.3. Substituting (4) into (3) we obtain

$$2(f_{\xi\eta}f - f_{\xi}f_{\eta} + g_{\xi\eta}g - g_{\xi}g_{\eta})fg - (f_{\xi\eta}g + fg_{\xi\eta} - f_{\xi}g_{\eta} - f_{\eta}g_{\xi} - fg)(f^2 + g^2) = 0.$$

Therefore, if f and g satisfy

$$f_{\xi\eta}f - f_{\xi}f_{\eta} + g_{\xi\eta}g - g_{\xi}g_{\eta} = 0, \tag{91}$$

$$f_{\xi\eta}g + fg_{\xi\eta} - f_{\xi}g_{\eta} - f_{\eta}g_{\xi} - fg = 0, \tag{92}$$

then u solves (3). We can check that

$$f = <\sigma(x)\sigma(0)> \ ,$$

$$g = <\mu(x)\mu(0)>,$$

satisfy (91), (92) if we set

$$\xi = -x^{-}, \quad \eta = x^{+}.$$

Let us consider (92). We have

$$\partial_{\xi}\sigma(x) = :\phi_{1}(x)e^{\rho(x)}: \ ,$$

$$\partial_{\eta}\sigma(x) = :\phi_{-1}(x)e^{\rho(x)}: \ ,$$

$$\partial_{\xi}\partial_{\eta}\sigma(x) = :(\phi_{0}(x)+i\phi_{1}(x)\phi_{-1}(x)\phi_{0}(x))e^{\rho(x)}: \ ,$$

$$\partial_{\xi}\mu(x) = -i:\phi_{1}(x)\phi_{0}(x)e^{\rho(x)}: \ ,$$

$$\partial_{\eta}\mu(x) = i:\phi_{-1}(x)\phi_{0}(x)e^{\rho(x)}: \ ,$$

$$\partial_{\xi}\partial_{\eta}\mu(x) = -i:\phi_{1}(x)\phi_{-1}(x)e^{\rho(x)}: \ .$$

Now the left hand side of (92) reads as

$$-i<:\phi_{1}(x)\phi_{-1}(x)e^{\rho(x)}:\mu(0)><:\phi_{0}(x)e^{\rho(x)}:\sigma(0)>$$

$$+ <:e^{\rho(x)}:\mu(0)><:(\phi_{0}(x)+i\phi_{1}(x)\phi_{-1}(x)\phi_{0}(x))e^{\rho(x)}:\sigma(0)>$$

$$+ i<:\phi_{1}(x)\phi_{0}(x)e^{\rho(x)}:\mu(0)><:\phi_{-1}(x)e^{\rho(x)}:\sigma(0)>$$

$$- i < :\phi_{-1}(x)\phi_0(x)e^{\rho(x)}:\mu(0)> < :\phi_1(x)e^{\rho(x)}:\sigma(0)>$$

$$- < :e^{\rho(x)}:\mu(0)> < :\phi_0(x)e^{\rho(x)}:\sigma(0)> \ . \tag{93}$$

Wick's theorem gives the identity

$$< :e^{\rho(x)}:\mu(0)> < :\phi_1(x)\phi_{-1}(x)\phi_0(x)e^{\rho(x)}:\sigma(0)>$$

$$= < :\phi_{-1}(x)\phi_0(x)e^{\rho(x)}:\mu(0) > < :\phi_1(x)e^{\rho(x)}:\sigma(0)>$$

$$- < :\phi_1(x)\phi_0(x)e^{\rho(x)}:\mu(0)> < :\phi_{-1}(x)e^{\rho(x)}:\sigma(0)>$$

$$+ < :\phi_1(x)\phi_{-1}(x)e^{\rho(x)}:\mu(0)> < :\phi_0(x)e^{\rho(x)}:\sigma(0)> \ . \tag{94}$$

It is easy to see that (94) implies that (93) vanishes. Note that
the above argument is valid even if instead of $\mu(0)$, $\sigma(0)$ we use

$$\sigma = :\lambda e^{\mu}: \ , \qquad \mu = :e^{\mu}: \ ,$$

with any $\lambda \in W$ and $\mu \in \Lambda^2 W$. In this way we may have rather general
representation of solutions to (3) in terms of free fermions. In
what follows we elaborate on details in the case of the KdV and the
KP equation.

5.2   The KP equation

$$\frac{3}{4}\frac{\partial^2 u}{\partial y^2} = \frac{\partial}{\partial x}\left(\frac{\partial u}{\partial t} - \frac{3}{2}u\frac{\partial u}{\partial x} - \frac{1}{4}\frac{\partial^3 u}{\partial x^3}\right) \tag{95}$$

is the consistency condition of the linear system

$$\frac{\partial w}{\partial y} = \left(\frac{\partial^2}{\partial x^2} + u\right)w, \tag{96}$$

$$\frac{\partial w}{\partial t} = \left(\frac{\partial^3}{\partial x^3} + \frac{3}{2}u\frac{\partial}{\partial x} + v\right)w. \tag{97}$$

From (96) and (97) we have

$$\left(\frac{\partial u}{\partial t} - \left(\frac{3}{2}\frac{\partial u}{\partial y}\frac{\partial}{\partial x} + \frac{\partial v}{\partial y}\right) + \left[\frac{\partial^2}{\partial x^2} + u, \frac{\partial^3}{\partial x^3} + \frac{3}{2}u\frac{\partial}{\partial x} + v\right]\right)w = 0. \tag{98}$$

The consistency means that (98), which is an equation for w at fixed value of y and t, should be void. Therefore, we have

$$\frac{\partial v}{\partial x} = \frac{3}{4}(\frac{\partial u}{\partial y} + \frac{\partial^2 u}{\partial x^2}) \ ,$$

and the KP equation (95).

If we impose the condition

$$\frac{\partial u}{\partial y} = 0 \ ,$$

then (95) reduces to a simpler equation known as the KdV equation

$$\frac{\partial u}{\partial t} = \frac{3}{2}u\frac{\partial u}{\partial t} + \frac{1}{4}\frac{\partial^3 u}{\partial x^3} \ .$$

<u>5.3</u> Hirota's transformation for the KP equation is

$$u = - \frac{1}{2}\frac{\partial^2}{\partial x^2}\log \tau \ . \tag{99}$$

Then (95) is rewritten as

$$(D_x^4 + 3D_y 2 - 4D_x D_z)\tau \cdot \tau = 0,$$

where we used Hirota's bilinear notation which we explain below. Let
f and g be functions in x, y, etc., and let $P(D_x,D_y,\cdots)$ be a
polynomial in D_x, D_y, etc. Then we define

$$P(D_x,D_y,\cdots)f \cdot g$$

$$= P(\partial_a,\partial_b,\cdots)(f(x+a,y+b,\cdots)g(x-a,y-b,\cdots))\Big|_{a=b=\cdots=0},$$

where $\partial_a = \partial/\partial a$, $\partial_b = \partial/\partial b$, etc. Note that

$$P(D_x,D_y,\cdots)f \cdot g = P(-D_x,-D_y,\cdots)g \cdot f.$$

Therefore, if f = g as in the KP equation then only the even part
of P is meaningful. Consider Hirota's equation

$$P(D_x,D_y,\cdots)\tau \cdot \tau = 0, \tag{100}$$

with an even polynomial P. Hirota's ansatz for N soliton solutions
to (100) goes as follows. We choose linear phases

$$\xi_j = c_j + \lambda_j x + \mu_j y + \cdots , \quad (j = 1,\cdots,N) \tag{101}$$

such that

$$P(\lambda_j,\mu_j,\cdots) = 0.$$

For a subset $J \subset \{1,\cdots,n\}$ we define

$$\xi_J = \sum_{j \in J} \xi_j + \sum_{\substack{j,j' \in J \\ j<j'}} c_{jj'} ,$$

where the phase shift $c_{jj'}$ is given by

$$e^{c_{jj'}} = - \frac{P(\lambda_1-\lambda_{j'},\mu_j-\mu_{j'},\cdots)}{P(\lambda_j+\lambda_{j'},\mu_j+\mu_{j'},\cdots)} . \tag{102}$$

Then the candidate for solution is

$$\tau_N = \sum_{J \subset \{1,\cdots,n\}} e^{\xi_J} . \tag{103}$$

In fact, for any even P, τ_N solves (100) if $N \leq 2$. For the KP equation (101) and (102) read as

$$\xi_j = c_j + (p_j-q_j)x + (p_j^2-q_j^2)y + (p_j^3-q_j^3)t, \tag{104}$$

$$e^{c_{jj'}} = \frac{(p_j-p_{j'})(q_j-q_{j'})}{(p_j-q_{j'})(p_{j'}-q_j)} .$$

Then, the fact is that (103) gives a solution for arbitrary N.

$\underline{5.4}$ From (104) we naturally count degree of x,y,t as $1,2,3$, respectively. In general, we introduce a variable x_n of degree n. We change the notation so that we use x_1, x_2, x_3 instead of x,y,t, and we denote by x the totality of variables,

$$x = (x_1,x_2,x_3,\cdots).$$

We set

$$\xi(x,k) = \sum_{n=1}^{\infty} k^n x_n.$$

By modifying (101) into

$$\xi_j = c_j + \xi(x,p_j) - \xi(x,q_j),$$

we obtain the N soliton formula (103) containing infinitely many variables x. The fact is that this τ_N satisfies infinitely many Hirota's bilinear equations among which the following are the lowest and the next lowest.

$$(D_1^{\ 4} + 3D_2^{\ 2} - 4D_1D_3)\tau\cdot\tau = 0, \tag{105}$$

$$(D_1^{\ 3}D_2 + 2D_2D_3 - 3D_1D_4)\tau\cdot\tau = 0. \tag{106}$$

We call them the KP hierarchy of bilinear equations.

5.5 The KP hierarchy can be also introduced as the consistency conditions for a system of infinitely many linear equations among which (105) and (106) are the lowest and the next lowest. This is called the KP hierarchy of linear equations. In order to describe it neatly we prepare a formal calculus of pseudo differential operators in

$$\partial = \frac{\partial}{\partial x_1} \,.$$

A pseudo differential operator is a formal series of the form

$$P(x,\partial) = \sum_{-\infty < j \leq m} a_j(x)\partial^j.$$

The product of $P(x,\partial)$ and

$$Q(x,\partial) = \sum_{-\infty < j \leq n} b_j(x)\partial^j$$

is defined as

$$P(x,\partial)Q(x,\partial) = \sum_{j=0}^{\infty} \sum_{k,\ell} \binom{k}{j} a_k(x)\partial^j b_\ell(x)\partial^{k+\ell-j}.$$

The adjoint of $P(x,\partial)$ is defined as

$$P^*(x,\partial) = \sum_j (-\partial)^j a_j(x).$$

We use the following notation:

$$[P(x,\partial)]_+ = \sum_{j=0}^{m} a_j(x)\partial^j,$$

$$[P(x,\partial)]_- = \sum_{j=-\infty}^{-1} a_j(x)\partial^j.$$

Consider a formal series of the form

$$w(x,k) = (\sum_{-\infty<j\leq m} a_j(x)k^j)e^{kx_1}.$$

We use the convension to write

$$w(x,k) = P(x,\partial)e^{kx_1}.$$

Then, we can show that

$$\int \frac{dk}{2\pi i} P(x,\partial)e^{kx_1} \cdot Q(x',\partial')e^{-kx_1'} = 0, \tag{107}$$

if and only if

$$[P(x,\partial)Q^*(x,\partial)]_- = 0. \tag{108}$$

In (107) $\int \frac{dk}{2\pi i}$ means to take the coefficient of k^{-1}.

<u>5.6</u> Consider a pseudo differential operator

$$L(x,\partial) = \partial + u_2(x)\partial^{-1} + u_3(x)\partial^{-2} + \cdots ,$$

and the system of equations (110) and (111) for

$$w(x,k) = P(x,\partial)e^{\xi(x,k)} = (1 + \frac{w_1(x)}{k} + \frac{w_2(x)}{k^2} + \cdots)e^{\xi(x,k)}. \tag{109}$$

$$L(x,\partial)w(x,k) = kw(x,k). \tag{110}$$

$$\frac{\partial}{\partial x_n} w(x,k) = B_n(x,\partial)w(x,k), \tag{111}$$

where

$$B_n(x,\partial) = [L(x,\partial)^n]_+. \tag{112}$$

The consistency conditions of (110) and (111) read as

$$\frac{\partial L}{\partial x_n} = [B_n, L] \, , \tag{113}$$

$$\frac{\partial B_m}{\partial x_n} - \frac{\partial B_n}{\partial x_m} + [B_m, B_n] = 0. \tag{114}$$

For $n = 2,3$ (113) gives (96), (97). The equation (110) merely means

$$L(x,\partial) = P(x,\partial) \cdot \partial \cdot P(x,\partial)^{-1}.$$

The choice of $B_n(x,\partial)$ as in (112) is crucial. With this choice one can show that (114) follows from (113), and that

$$[[B_n, L]]_+ = 0,$$

which is a part of (113). Therefore, the net content of consistency is

$$\frac{\partial L}{\partial x_n} = [[B_n, L]]_- .$$

These are non linear evolution equations for the coefficients of L.

<u>5.7</u> We call $w(x,k)$ in (109) the wave function. We define the adjoint wave function $w^*(x,k)$ by

$$w^*(x,k) = Q(x,\partial)e^{-\xi(x,k)} = (1 + \frac{w_1^*(x)}{k} + \frac{w_2^*(x)}{k^2} + \cdots)e^{-\xi(x,k)}, \tag{115}$$

where

$$Q(x,\partial) = P^*(x,\partial)^{-1}. \tag{116}$$

Now we show that the KP hierarchy of linear differential equations (110) and (111) is rewritten into single bilinear identity.

Assume that $w(x,k)$ of the form (109) satisfies (111). Then $w(x,k)$ and $w^*(x,k)$ defined by (115) and (116) satisfies

$$\int \frac{dk}{2\pi i} w(x,k)w^*(x',k) = 0. \tag{117}$$

We call (117) the bilinear identity for the KP hierarchy.

The proof goes as follows. The equivalence of (107) and (108) shows that (117) is valied if $x_n = x_n'$ for $n = 2,3,\cdots$. Therefore, we have

$$\int \frac{dk}{2\pi i} \frac{\partial^\ell}{\partial x_1^\ell} w(x,k) \cdot w^*(x,k) = 0,$$

for any ℓ. Since $\frac{\partial}{\partial x_n} w(x,k)$ can be replaced by $B_n(x,\partial)w(x,k)$ we can conclude

$$\int \frac{dk}{2\pi i} \frac{\partial^{\ell_1}}{\partial x_1^{\ell_1}} \cdots \frac{\partial^{\ell_n}}{\partial x_n^{\ell_n}} w(x,k) \cdot w^*(x,k) = 0.$$

Thus we derived (117).

The converse statement is also true. Assume that $w(x,k)$ of the form (109) and $w^*(x,k)$ of the form (115) satisfy (117). Then, they are connected by (116) and $w(x,k)$ satisfies (111). Here is the proof. The identity (116) follows from the equivalence of (107) and (108). The bilinear identity implies

$$\int \frac{dk}{2\pi i} (\frac{\partial}{\partial x_n} - B_n(x,\partial)) w(x,k) \cdot w^*(x',k) = 0.$$

Now we have

$$(\frac{\partial}{\partial x_n} - B_n)w = (\frac{\partial}{\partial x_n} - B_n)Pe^\xi = (\frac{\partial P}{\partial x_n} - B_n P + k^n P)e^\xi$$

$$= (\frac{\partial P}{\partial x_n} - B_n P + P\partial^n)e^\xi$$

$$= (\frac{\partial P}{\partial x_n} - B_n P + L^n P)e^\xi$$

$$= (\frac{\partial P}{\partial x_n} + (L^n)_- P)e^\xi.$$

Namely, we have

$$(\frac{\partial}{\partial x_n} - B_n)w = R(x,\partial)w,$$

with such $R(x,\partial)$ that

$$[R(x,\partial)]_+ = 0.$$

By using the equivalence of (107) and (108) again we conclude $R(x,\partial)$ = 0. Thus we derived (111).

5.8   We define vertex operators by

$$X(k) = e^{\xi(x,k)}e^{-\xi(\tilde{\partial},k^{-1})},\tag{118}$$

$$X^*(k) = e^{-\xi(x,k)}e^{\xi(\tilde{\partial},k^{-1})},\tag{119}$$

where

$$\tilde{\partial} = (\frac{\partial}{\partial x_1}, \frac{1}{2}\frac{\partial}{\partial x_2}, \frac{1}{3}\frac{\partial}{\partial x_3}, \cdots).$$

Let $w(x,k)$ and $w^*(x,k)$ satisfy the bilinear identity. Then we can show that there exists a function $\tau(x)$ satisfying

$$w(x,k) = X(k)\tau(x)/\tau(x),\tag{120}$$

$$w^*(x,k) = X^*(k)\tau(x)/\tau(x).\tag{121}$$

Since we have

$$u(x) = -\frac{1}{2}\frac{\partial}{\partial x_1}w_1(x),$$

the formula (120) contains Hirota's transformation (99). In terms of $\tau(x)$ the bilinear identity (117) is written as

$$\int\frac{dk}{2\pi i}e^{\xi(\tilde{\partial}_y,k^{-1})}(\tau(x+y)\tau(x-y))e^{-2\xi(y,k)} = 0.\tag{122}$$

We introduce polynomials $p_j(x)$ by expanding $e^{\xi(x,k)}$,

$$e^{\xi(x,k)} = \sum_{j=0}^{\infty}k^j p_j(x).\tag{123}$$

For example,

$$p_0(x) = 1, \quad p_1(x) = x_1, \quad p_2(x) = x_2 + \frac{x_1^2}{2}.$$

Now we rewrite (122) as follows.

$$0 = \sum_{j=0}^{\infty} p_j(-2y)p_{j+1}(\tilde{\partial}_y)(\tau(x+y)\tau(x-y)) = 0.$$

$$= \sum_{j=0}^{\infty} p_j(-2y)[p_{j+1}(\tilde{\partial}_z)e^{\sum_{\ell=1}^{\infty} y_\ell \partial z_\ell}\tau(x+z)\tau(x-z)]\Big|_{z=0},$$

$$= \sum_{j=0}^{\infty} p_j(-2y)p_{j+1}(\tilde{D}_x)e^{\sum_{\ell=1}^{\infty} y_\ell D x_\ell}\tau(x)\cdot\tau(x). \tag{124}$$

By expanding (124) we get the KP hierarchy of bilinear equations. In this way, the KP hierarchy is characterized by the bilinear identity (117) and the wave function and the τ function are connected by the vertex operator (118).

<u>6.1</u> Consider free fermions ψ_n, ψ_n^* indexed by an integer n. We impose the anti-commutation relations

$$[\psi_m,\psi_n]_+ = [\psi_m^*,\psi_n^*]_+ = 0, \quad [\psi_m^*,\psi_n]_+ = \delta_{mn}.$$

We denote by A the free fermion algebra generated by them. The Fock space $\mathcal{F}$ is the left A module generated by the vacuum vector $|0>$ characterized by

$$\psi_n|0> = 0 \qquad n < 0,$$

$$\psi_n^*|0> = 0 \qquad n \geq 0.$$

We dente by ι the auto-morphism of A such that

$$\iota(\psi_n) = \psi_{n+1},$$

$$\iota(\psi_n^*) = \psi_{n+1}^*.$$

We denote also by ι the isomorphism of $\mathcal{F}$ such that

$$\iota(|0>) = \psi_0|0>,$$

$$\iota(a)\iota(|v>) = \iota(a|v>),$$

for $a \in A$ and $|v> \in \mathcal{F}$. We set

$$|\ell\rangle = \iota^{\ell}(|0\rangle).$$

Now we denote by $\mathcal{O}\!f$ the infinite dimensional Lie algebra in A spanned by $\psi_m \psi_n^*$ $(m, n \in \mathbf{Z})$ and 1. We denote by G the infinite dimensional group

$$G = \{e^{X_1} \cdots e^{X_n} \mid X_1, \cdots, X_n \in \mathcal{O}\!f\}.$$

Consider the following bilinear identity for an element $|v\rangle$ in $\mathcal{F}$:

$$\sum_{n \in \mathbf{Z}} \psi_n |v\rangle \otimes \psi_n^* |v\rangle = 0. \tag{125}$$

The fact is that (125) is valid if and only if

$$|v\rangle \in \bigsqcup_{\ell \in \mathbf{Z}} G|\ell\rangle. \tag{126}$$

Our main goal is to establish the equivalence of (117) and (125).

<u>6.2</u>   The dual Fock space $\mathcal{F}^+$ is the right A module generated by the vacuum vector $\langle 0|$ characterized by

$$\langle 0|\psi_n = 0 \qquad n \geq 0,$$

$$\langle 0|\psi_n^* = 0 \qquad n < 0.$$

We define $\langle \ell|$ in the same way as in 6.1. There exists a unique map

$$\mathcal{F}^* \otimes \mathcal{F} \xrightarrow{\quad} \mathbb{C}$$
$$A$$

$$(\langle a|, |b\rangle) \longmapsto \langle a|b\rangle$$

normalized by $\langle 0|0\rangle = 1$. Then the expectation value on A is defined by

$$A \xrightarrow{\quad} \mathbb{C}$$
$$\Psi \qquad\qquad \Psi$$
$$a \longmapsto \langle 0|a|0\rangle$$

We denote by $:\ :$ the corresponding normal ordering.

<u>6.3</u> We define a larger Lie algebra $\mathfrak{gl}(\infty)$ containing $\mathfrak{g}$ as follows. An element in $\mathfrak{gl}(\infty)$ is of the form

$$X = \sum a_{jk} {:} \psi_j \psi_k^* {:} + c,$$

where we allow the range of summation to be infinite with the restriction that there exists $N = N(X)$ such that $a_{jk} = 0$ if $|j-k| > N$. (Note that only a finite sum is allowed in $\mathfrak{g}$.) By this restriction the Lie bracket in $\mathfrak{gl}(\infty)$ is well-defined so that $\mathfrak{g}$ is contained in it as a sub algebra. We can also define the action of $\mathfrak{gl}(\infty)$ on $\mathcal{F}$ by extending the action of $\mathfrak{g}$.

The center of $\mathfrak{gl}(\infty)$ is two dimensional and contains 1 and

$$H_0 = \sum_{n \in \mathbf{Z}} {:} \psi_n \psi_n^* {:} \; .$$

The operator 1 acts on $\mathcal{F}$ as the identity operator. The action of H_0 is not constant. In fact, $\mathcal{F}$ splits into the direct sum of the eigen spaces of H_0.

$$\mathcal{F} = \bigoplus_{\ell \in \mathbf{Z}} \mathcal{F}_\ell \, ,$$

$$\mathcal{F}_\ell = \{ |a\rangle \in \mathcal{F} \,|\, H_0 |a\rangle = \ell |a\rangle \}.$$

It is easy to show that $\mathcal{F}_\ell$ is an irreducible $\mathfrak{gl}(\infty)$ module and spanned by

$$\psi_{m_1}^* \cdots \psi_{m_s}^* \psi_{n_t} \cdots \psi_{n_1} |0\rangle, \tag{127}$$

where

$$m_1 < \cdots < m_s < 0 \leqq n_t < \cdots < n_1,$$

and

$$t = s + \ell.$$

We set

$$D = \sum_{n \in \mathbf{Z}} n {:} \psi_n \psi_n^* {:} \; .$$

The action of D on $\mathcal{F}$ defines degree in such a way that

$$\deg \psi_n = \deg \psi^*_{-n} = n.$$

Then $|\ell\rangle$ is the lowest vector in $\mathcal{F}_\ell$.

<u>6.4</u> We define

$$H_n = \sum_{m \in \mathbb{Z}} : \psi_m \psi^*_{m+n} : \ .$$

Note that

$$[D, H_n] = -n H_n ,$$

$$[H_m, H_n] = m \delta_{m+n} .$$

These commutation relations are realized by

$$D = \sum_{n=1}^{\infty} n x_n \frac{\partial}{\partial x_n} , \tag{128}$$

$$H_n = \begin{cases} \dfrac{\partial}{\partial x_n} & n > 0. \\[2mm] -n x_{-n} & n < 0 \end{cases} \tag{129}$$

Consider a formal element

$$H(x) = \sum_{n=1}^{\infty} x_n H_n .$$

Since $H(x)$ lower the degree, its action on $\mathcal{F}$ is well-defined. Now we set

$$V_\ell = \mathbb{C}[x_1, x_2, x_3, \cdots] \otimes \mathbb{C}u^\ell ,$$

and consider the map

$$\begin{array}{ccc}
\rho_\ell : \mathcal{F}_\ell & \longrightarrow & V_\ell \\
\quad\ \text{\rotatebox{90}{$\in$}} & & \text{\rotatebox{90}{$\in$}} \\
|v\rangle & \longmapsto & \langle \ell | e^{H(x)} | v\rangle u^\ell \ \ .
\end{array} \tag{130}$$

This is an isomorphism and consistent with the identification (128),

(129). Moreover we can realize H_0 by

$$H_0 = u\frac{\partial}{\partial u} .$$

The isomorphism (130) is explicitly given as follows. Let Y be a Young diagram of the form

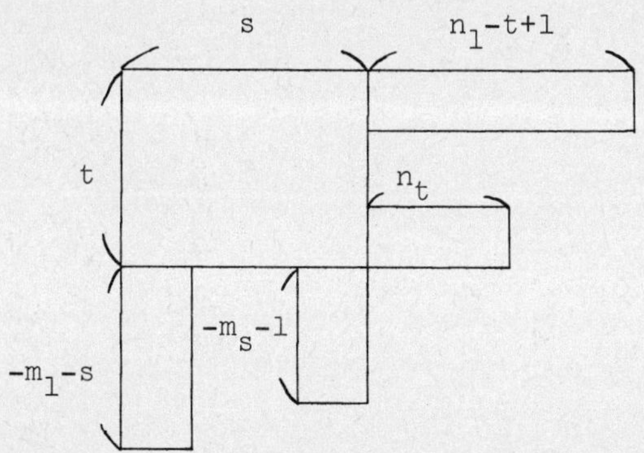

,

and denote by ρ_Y the representation of $GL(N,\mathbb{C})$ with sufficiently large N. For $g \in GL(N,\mathbb{C})$ we define $x_1, x_2, x_3, \cdots$ by

$$x_n = \frac{1}{n} \text{ trace } g^n.$$

The polynomial $\chi_Y(x)$ given by

$$\chi_Y(x) = \text{trace } \rho_Y(g),$$

is called Schur's polynomial. The polynomials $p_j(x)$ of (123) are Schur's polynomials corresponding to the Young diagram of single row of length j.

Let $|v\rangle$ be the element (130). Then we have

$$\rho_\ell(|v\rangle) = (-)^{m_1+\cdots+m_s+(t-s)(t-s-1)/2} \chi_Y(x)u^\ell.$$

<u>6.5</u> Let us define

$$\psi(k) = \sum_{n \in \mathbb{Z}} k^n \psi_n,$$

$$\psi^*(k) = \sum_{n\in\mathbf{Z}} k^{-n}\psi_n^*.$$

Then we have (see (117), (118))

$$\langle\ell+1|e^{H(x)}\psi(k)|v\rangle = k^\ell X(k)\langle\ell|e^{H(x)}|v\rangle, \tag{131}$$

$$\langle\ell-1|e^{H(x)}\psi^*(k)|v\rangle = k^{1-\ell}X^*(k)\langle\ell|e^{H(x)}|v\rangle. \tag{132}$$

Now, take

$$|v\rangle \in G|0\rangle.$$

Then (125) implies

$$\int\frac{dk}{2\pi ik}\,\psi(k)|v\rangle \otimes \psi^*(k)|v\rangle = 0.$$

Therefore we have

$$\int\frac{dk}{2\pi ik}\,\langle1|e^{H(x)}\psi(k)|v\rangle\langle-1|e^{H(x')}\psi^*(k)|v\rangle = 0. \tag{133}$$

Now we set

$$\tau(x) = \langle0|e^{H(x)}|v\rangle,$$

$$w(x,k) = \frac{\langle1|e^{H(x)}\psi(k)|v\rangle}{\tau(x)},$$

$$w^*(x,k) = \frac{\langle-1|e^{H(x)}\psi^*(k)|v\rangle}{k\tau(x)}.$$

The formula (131), (132) implies (120), (121), and (133) then reduces to (117).

Thus, the bilinear identity (125) characterizing the group orbit (126) is identified with the bilinear identity (117) characterizing the KP hierarchy.

Finally, we note that the N soliton formula (103) is recovered by the choice

$$|v\rangle = e^{\sum_{j=1}^{N} a_j \psi(p_j)\psi^*(q_j)} |0\rangle.$$

If we take $p_j = -q_j$, then we get the τ function for the KdV hierarchy. The sub algebra in $g\ell(\infty)$ generated by $\psi(p)\psi^*(-p)$ is isomorphic to the Kac-Moody algebra $A_1^{(1)}$. Therefore the solution space to KdV equation is identified with the group orbit of $A_1^{(1)}$ in the vertex representation constructed as above.

References

[1] M. Sato, T. Miwa and M. Jimbo, Publ. RIMS 14 (1978) 223, 15 (1979) 201, 577, 871, 16 (1980) 531.

[2] E. Date, M. Kashiwara, M. Jimbo and T. Miwa, Non-linear Integrable Systems – Classical Theory and Quantum Theory (ed. M. Jimbo and T. Miwa), World Science Publishing Co., Singapore, 1983.

[3] T. T. Wu, B. McCoy, C. Tracy and E. Barouch, Phys. Rev. B13 (1976) 316.

[4] R. Hirota, Solitons (ed. R. K. Bullough and P. J. Candrey), Springer, 1980.

ANTIFERROMAGNETS

Chanchal Kumar Majumdar
Magnetism Department
Indian Association for the Cultivation
of Science, Jadavpur, Calcutta
700032.

1. Introduction

The Bethe ansatz [1] first provided the exact ground state of
the antiferromagnetic Heisenberg linear chain. Recent work stimulated
by the connection of the ansatz to the inverse scattering methods
clarified some questions about the low-lying excited states. We shall
present a brief account of these results. Our aim is to extract from
these results some information that may be useful for the three
dimensional antiferromagnets which are not generally solvable. The
antiferromagnets have been discussed, faute de mieux, in terms of the
Neel states with broken symmetry and the antiferromagnetic spin waves.
We shall show that the broken symmetry in antiferromagnets is not
properly understood.

The Heisenberg Hamiltonian is written as

$$H = J \sum_{<i,j>} \vec{S}_i \cdot \vec{S}_j . \tag{1.1}$$

The exchange constant J is positive for antiferromagnets. The spins
S_i are distributed over a lattice. The angular brackets $< >$ indicate
that the sum goes over the nearest neighbours only. The linear chain
is represented by

$$H = 2J \sum_{i=1}^{N} \vec{S}_i \cdot \vec{S}_{i+1} , \tag{1.2}$$

with periodic boundary conditions, $N+1 \equiv 1$. Each spin has magnitude
1/2. This is the extreme quantum limit.

If we had a ferromagnetic situation, we would have

$$H_F = -2J \sum_{i=1}^{N} \vec{S}_i \cdot \vec{S}_{i+1} , \tag{1.3}$$

and the ground state wavefunction can be taken to be the state with
all spins down

$$\phi_o = \beta_1 \beta_2 \ldots \beta_N . \tag{1.4}$$

α and β are the up and down spin functions, respectively.

The ground state energy $E_o = -\frac{1}{2} NJ$. The first excited states can be written as [2]

$$\Phi_k = \sum_{m=1}^{N} e^{ikm} S_m^+ \Phi_o ; \tag{1.5}$$

the wave vectors are determined by the periodic boundary conditions

$$k = \frac{2\pi}{N} q, \quad q = 0,1,2,\ldots,N-1. \tag{1.6}$$

The excitation spectrum is given by

$$\varepsilon/2J = 1 - \cos k. \tag{1.7}$$

For small k, $\varepsilon \sim k^2$. A similar result holds in three dimensions. This is the Goldstone boson mode [3] in the ferromagnet. Note that the ground state is degenerate and the choice of (1.6) out of the degenerate set (for instance, by the application of a vanishingly small magnetic field) breaks the rotational symmetry. As a matter of fact, condensed matter physics provides many variations of broken symmetry and their associated Goldstone bosons [4].

The ground state wavefunction of the antiferromagnet, Eq.(1.1), in three dimensions is not known. The partition function and the free energy cannot be explicitly calculated. Approximate treatments are based on the molecular field approximations suggested by Neel [5]. The molecular field is not uniform but oscillatory in antiferromagnets. In the one dimensional linear chain the spins on the even sites point up and those on the odds sites point down or vice versa. The Neel states are

$$\Psi_1 = \alpha_1 \beta_2 \alpha_3 \beta_4 \cdots \alpha_{N-1} \beta_N,$$

$$\Psi_2 = \beta_1 \alpha_2 \beta_3 \alpha_4 \cdots \beta_{N-1} \alpha_N. \tag{1.8}$$

In three dimensions, if we consider a bipartite lattice, the spins on one sublattice point up and those on the other point down, or vice versa. Individually, each Neel state breaks the translational as well as rotational symmetry. In fact they are not even eigenstates of H and cannot serve as 'reference states' in the Bethe ansatz calculation. Nevertheless, Van Vleck's calculation of the susceptibility of an antiferromagnet based on the Neel states was successful in explaining the qualitative features: the 1/T decrease at high temperatures; a maximum with a cusp at the transition temperature T_N; vanishing parallel susceptibility and constant perpendicular susceptibility at low temperature. Accurate experiments reveal that the parallel susceptibility goes through a maximum at a temperature T_{max} and the transition occurs at T_N slightly below T_{max} with a divergence in the derivative of the parallel susceptibility [6]. From the neutron diffraction experiments on antiferromagnets, Neel-type spin patterns are deduced. Thus in a typical example, MnO, the spins in a single [111] plane are parallel but in adjacent [111] planes they are anti-parallel.

The antiferromagnetic spin wave theory is based on the Neel states. The Holstein-Primakoff transformations [7] are applied to each sublattice to get the boson excitation spectrum. For low energy ω and small wave-vector k it is given by the linear relation

$$\omega \sim k. \tag{1.9}$$

Two arguments are advanced to justify the broken symmetry [8]. First, one may assume that a small amount of anisotropy energy is present to hold the spins constant in the z direction. Secondly, one may suppose that the spin wave theory starts from 'wavepackets' of states chosen to make the z-component of the total spin roughly constant and that the time required for the total spin to drift around from one orientation to another essentially different one is extremely large. Neutron diffraction pictures are possible because the time a neutron takes to sample the spins is small compared with the turn-over time.

We shall now try to answer the following questions: What can we learn from the exact solutions about the broken symmetry and Goldstone bosons in antiferromagnets? Is it possible that the correlations seen in the neutron experiments have the Neel structure even though the ground state wavefunction is very complicated? What is the role of anisotropy?

2. Bethe Ansatz in the Linear Chain

For the antiferromagnetic Hamiltonian (1.2) the state Φ_0 is an eigenstate belonging to the eigenvalue $\frac{1}{2}$ NJ and serves as a reference state. If one spin is turned over, we generate a set of eigenstates given by (1.5), with k given by (1.6) and energy

$$\varepsilon/2J = -1 + \cos k \tag{2.1}$$

measured with respect to the energy $\frac{1}{2}$ NJ of the reference state. If two spins are turned over, we construct

$$\Psi_2 = \sum_{\substack{m_1,m_2 \\ m_1 < m_2}} a(m_1,m_2) S^+_{m_1} S^+_{m_2} \Phi_0 \tag{2.2}$$

The Bethe ansatz for $a(m_1,m_2)$ is

$$a(m_1,m_2) = \exp[i(f_1 m_1 + f_2 m_2 + \tfrac{1}{2}\phi)] + \exp[i(f_2 m_1 + f_1 m_2 - \tfrac{1}{2}\phi)]. \tag{2.3}$$

The equations determining f_1, f_2 and ϕ are

$$2 \cot \tfrac{1}{2}\phi = \cot \tfrac{1}{2} f_1 - \cot \tfrac{1}{2} f_2 \tag{2.4}$$

and

$$Nf_1 = 2\pi q_1 + \phi, \quad Nf_2 = 2\pi q_2 - \phi. \tag{2.5}$$

The energy

$$\varepsilon_2 = -1 + \cos f_1 - 1 + \cos f_2. \tag{2.6}$$

The pair of equations in (2.5) comes from periodic boundary conditions, while (2.4) arises from the interaction of the two up-spins. Note that the sum of f_1 and f_2 is independent of ϕ. This is a result of the translational symmetry of the Hamiltonian.

The constants q_1 and q_2 cannot differ by 1. If $f_1 = f_2$, $q_1 = q_2 \pm 1$, and $\phi = \pm\pi$; this makes $a(m_1,m_2) = 0$. The allowed values of q_1 are

$$q_1 = 0,1,2,\ldots q_2-2, \; q_2+2,\ldots N-1. \tag{2.7}$$

Because of the symmetry of (2.3) under interchange, we can restrict to $q_1 < q_2$. These solutions with real f's give then $\tfrac{1}{2}(N-1)(N-2)$ states. The remaining $(N-1)$ states are given by complex values

$$f_1 = u + iv, \; f_2 = u - iv, \quad (v > 0). \tag{2.8}$$

From (2.5), ϕ is also complex

$$\phi = \xi + i\chi, \tag{2.9}$$

and from (2.5) we get

$$N(f_1-f_2) = 2\pi(q_1-q_2) + 2\phi$$

or

$$\tag{2.10}$$

$$\xi = \pi(q_2-q_1), \quad \chi = Nv.$$

The large value of χ localizes the wavefunction; the two up-spins stay sude-by-side while propagating along the chain. This forms a bound complex or a bound magnon state. The eigenvalue spectrum of the two spin deviations consists of a continuum of scattering states, Eq.(2.6) with real wavevectors, and a branch of the bound magnon state

$$\varepsilon_2^B = \frac{1}{2}(1 - \cos 2u). \tag{2.11}$$

Note that Eq.(2.8) gives a 'two string' with centre u [9].

The general Bethe ansatz for an ℓ-up-spin state can be written as

$$a(m_1,m_2,\ldots,m_\ell) = \sum_P \exp[i\{\sum_{k=1}^{\ell} f_{Pk} m_k + \frac{1}{2} \sum_{k<r}^{1,\ell} \phi_{Pk,Pr}\}]. \tag{2.12}$$

The equations determining the parameters f_i, ϕ_{ij} and

$$2 \cot \frac{1}{2} \phi_{ij} = \cot \frac{1}{2} f_i - \cot \frac{1}{2} f_j, \quad \pi \le \phi_{ij} \le \pi, \tag{2.13}$$

$$N_{fi} = 2\pi q_i + \sum_k{}' \phi_{ik}, \tag{2.14}$$

$$\varepsilon = - \sum_{i=1}^{\ell} (1-\cos f_i). \tag{2.15}$$

We have exactly the right number of equations to find all the parameters introduced in the wavefunction. The proof that the Bethe ansatz works for the ground state wavefunction is given by Yang and Yang [10].

Because of the permutation symmetry, we can choose an ordering in the numbers q_i. Also, it has been shown by Hulthen [11] that a state with some q's = 0 has the energy and total spin as the corresponding state in which the zeros have been eliminated. One can therefore order the accepted q's as

$$0 < q_1 \le q_2 \le q_3 \le \ldots . \tag{2.16}$$

Again, $q_{j+1} \ge q_j+2$, otherwise the wavefunction vanishes by the same argument as above. Thus with N even and N/2 spins up and N/2 spin down, we have a unique determination of the q's

$$q_1 = 1, \; q_2 = 3, \; q_3 = 5 \ldots, \; q_{\frac{1}{2}N} = N-1. \tag{2.17}$$

The state has clearly $S_z = 0$, total spin $S = 0$ and is non-degenerate. The ground state wavevector can be calculated as

$$P = \sum_{i=1}^{N/2} f_i = \frac{2\pi}{N} (1+3+\ldots+(N-1)) = (\pi/2)N. \tag{2.18}$$

If $N = 4j$, j positive, the wavevector is 0 (mod 2π); if $N = 4j+2$, it is π (mod 2π). The ground state energy is calculated by a passage to the continuum limit :

$$q_j = \frac{2j-1}{N} \rightarrow x, \qquad (2.19)$$

$$1/N \sum_{\ell=1}^{N/2} \phi_{j\ell} \rightarrow \frac{1}{2} \int_0^1 \phi(x,y)\,dy; \qquad (2.20)$$

Eqs.(2.13), (2.14) and (2.15) become

$$2 \cot \frac{1}{2} \phi(x,y) = \cot \frac{1}{2} f(x) - \cot \frac{1}{2} f(y), \qquad (2.21)$$

$$f(x) = 2\pi x + \frac{1}{2} \int_0^1 \phi(x,y)\,dy, \qquad (2.22)$$

$$\varepsilon = - \sum_{j=1}^{N/2} (1-\cos f_j) = -(N/2) \int_0^1 (1-\cos f(x))\,dx. \qquad (2.23)$$

Remembering that when $x = y$, ϕ jumps from π to $-\pi$, and differentiating (2.22) we get

$$\frac{df}{dx} = \pi + \cos ec^2 (\tfrac{1}{2} f)\frac{df}{dx} \int_0^1 \frac{dy}{4+\{\cot \frac{1}{2} f(x)-\cot \frac{1}{2} f(y)\}^2} \qquad (2.24)$$

Put

$$g = \cot \frac{1}{2} f, \qquad (2.25)$$

and

$$\rho_o(g) = - \frac{dx}{dy}. \qquad (2.26)$$

Thus we arrive that Hulthen's integral equation

$$\frac{2}{\pi(1+g^2)} = \rho_o(g) + \frac{2}{\pi} \int_{-\infty}^{\infty} \frac{\rho_o(g')dg'}{4+(g-g')^2}. \qquad (2.27)$$

The solution of the integral equation is

$$\rho_o(g) = \frac{1}{2} \operatorname{sech} \frac{1}{2} \pi g, \qquad (2.28)$$

and

$$\varepsilon = -N \int_{-\infty}^{\infty} \frac{\rho_o(g)}{1+g^2}\,dg = -N\ell n2. \qquad (2.29)$$

In absolute measure the energy is

$$E_o = \frac{1}{2} NJ - 2JN\ln 2. \tag{2.30}$$

3. The Low-lying Excited States

In their attempt to calculate the excited states of the linear chain des Cloizeaux and Pearson [12] made some plausible assumptions about the change in the distribution of q's in (2.17) and arrived at the spectrum

$$\varepsilon = \frac{1}{2} \pi \, |\sin k|, \quad -\pi \leq k \leq \pi, \tag{3.1}$$

for spin 1 states. The wave vector k is measured from that of the ground state. Notice the characteristic double periodicity of the spectrum. Experiments [13] on dichloro<u>bis</u> (pyridine) copper II (CPC) have seen a spin-1 branch of this type.

A rigorous calculation of the low-lying excited states is now available from the work of Takahasi [14], Woynarovich [15] and Faddeev and Takhtajan [16]. There are S=1 as well as S=0 states; the excitation spectrum has the form (3.1) but requires careful interpretation.

In (2.13) we introduce complex numbers λ_j by

$$2\lambda_j = - \cot \frac{1}{2} f_j, \quad j = 1, 2, \ldots, \tag{3.2}$$

so that

$$\exp(if_j) = \frac{\lambda_j - \frac{1}{2} i}{\lambda_j + \frac{1}{2} i}, \tag{3.3}$$

and from Eq. (2.3) we get

$$\exp(i\phi_{ij}) = \frac{\lambda_i - \lambda_j - i}{\lambda_i + \lambda_j + i}. \tag{3.4}$$

Exponentiating (2.14), and using (3.3) and (3.4) we get an algebraic relation for the $\ell \leq \frac{1}{2} N$ complex number λ_j

$$\left(\frac{\lambda_j - \frac{1}{2} i}{\lambda_j + \frac{1}{2} i} \right)^N = \prod_{\substack{k=1 \\ k \neq j}}^{\ell} \frac{\lambda_j - \lambda_k - i}{\lambda_j - \lambda_k + i}, \quad j = 1, 2, \ldots, \ell. \tag{3.5}$$

Eq. (2.15) and (2.18) for energy and momenta are

$$E = - \sum_{i=1}^{\ell} \frac{2}{1 + 4\lambda_i^2}, \tag{3.6}$$

$$P = \sum_{i=1}^{\ell} (2 \tan^{-1} 2\lambda_j - \pi) \pmod{2\pi}. \tag{3.7}$$

When $N \to \infty$, the numbers λ_j asymptotically cluster in various 'strings'. A string of length n - an n-string - has n numbers of the form

$$\lambda_{j,\alpha}^n = \lambda_j^n + \frac{1}{2} i(n+1-2\alpha), \quad \alpha = 1,2,\ldots,n. \tag{3.8}$$

λ_j^n is the 'centre of the string' and is a real number. A 1-string is just a real number. A configuration is defined by a partition of the numbers $\{\lambda_j\}$ into strings. Let ν_n be the number of n-strings present. Then a configuration is parametrized by the set $\{\nu_n\}$, $n = 1,2,\ldots,$ and clearly we have

$$\sum n\nu_n = \ell. \tag{3.9}$$

For a given configuration we must get equations for λ_j^n, $j = 1,2,\ldots,\nu_n$. These can be obtained from (3.5) in the form

$$N\Phi_n(\lambda_j^n) = 2\pi Q_j^n + \sum_m \sum_{k=1}^{\nu_m} \Phi_{nm}(\lambda_j^n-\lambda_k^m), \tag{3.10}$$

where

$$\Phi_n(x) = 2\tan^{-1}(\frac{2x}{n}), \quad -\frac{1}{2}\pi \le \tan^{-1}\theta \le \frac{1}{2}\pi, \tag{3.11}$$

$$\Phi_{nm}(x) = 2\sum_k{}' (\tan^{-1}\frac{2x}{k} + \tan^{-1}\frac{2x}{k+2}). \tag{3.12}$$

In (3.12) the summation extends over $k = |n-m|, |n-m| + 2, |n-m| + 4, \ldots, n+m-2$ and for n=m the term with vanishing denominator is omitted. The numbers Q_j^n vary in the interval $|Q_j^n| \le Q_{max}^n$ where

$$Q_{max}^n = \frac{1}{2} N - \frac{1}{2\pi} \sum_m \Phi_{nm}(\infty)\nu_m - \frac{1}{2}, \tag{3.13}$$

and Q_j^n are integers or half integers according as Q_{max}^n is. Any array of Q_j^n, such that $Q_j^n \ne Q_{j'}^n$, for $j \ne j'$ and each n, defines a unique Bethe vector. The total number of state generated in this way is, of course, 2^N.

For the ground state, $\ell = \frac{1}{2} N$ and we have only 1-strings (no complex numbers); hence from (3.9) $\nu_1 = \frac{1}{2} N$; $\nu_n = 0$, $n > 1$. $\Phi_{11}(\infty) = \pi$, and (3.13) gives $Q_{max}^1 = \frac{1}{4} N - \frac{1}{2}$. The number of vacancies $2Q_{max}^1 +1$ for the quantities Q_j is equal to $\frac{1}{2} N$, so that there is no freedom of choice left. This shows that the ground state is non-degenerate.

We now consider a class of configurations for which $\frac{1}{2} N-\nu_1$ is fixed and finite when $N \to \infty$; all ν_n, $n > 1$, are finite. This class can be described as a Dirac sea of strings of length 1 (magnons), with a finite number of embedded strings of length 2, 3, etc. (magnon bound states). This class contains all two lying excitations.

The states of our interest are generated as follows:

(1) $\ell = \frac{1}{2} N-1$, the spin of the state is 1. As solution of (3.8) we consider $\nu_1 = \frac{1}{2} N-1$; $\nu_n = 0$, $n > 1$. $\Phi_{11}(\infty) = \pi$, and $Q^1_{max} = \frac{1}{4} N$, and the number of vacancies for Q^1_j is $2Q^1_{max} + 1 = \frac{1}{2} N+1$. So there are two free parameters characterizing the states. This corresponds to the des Cloizeaux–Pearson branch.

(2) $\ell = \frac{1}{2} N$, the spin of the state is 0. From (3.8) we take $\nu_1 = \frac{1}{2} N-2$, $\nu_2 = 1$; $\nu_n = 0$, $n > 2$. $\Phi_{11}(\infty) = \pi$, $\Phi_{12}(\infty) = 2\pi$, $\Phi_{21} = 2\pi$, $\Phi_{22}(\infty) = 3\pi$. Hence $Q^1_{max} = \frac{1}{4} N - \frac{1}{2}$, $Q^2_{max} = 0$. The number of vacancies for the parameters Q^1_j is $\frac{1}{2} N$ and the only possible value for Q^2_j is 0. This means that the number of vacancies exceeds the number of Q^1_j by 2. So this configuration is also characterized by two parameters – two holes in the distribution of Q^1_j. This branch is the one guessed by Fazekas and Suto [17] and computed by Woynarovich.

The numbers λ^1_j are distributed on the real axis when $N \to \infty$ with a density $\rho(\lambda)$ and the λ^n_j, $n > 1$, run through a finite number of values depending on the configuration. The system of Eqs. (3.10) to (3.12) turns into a linear integral equation for $\rho(\lambda)$ and a finite number of equations for λ^n_j, $n > 1$. The equations, together with (3.6) and (3.7), show that the energy and momentum of these states, with respect to those of the ground state, have the form

$$\epsilon(\lambda_1, \lambda_2) = \epsilon(\lambda_1) + \epsilon(\lambda_2),$$

$$k(\lambda_1, \lambda_2) = k(\lambda_1) + k(\lambda_2), \quad -\infty < \lambda_1, \lambda_2 < \infty, \tag{3.14}$$

and

$$\epsilon(\lambda) = \frac{\pi}{2\cosh\pi\lambda}, \tag{3.15}$$

$$k(\lambda) = \frac{\pi}{2} - \tan^{-1}\sinh\pi\lambda. \tag{3.16}$$

Hence the excitation spectrum has the dispersion law

$$\epsilon(k) = \frac{\pi}{2} \sin k, \quad 0 \le k \le \pi. \tag{3.17}$$

The parameters λ_1 and λ_2 characterize the holes in the distribution of Q^1_j. For the spin 0 excitation, the sum of the parameters $\lambda_s = \frac{1}{2}(\lambda_1 + \lambda_2)$ is uniquely determined and distinguishes this spin 0 state, but the contributions to energy and momentum of the 2-string are zero. The excitation is a kink rather than an ordinary particle. Note also that

that k goes over half the Brillouin zone; the double periodicity is
recovered because the k is measured from that of the ground state, 0
or π.

All the low lying configurations are described by the holes in
the distribution of Q^1_j and a finite number of strings of higher length.
The number of holes is always even. The energy and momentum are given
additively by the energies $\varepsilon(\lambda_j)$ and momenta $k(\lambda_j)$ of the individual
holes. The contribution of strings of higher length is always zero,
as in the singlet case.

The only excitation uniquely parametrized by momentum is a spin
wave of spin 1/2. All physical states, however, contain an even number
of spin waves and have integer spin. The fermion character of the
excitation is suppressed and the bosonic character is seen. This
picture of excitations is similar to the phenomenon of decolouring in
relativistic quantum field theory of coloured objects.

4. Antiferromagnets in Three Dimension

For the three dimensional Hamiltonian (1.6) Peierls, Marshall
[18], and Lieb, Schultz and Mattis [19] have shown that the ground
state is unique and has spin 0. Consider the relation

$$\vec{S}^2 = (\sum_{i=1}^{N} \vec{S}_i)^2 = \sum_{i=1}^{N} \vec{S}_i^2 + 2 \sum_{i<j}^{1,N} \vec{S}_i \cdot \vec{S}_j \tag{4.1}$$

or

$$2 \sum_{i<j}^{1,N} \vec{S}_i \cdot \vec{S}_j = \vec{S}^2 - \sum_{i=1}^{N} \vec{S}_i^2 . \tag{4.2}$$

Since $\vec{S}^2 = S(S+1)$ has values $S = 0,1,2,...$ (N even), the eigenvalues
of the left hand side of (4.2) are arranged in an increasing sequences.
The ground state belongs to S=0. This ground state wavefunction is
shown to be <u>not</u> orthogonal to that of the Hamiltonian (1.1), which
then must also have S=0. Further, all the $C^N_{N/2}$ states in the $S_z=0$
subspace occur in the ground state, and their coefficients, with a
factor taken out, are all positive. Hence the ground state is unique.
There is no 'degeneracy of the vacuum'.

5. Antiferromagnets with known Ground States

We must understand the results of neutron scattering
experiments, which are determined by correlation function [20]. The
pictures of the spin pattern are pictorial representations of the

correlation functions. We should explore the possibility that the correlation functions have the Neel structure even though the ground state wavefunction is very complicated.

The space-time correlation functions $\langle S^z_{i+n}(t) S^z_i(0) \rangle$ have only been numerically studied [21]. Because of the difficulty in getting the normalization of the wavefunction in the Bethe ansatz, the static correlation function $\langle S^z_{i+1} S^z_i \rangle$ has not been calculated for the linear chain. Only $\langle S^z_{i+1} S^z_i \rangle$ is exactly known, as it is related to the ground state energy. From numerical computation [22], it is found that $\langle S^z_{i+n} S^z_i \rangle$ oscillates in sign, as the Neel function suggests, but the amplitude diminishes with increasing n and there is no long range order (Table 1).

An even clearer picture emerges from the study of a Hamiltonian for which the ground state can be explicitly written down [23]. This is

$$H_A = 2J \sum_{i=1}^{N} \vec{S}_i \cdot \vec{S}_{i+1} + J \sum_{i=1}^{N} \vec{S}_i \cdot \vec{S}_{i+2} \tag{5.1}$$

with periodic boundary conditions (N+1 $\equiv$ 1, N+2 $\equiv$ 2). The ground state energy is $-\frac{3}{4} NJ$ and is found by proving that this is both an upper and a lower bound for the ground state energy [24]. Let

$$[\ell,m] = \alpha(\ell)\beta(m) - \beta(\ell)\alpha(m) \tag{5.2}$$

be the usual singlet combination. We then construct the states

$$\phi_1 = [1,2][3,4][5,6] \;\ldots\; [N-1,N],$$

$$\phi_2 = [2,3][4,5][6,7] \;\ldots\; [N,1]. \tag{5.3}$$

Both then belong to the eigenvalue $-\frac{3}{4} NJ$. We have a degenerate ground state. If T denotes the translation operator for unit displacement, we have

$$T\phi_1 = \phi_2, \quad T\phi_2 = \phi_1. \tag{5.4}$$

The states ϕ_1 and ϕ_2 break the discrete translation symmetry. However, we can construct two ground state functions which diagonalize the translation operator T:

$$\phi^+ = \frac{1}{\sqrt{2}}(\phi_1 + \phi_2), \quad \phi^- = \frac{1}{\sqrt{2}}(\phi_1 - \phi_2), \tag{5.5}$$

$$T\phi^+ = \phi^+, \quad T\phi^- = -\phi^-. \tag{5.6}$$

The correlation functions can be calculated exactly [25] and are given in Table 1. Notice that for finite N the order quickly settles down to an alternating function - exactly what is expected from the Neel states - although the ground state wavefunction is very different. The long range order is evanescent and as $N \rightarrow \infty$ only the nearest neighbour order remains non-vanishing and antiferromagnetic. The model in that limit represents a quantum spin liquid. Though there is no long range order in spin correlation function

$$K^2(i,j) = \langle S_i^z S_j^z \rangle = \frac{1}{4}\, \delta_{ij} - \frac{1}{8}\, \delta|i-j|,1, \qquad (5.7)$$

the four spin correlation function have off-diagonal long range order [26]:

$$K^4(ij;\ell m) = \langle S_i^x S_j^x S_\ell^y S_m^y \rangle = K^2(ij)K^2(\ell m) + \frac{1}{64}\, \delta|i-j|,1\, \delta|\ell-m|,1$$
$$x\, \exp(i\pi(\frac{i+j}{2} - \frac{\ell+m}{2})). \qquad (5.8)$$

Table 1. Spin correlations for $\Phi^{\pm}$ and the linear chain

No.of spins	12		∞		∞ (linear chain)
	+	-	+	-	
$\langle \sigma_1^z \sigma_2^z \rangle$	-0.516	-0.484	-0.5	-0.5	-0.591
$\langle \sigma_1^z \sigma_3^z \rangle$	0.030	-0.032	0	0	0.25
$\langle \sigma_1^z \sigma_4^z \rangle$	-0.030	0.032	0	0	-0.19
$\langle \sigma_1^z \sigma_5^z \rangle$	0.030	-0.032	0	0	0.15
$\langle \sigma_1^z \sigma_6^z \rangle$	-0.030	0.032	0	0	
$\langle \sigma_1^z \sigma_7^z \rangle$	0.030	-0.032	0	0	

The excitation spectrum of the Hamiltonian (5.1) is not exactly known. Approximate calculations by Majumdar, Krishan and Mubayi [27] indicated the existence of a spin 1 and a spin 0 branches. Shastry and Sutherland [26] construct a spectrum of 'defect' states. Since the two ground states Φ_1 and Φ_2 have two different modes of spin pairing the excitations are thought of as mixtures of these two modes separated by a defect. As an example, the wavefunction

$$\Psi(p,m) = [2p-3,2p-2]\alpha_{2p-1}[2p,2p+1] \ldots [2m-2,2m-i]\alpha_{2m}[2m+1,2m+2]$$

$$\ldots \tag{5.9}$$

is an excitation function with two up-spins at the defects. This state has spin 1. In all taking $N=2M$, we can form $4M^2$ defect states which are mixtures of spin 1 and spin 0 states. The spectrum of these defects states has a continuum with a lower edge at $J(\frac{5}{2} - 2 |\cos k|)$. There is a bound state branch emerging out of the continuum at 0.36 and re-entering the continuum at 0.64 . The excitation spectrum may have a gap $\frac{1}{2} J$. Exact information is available for chains containing up to 12 spins [27], but this is not long enough to check the existence of a gap. The low-lying spin 1 and spin 0 states are so intermingled that they may be asymptotically degenerate. Perhaps there is decolouring here also.

Other antiferromagnetic models where some eigenstates are soluble are discussed by Shastry and Sutherland [28], Van der Broek [29], Caspers Magnus [30], Bader and Schilling [31], and D.J. Klein [32].

6. Hamiltonian with Neel states as ground states

We now discuss the question of anisotropy. The Neel states are indeed the lowest states of the Ising model, an extremely anisotropic limit. A more general Hamiltonian with Neel states a ground states can be generated [33]. Start from the ferromagnetic XXZ Hamiltonian, which can be solved by the Bethe ansatz:

$$H_F(-J,-\Delta) = \sum_{i=1}^{N} [-JS_i^z S_{i+1}^z - \frac{1}{2} \Delta(S_i^+ S_{i+1}^- + S_i^- S_{i+1}^+)], \tag{6.1}$$

$(J > 0, \Delta > 0)$. The ground state is the fully aligned state. First, we rotate every alternate spin by an angle π about the z-axis. $S_i^x \to -S_i^x$, $S_i^y \to -S_i^y$, $S_i^z \to S_i^z$. We get

$$H_F(-J,\Delta) = \sum_{i=1}^{N} [-JS_i^z S_{i+1}^z + \frac{1}{2} \Delta(S_i^+ S_{i+1}^- + S_i^- S_{i+1}^+)]. \tag{6.2}$$

Next we rotate every alternate spin by an angle π about the x-axis $S_i^x \to S_i^x$, $S_i^y \to -S_i^y$, $S_i^z \to -S_i^z$. We obtain the antiferromagnetic Hamiltonian

$$H_A(J,\Delta) = \sum_{i=1}^{N} [JS_i^z S_i^z + \frac{1}{2} \Delta(S_i^+ S_{i+1}^+ + S_i^- S_{i+1}^-)]. \tag{6.3}$$

Since either the odd or the even spins can be rotated, we generate the two Neel states by the above rotations. The Neel states are not affected by the S^+S^+ or S^-S^- terms; they are stable with regard to the addition of these terms. Eq.(6.3) is also a special case of the XYZ model of Baxter [34] where only the transverse anisotropy has been retained.

Each Neel state breaks the translational symmetry, but we can restore translational symmetry by taking their linear combination. The ground state has oscillating long range order. For $\Delta = J$, the spectrum has no gap; in fact the dispersion law is $\varepsilon \sim k^2$ as $k \to 0$, $\varepsilon \to 0$. This is, of course, the ferromagnetic spectrum appearing without change under canonical transformation.

These results can be generalized to three dimensions. Consider a bipartite lattice and start with the ferromagnetic Hamiltonian $(J > 0)$

$$H_F = \frac{1}{2} \sum_{<ij>} [-JS_i^z S_j^z + \frac{1}{2} \Delta (S_i^+ S_j^- + S_i^- S_j^+)]. \tag{6.4}$$

Now we rotate the spins of one sublattice about the x-axis by an angle π and obtain the antiferromagnetic Hamiltonian

$$H_A = \frac{1}{2} \sum_{<ij>} [JS_i^z S_j^z + \frac{1}{2} \Delta (S_i^+ S_j^+ + S_i^- S_j^-)]. \tag{6.5}$$

We generate the Neel states as before. The spectrum for $J = \Delta$ is again $\varepsilon \sim q^2$ as $\varepsilon \to 0$, $q \to 0$.

Dyson, Lieb and Simon [35] prove that the two dimensional version of (6.5) has long range order at finite temperature. So we expect that the three dimensional one has an antiferromagnetic transition too. The molecular field results can be reproduced from (6.5) as they depend on the properties of the Neel states.

7. Conclusion

We have concentrated on the spin 1/2 antiferromagnetic models. This is where the Bethe ansatz works. Some results for higher spins are also known. Fisher [36] has shown that the partition function for the infinite spin linear chain can be calculated. Each spin $\vec{S}_j$ can be scaled by its magnitude S and the spins then become commuting variables in the limit $S \to \infty$.

$$\vec{S}_j = \vec{s}_j \, s, \tag{7.1}$$

$$s_j^x s_j^y - s_j^y s_j^x = is_j^z / S \to 0. \tag{7.2}$$

$$S \to \infty$$

The correlation function is

$$\langle S^z_i S^z_{i+\ell}\rangle = \frac{1}{3}(-1)^{|\ell|}[u(T)]^{|\ell|} \qquad (7.3)$$

with

$$u(T) = \coth(J/2k_b T) - \frac{2k_B T}{J}. \qquad (7.4)$$

As $T \to 0$, this gives the Neel picture. It is also known from numerical computations on short chains of higher spin particles [37] that the excitation spectrum approaches the spectrum calculated from the spin wave theory. In practice, the spin 5/2 (Mn ion) or 7/2 (Eu ion) may differ by about 10 percent from the infinite spin results.

In the extreme quantum limit (spin 1/2), there is no evidence for broken symmetry or Goldstone modes in antiferromagnets. However, the correlation functions have not been calculated, and the nature of long range order for the Hamiltonian (11) in three dimensions remains unclear. For larger spins, the conventional treatments provide resonable physical descriptions of the experimental phenomena.

References

1. H. Bethe, Z. Physik 71, 205 (1931).
2. F. Bloch, Z. Physik 61, 206 (1930); 74, 295 (1932).
3. J. Goldstone, Nuovo Cimento 19, 154 (1961).
4. P.W. Anderson, Phys. Rev. 130, 439 (1963) and 'Concepts in Solids', pp. 175-182 (Benjamin, New York, 1964). H. Wagner, Z. Physik 195, 273 (1966). R.V. Lange, Phys. Rev. 146, 381 (1968).
5. J. Van Vleck, J. Chem. Phys. 9, 85 (1941).
6. M.E. Fisher, Phil. Mag. 7, 1731 (1962).
7. F. Keffer, Handbuch der Physik XVIII/2, pp.1-273 (1966).
8. P.W. Anderson, Phys. Rev. 86, 694 (1952).
9. The two spin deviation problem in higher dimensions was solved by M. Wortis, Phys. Rev. 132, 85 (1963). The three spin deviation problem has been discussed in C.K. Majumdar, Phys. Rev. B1, 287 (1970); C.K. Majumdar and I. Bose, J. Math. Phys. 19, 2189 (1978); C.K. Majumdar and S. Banerjee, J. Math. Phys. 23, 71 (1982) and S. Banerjee and C.K. Majumdar, Physica 107A, 212 (1981). From these articles other references can be traced.
10. C.N. Yang and C.P. Yang, Phys. Rev. 147, 303 (1966); 150, 321 (1966); 150, 327 (1966).

11. L. Hulthen, Arkiv, Mat. Astron. Fysik 26A, No.11 (1938).

12. J. des Cloizeaux and J.J. Pearson, Phys. Rev. 128, 2131 (1967).

13. Y. Endoh, G. Shirane, R.J. Birgeneau, P.M. Richards and S.L. Holt, Phys. Rev. Lett. 32, 170 (1974).

14. M. Takahasi, Prog. Theor. Phys. 46, 401 (1971).

15. F. Woynarovich, J. Phys. A15, 2985 (1982).

16. L.D. Faddeev and L.A. Takhtajan, Phys. Lett. 85A, 375 (1981); L.A. Takhtajan and L.D. Faddeev, Zap. Nauk. Semin. LOMI 109, 134 (1984) (in Russian).

17. P. Fazekas and A. Sueto, Sol. St. Commun. 19, 1045 (1976); A. Sueto, Sol. St. Commun. 20, 681 (1976).

18. W. Marshall, Proc. Roy. Soc. A232, 48 (1955), A232, 69 (1955).

19. E.H. Lieb, T.D. Schultz and D.C. Mattis, Ann. Phys. (New York) 16, 417 (1961).

20. L. Van Hove, Phys. Rev. 95, 1374 (1954).

21. F. Carboni and P.M. Richards, Phys. Rev. 177, 889 (1969).

22. J. Bonner and M.E. Fisher, Phys. Rev. 135A, 64 (1967).

23. C.K. Majumdar and D.K. Ghosh, J. Math. Phys. 10, 1388, 1399 (1969).

24. C.K. Majumdar, J. Phys. C3, 911 (1970). The lower bound was found by J. Pasupathy; C.K. Majumdar 'Solid State Physics', Vol.II (Ed. by F.C. Auluck, Thomson Press (India) Ltd., New Delhi 1972). The same proof is found in P.M. Van der Broek, Phys. Lett. 77A, 261 (1980).

25. B.S. Shastry and C.K. Majumdar, Phys. Rev. B12, 2856 (1975).

26. B.S. Shastry and B. Sutherland, Phys. Rev. Lett. 47, 964 (1981).

27. C.K. Majumdar, K. Krishan and V. Mubayi, J. Phys. C5, 2846 (1972).

28. B.S. Shastry and B. Sutherland, Physica 108B, 1069 (1981).

29. P.M. Van der Broek, W.J. Caspers and M.W.M. Willemse, Physica 104A, 298 (1980).

30. W.J. Caspers, Phys. Rep. 63, 225 (1980); W.J. Caspers, Physica 115A, 275 (1982). W.J. Caspers and W. Magnus, Phys. Lett. 88A, 103 (1982). W.J. Caspers and W. Magnus, Physica 119A, 291 (1983).

31. H.P. Bader and R. Schilling, Phys. Rev. B19, 3556 (1979).

32. D.J. Klein, J. Phys. A15, 661 (1982).

33. I. Bose, S. Chatterjee and C.K. Majumdar, Phys. Rev. B29, 2741 (1984).

34. R. Baxter, Ann. Phys. (New York) 70, 323 (1972).

35. F.J. Dyson, E. Lieb and B. Simon, J. Stat. Phys. 18, 335 (1978).

36. M.E. Fisher, Am. J. Phys. 32, 343 (1964).

37. C.S. Jain, K. Krishan, C.K. Majumdar and V. Mubayi, Phys. Rev. B12, 5235 (1975).

CLASSICAL AND QUANTUM L-MATRICES

L.D. Faddeev

Contents

Lecture 1.   General features of the R-matrix

Lecture 2.   Fundamental L operators, local Quantum hamiltonians

Lecture 3.   Classification of FPR's

Lecture 4.   The Fundamental Poisson brackets (cont'd) The Classical Lattice Case.

[*] Notes by T.R. Ramadas. Professor Faddeev also gave two lectures each on σ-models and anomalies. These are however not included here.

Lecture 1 : <u>General features of the R-matrix method</u>

In the lectures of Professor Takhtadjan the decisive role of the
local L-operator and corresponding R-matrix in the construction of
quantum models on the lattice soluble by means of the Algebraic
Bethe Ansatz was illustrated. In these lectures I shall discuss
some general properties of L-operators and outline a path towards
their possible classification. In particular the relation of the
L-operator with the Lax operator of the inverse scattering method
will become clear. This is why the Algebraic Bethe Ansatz is some-
times called the quantum inverse scattering method.

1. Recall from Professor L. Takhtadjan's lectures the Fundamental
   Commutation Relations (FCRs from now on)

$$R(\lambda,\mu)(L_n(\lambda) \otimes L_n(\mu)) = (L_n(\mu) \otimes L_n(\lambda))R(\lambda,\mu) \quad \lambda,\mu \in \mathbb{C}.$$

Here $L_n(\lambda)$ acts on $h_n \otimes V$, where h_n is the 'quantum space'
associated to site 'n' of a one-dimensional lattice, V the 'auxiliary'
space, and R acts on $V \otimes V$. A partial contraction is implicit in
the notation $L_n(\lambda) \otimes L_n(\mu)$ - this operator acts on $h_n \otimes V \otimes V$. When
no confusion is possible, the subscript 'n' will be dropped from now
on.

Consider a representative example of the FCR in the 'rational'
case, so called because the operators L, R are rational functions of
λ,μ. Let $\{\sigma^a\}_{a=1,2,3}$ be the Pauli matrices acting on $\mathbb{C}^2 \simeq V$, and S^a
the spin-matrices in a spin S representation ($S = \frac{m}{2}$, m integer) acting
on $\mathbb{C}^{2S+1} \simeq h$. We shall verify that the FCR's are satisfied with the
following choice: (repeated indices are summed over whenever they
appear)

$$L(\lambda) = I_h \otimes I_V + \frac{i}{\lambda} S^a \otimes \sigma^a$$

$$R(\lambda,\mu) = P + \frac{i}{\lambda-\mu} I_{V \otimes V}$$

where P permutes the two factors in $V \otimes V$, that is, $P(a \otimes b) = b \otimes a$,
$a,b \in V$. If we write out $L(\lambda) \otimes L(\mu)$ we get

$$L(\lambda) \otimes L(\mu) = I_h \otimes I_V \otimes I_V + \frac{i}{\lambda} S^a \otimes \sigma^a \otimes I_V + \frac{i}{\mu} S^a \otimes I_V \otimes \sigma^a$$
$$- \frac{1}{\lambda\mu} S^a S^b \otimes \sigma^a \otimes \sigma^b$$

and one can check, using $P(A \otimes B) = (B \otimes A)P$, that the FCR will be
satisfied if

$$([S^a, S^b] \sigma^a \otimes \sigma^b) P = S^c (1 \otimes \sigma^c - \sigma^c \otimes 1)$$

This in turn can be checked using the relations $[S^a, S^b] = i\epsilon^{abc} S^c$, $\sigma^a \sigma^b = \delta^{ab} I_V + i\epsilon^{abc} \sigma^c$ and the identity

$$P = \frac{1}{2} (I_V \otimes I_V + \sigma^a \otimes \sigma^a)$$

Note that the FCR is satisfied independent of the quantum spin S, and in fact this enables one to have different quantum spins at different sites.

2. We shall return to the 'quantum mechanical lattice' case in Lecture §2. Let us now take a "classical limit" of the above algebra. Thus we suppose that associated to each lattice point is some symplectic manifold, i.e., a 'phase-space' where a Poisson bracket is defined between functions. These structures are to be in natural correspondence with the quantum system defined on h. Also we think of the $L_n(\lambda)$ as dim V x dim V matrices with entries which are operators on h_n to start with and which become functions on the local phase-space in the classical limit. Then the tensor product $L_n(\lambda) \otimes L_n(\mu)$ will be a Kronecker product of matrices $[(A \otimes B)ij, k\ell = A_{ik} B_{j\ell}]$ with operator - (respectively functions-on-phasespace-) valued entries. In our example dim V=2, and the local phase-spaces will be spheres in the Lie algebra of SU(2). More of this in Lecture §3, but we now proceed with the classical limit. Reintroduce the Planck's constant, and write

$$R(\lambda) = P(1 + \frac{i\hbar P}{\lambda})$$

One can check that the FCRs are satisfied once more with this choice of R provided we remember to reintroduce $\hbar$ in the commutation relations of the S^a as well. We can now rewrite the FCR as follows (we use $P^2 = I$)

$$L(\lambda) \otimes L(\mu) - PL(\mu) \otimes L(\lambda)P = \frac{i\hbar}{\lambda - \mu} P(L(\mu) \otimes L(\lambda) - L(\lambda) \otimes L(\mu))$$

$$= [\frac{-i\hbar}{\lambda - \mu} P, \ L(\lambda) \otimes L(\mu)]$$

$$-[L(\lambda) \otimes L(\mu) - PL(\mu) \otimes L(\lambda)P]$$

$$\times \frac{i\hbar P}{\lambda - \mu}$$

We introduce the notation $[A \otimes B]_{ij, k\ell} = [A_{ik}, B_{j\ell}]$ where on the right

we mean the operator commutator. Then the above equations can be
written

$$\frac{[L(\lambda) \otimes L(\mu)]}{i\hbar} \, (1 + \frac{i\hbar P}{\lambda - \mu}) = [r, L(\lambda) \otimes L(\mu)]$$

where $r(\lambda) = -\frac{P}{\lambda}$. As $\hbar \to 0$ we get

$$(*) \qquad \{L(\lambda) \otimes L(\mu)\} = [r(\lambda - \mu), L(\lambda) \otimes L(\mu)]$$

where now $\{A \otimes B\}_{ij,k\ell} = \{A_{ik}, B_{j\ell}\}$ and $\{ \ \}$ denotes Poisson Brackets.
Thus spin operators have become phase-space functions satisfying
$\{S^a, S^b\} = \varepsilon^{abc} S^c$, and of course, $(*)$ can be verified directly with
this choice of local phase-space functions.

3. We now write down the continuum limit of the above algebra. We
suppose that the dynamical variables are (perhaps 'nonlinear')
fields defined on one-dimensional space, and that a Poisson-bracket
is defined on the space of functions of the fields. Thus, for
example, one would have in our 'magnetic' example, a field
$\vec{S}(x) = \{S^a(x), \ a=1,2,3\}$ with $\vec{S}^2 \equiv \sum (S^a)^2 = 1$ and the Poisson brackets

$$\{S^a(x), S^b(x)\} = \varepsilon^{abc} S^c(x) \ \delta(x-y)$$

If we define $L(x,\lambda) = \frac{1}{\lambda} \vec{S}(x) \cdot \vec{\sigma}$ one can check

$$\{L(x,\lambda) \otimes L(y,\mu)\} = \{r(\lambda - \mu), \ L(x,\lambda) \otimes 1 + 1 \otimes L(y,\mu)\} \ \delta(x-y)$$

This will be called the Fundamental Poisson bracket Relations (FPR),
and the matrix r will be called the classical r-matrix. There are
some very important remarks to be made at this point (see the Les
Houches Lectures [LDF] for details).
a) Consider the linear operator

$$\tilde{L} = \frac{1}{i} \frac{\partial}{\partial x} + L(x,\lambda)$$

and let $T(x,y|\lambda)$ $(x \geq y)$ be the matrix solution of

$$\tilde{L} T = 0 \qquad T(y,y|\lambda) = \text{Identity}.$$

Then FPRs for L give:

$$\{T(\tau) \otimes T(\mu)\} = [r(\lambda - \mu), \ T(\lambda) \otimes T(\mu)]$$

so that indeed the relations $(*)$ are the integrated version of the
FPR. Note that $T = \overrightarrow{\exp} - i \int_y^x L \, dx$. (cf. the definition of the

monodromy matrix in Professor L. Takhtadjan's lectures). The
discretization of the FPR's is therefore done by defining $L_n = 1 +$
$\Delta L(n\Delta,\lambda)$ where Δ is the lattice spacing.
b) Let us now impose periodic boundary conditions on the fields and
let $-L \leq x \leq L$ be a corresponding fundamental domain. If we let
$T_L(\lambda) = T(L;-L|\lambda)$ we have

$$\{tr\ T_L(\lambda),\ tr\ T_L(\mu)\} = 0$$

which gives an infinite family of observables in involution, i.e.,
a commuting family of Hamiltonians for the system in a box. In fact
the kinematical structure represented by the FPR's is present in most
1+1 dimensional integrable field theories, and the $\tilde{L}$ operator
constructed above is the associated linear operator that occurs in
the Lax pair or more generally in the zero curvature representation
of the field equations [LDF]. This establishes the connection
between the Algebraic Bethe Ansatz and the Inverse Scattering
Methods. Both are based on the generating object - the operators L_n
or $\tilde{L}$.. The latter can be obtained by the contraction limits of the
former : the quasiclassical limit as $\hbar \to 0$ and **continuous limit as**
$\Delta \to 0$.
c) The algebraic structure involved in the FPR is that of a Lie
algebra - they show that the linear space spanned by the phase-space
functions which occur as the entries of the $L(x,\lambda)$ matrix is a Lie
algebra under Poisson-Bracket. The Jacobi identity is the following
equation satisfied by the 'structure constants' r: consider the
triple tensor product $V \otimes V \otimes V$ of the auxiliary space and let
$r_{ij} = -\dfrac{P_{ij}}{\lambda}$ where the operator P_{ij} permutes the factors i and j
(i,j = 1,2,3). Then

$$[r_{12}(\lambda-\mu),\ r_{13}(\lambda-\sigma)] + [r_{12}(\lambda-\mu),\ r_{23}(\mu-\sigma)]$$

$$+ [r_{13}(\lambda-\sigma),\ r_{23}(\mu-\sigma)] = 0$$

The classification of FPRs will be the subject of Lecture §3. We
shall introduce three rather rich classes of classical r-matrix and
classify the corresponding matrices $L(x,\lambda)$. Then we shall discuss
how to classify quantum operators $L_n(\lambda)$ via suitable deformation
opposite to the contraction mentioned above. The first step in this
direction-classifying quantum R-matrices, satisfying the Yang-Baxter
relations - is a subject widely discussed but not finished yet.
Main Reference: [LDF] L.D. Faddeev in Proceedings, Les Houches XXXIX,
1982. Elsevier, 1984.

Lecture 2 : Fundamental L operators, Local quantum Hamiltonians

1. We return to the lattice, and the quantum mechanical commutation relations $R(L \otimes L) = (L \otimes L)R$ where L is an operator on $D_S \otimes D_{1/2}$ (where D_S is 2S+1 dimensional and the spin operator S acts on it). Let $\tau(\lambda)$ be the corresponding monodromy matrix

$$\tau(\lambda) = \overset{N}{\underset{n=1}{\overleftarrow{\pi}}} \ L_n(\lambda)$$

where we impose periodic boundary conditions $h_n = h_n + N$. The above commutation relations ensure that $[\text{tr } \tau(\lambda), \text{tr } \tau(\lambda')] = 0$ and this furnishes us with a family of commuting operators. However we require operators which are <u>local</u>. In the case of the passage from the 8-vertex to the XYZ model [cf. the lectures of L. Takhtadjan] this was achieved due to the following reasons: i) the quantum space and the auxiliary space have the same dimension (S = 1/2) and ii) for a certain value of the parameter λ the L matrix becames a permutation. Such L matrices we call <u>fundamental</u>. The aim of this lecture is to show how we can find local hamiltonians for systems with L matrices which are not fundamental. The idea is to substitute such an L by an associated fundamental one.

2. Suppose we are given L acting on $D_S \otimes D_{1/2}$ as above, satisfying $R(L \otimes L) = (L \otimes L)R$. Denote by $\hat{L}$ the same operator, but with the understanding that D_S is now to be regarded as the auxiliary space and $D_{1/2}$ as the quantum. Suppose we can find $\hat{R}(\lambda)$ such that $\hat{R}(0) = $ Identity, and

$$\hat{R}(\lambda-\mu) \ \hat{L}(\lambda) \otimes \hat{L}(\mu) = \hat{L}(\mu) \otimes \hat{L}(\lambda)\hat{R}(\lambda-\mu)$$

Note that now $\hat{L} \otimes \hat{L}$ acts on $D_S \otimes D_S \otimes D_{1/2}$ and $\hat{R}$ on $D_S \otimes D_S$. Let $\tilde{L} = P\hat{R}$. Then the above equation can be represented by the diagram (double lines representing spin 1/2 lines):

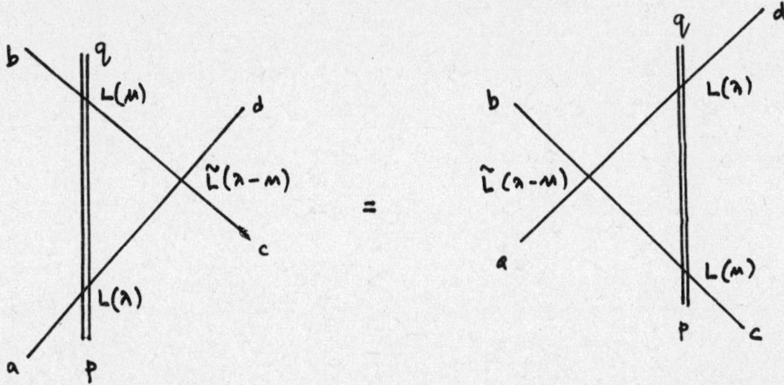

Consider now the monodromy matrices $\tau(\lambda) = \overleftarrow{\pi}\, L_n(\lambda)$, $\tilde{\tau}(\lambda) = \overleftarrow{\pi}\, \tilde{L}_n(\lambda)$.
Then $\tau(\lambda)$ acts on $\bigotimes_n (D_S)_n \otimes D_{1/2}$, $\tilde{\tau}(\lambda)$ on $\bigotimes_n (D_S)_n \otimes D_S$, and we have

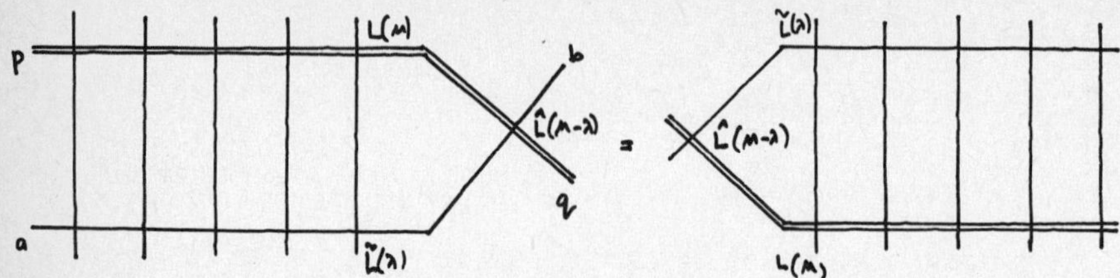

which gives $\hat{L}(\mu-\lambda)\,[\tau(\lambda) \otimes \tilde{\tau}(\mu)] = [\tilde{\tau}(\mu) \otimes \tau(\lambda)]\,\hat{L}(\mu-\lambda)$ and this
yields $[\mathrm{tr}_{D_{1/2}}\tau(\lambda),\ \mathrm{tr}_{D_S}\tilde{\tau}(\mu)] = 0$. We also have $[\mathrm{tr}_{D_S}\tilde{\tau}(\lambda),\ \mathrm{tr}_{D_S}\tilde{\tau}(\mu)]$
$= 0$. This shows that the commuting families generated by $\tilde{\tau}(\lambda)$
and $\tau(\lambda)$ could coincide. More work shows that this is indeed true.
Now $\tilde{\tau}(\lambda)$ can be used to produce local Hamiltonians exactly in the
way it was done for the XYZ model in the lectures of L. Takhtadjan.

3. We now proceed to give the construction of $\hat{R}$. Note that it acts
on $D_S \otimes D_S \simeq \overset{2S}{\underset{j=0}{+}} D_j$. Let $\Sigma^a = S^a \otimes I + I \otimes S^a$ be the spin operator
for the composite system. Then $\Sigma^a \Sigma^a$ (the Casimir, or Laplacian
operator) acts by the scalar $j(j+1)$ on D_j. We shall let J be the
operator which acts on D_j as multiplication by j; thus $(\Sigma^a)^2 = J(J+1)$.
Note the identity $2S(S+1) + 2S^a \otimes S^a = J(J+1)$. We write out
$\hat{R}(\hat{L} \otimes \hat{L}) = (\hat{L} \otimes \hat{L})\hat{R}$ using

$$\hat{L}(\lambda) \otimes \hat{L}(\mu) = I_{1/2} \otimes I_S \otimes I_S + i\sigma^a \otimes$$
$$\left(\frac{S^a \otimes I_S}{\lambda} + \frac{I_S \otimes S^a}{\mu} \right) - \frac{1}{\lambda\mu}\,\sigma^a\sigma^b \otimes S^a \otimes S^b$$
$$= I_{1/2} \otimes (I_S \otimes I_S - \frac{1}{\lambda\mu}\,S^a \otimes S^a)$$
$$+ i\sigma^c \otimes \left(\frac{-\varepsilon_{abc}\,S^a \otimes S^b}{\lambda\mu} + \frac{S^c \otimes I_S}{\lambda} + \frac{I_S \otimes S^c}{\mu} \right)$$

and let (A') be the resulting expression. Equating the coeffieicnets
of $I_{1/2}$ in A' we get

$$\hat{R}(\lambda-\mu)\,(S^a \otimes S^a) = (S^a \otimes S^a)\,\hat{R}(\lambda-\mu)$$

which implies $\hat{R}$ commutes with $(\Sigma^a)^2$ and leaves each D_j invariant.
We make the ansatz: $\hat{R}$ is a function of $\Sigma^a\Sigma^a$, i.e., acts by a scalar

on D_j. In particular it commutes with $\sum^a$.

We now equate terms in (A') proportional to σ^c, and get:

$$\hat{R}\left\{\frac{S^c \otimes I_S}{\lambda} + \frac{I_S \otimes S^c}{\mu} - \frac{\varepsilon_{abc}\, S^a \otimes S^b}{\lambda\mu}\right\} = \left\{\frac{S^c \otimes I_S}{\mu} + \frac{I_S \otimes S^c}{\lambda}\right.$$

$$\left. - \frac{\varepsilon_{abc}\, S^a \otimes S^b}{\lambda\mu}\right\}\hat{R}$$

This can be rewritten, using $[\hat{R},\, S^c \otimes I_S + I_S \otimes S^c] = [R,\, \sum^c] = 0$, and replacing $\lambda-\mu$ by λ

$$\frac{\lambda}{2}\left\{\hat{R}(\lambda)\,(I \otimes S^c - S^c \otimes I) + (I \otimes S^c - S^c \otimes I)\,\hat{R}(\lambda)\right\}$$

$$= [\hat{R}(\lambda),\, \varepsilon_{abc}\, S^a \otimes S^b]$$

Let us define now

$$X_+^a = -\frac{1}{2}\,(I \otimes S^a - S^a \otimes I)J + i\varepsilon^{abc}S^b \otimes S^c$$

$$X_-^a = \frac{1}{2}\,(I \otimes S^a - S^a \otimes I)\,(J+1) + i\varepsilon^{abc}S^b \otimes S^c$$

We have then $[J,X_\pm^a] = \pm X_\pm^a$ [TTF] and the equation to be satisfied by $\hat{R}$ becomes

$$\hat{R}(X_-^a(i\lambda-J) - X_+^a(i\lambda+J+1) = (-X_-^a(i\lambda+J) - X_+^a(J+1-i\lambda))\hat{R}$$

This is satisfied, if, with $f(i,\lambda) \underset{\text{def}}{\equiv} \hat{R}(\lambda)\big|_{D_j}$ we have

$$f(j,\lambda) = \frac{j-i\lambda}{j+i\lambda}\, f(j-1,\lambda)$$

The solution of this recurrence relation completes the construction of $\hat{R}(\lambda)$ and $\hat{L}_n(\lambda)$. In particular the hamiltonian (density) can be found in the form.

$$H = \sum_S f_S(S_n.\,S_{n+1})$$

where f_S is a polynomial of degree $2S$. Note now that eigenstates of $T(\lambda) = \text{tr } \tau(\lambda)$ are eigenstates of the local Hamiltonian as well, and thus the latter can be diagonalised by the Bethe ansatz. In fact one can prove that the energy $h(\lambda)$ (with respect to H_1) and the quasimomentum $p(\lambda)$ of one-particle states are related by

$$h(\lambda) = \frac{dp(\lambda)}{d\lambda}$$

We now make a few remarks on the lattice version of the (quantum) nonlinear Schrodinger model (NLS). An L matrix for the model can be written down:

$$L^{(NS)}(\lambda) = \frac{1}{is}\, \sigma_3\, L^{xxx}\left(\frac{-\lambda}{\kappa}\right)$$

where κ is the coupling constant, $s = -2/\kappa\Delta$, Δ the lattice constant, and L^{xxx} is the rational L matrix of the first lecture, with the quantum space $= L^2(\mathbb{R})$ and carrying an infinite-dimensional irreducible representation of SU(2) - the so-called Holstein-Primakoff representation. This representation is obtained as follows. Let Ψ^*, Ψ be the canonical creation and annihilation operators on $L^2(\mathbb{R})$, satisfying $[\Psi,\Psi^*] = 1$, and $\rho(t) = (1 + \frac{\kappa\Delta}{4}\, t)^{1/2}$. Then

$$s^1 = \frac{i}{\sqrt{\kappa\Delta}}\, (\Psi^*\rho + \rho\Psi)$$

$$s^2 = \frac{1}{\sqrt{\kappa\Delta}}\, (\rho\Psi - \Psi^*\rho)$$

$$s^3 = \frac{-2}{\kappa\Delta}\, (1 + \frac{\kappa\Delta}{2}\, \Psi^*\Psi)$$

If we consider the monodromy matrix we have (assuming an even number N of lattice sites)

$$\tau_N^{(NS)}(\lambda) = \frac{1}{(is)^N}\, \overset{N}{\underset{n=1}{\overleftarrow{\pi}}}\, \tilde{L}_n^{xxx}\left(\frac{-\lambda}{x}\right)$$

where

$$\tilde{L}_n^{xxx} = \begin{cases} L_n^{xxx}, & n \text{ even} \\ \sigma_3 L_n^{xxx}\sigma_3, & n \text{ odd} \end{cases}.$$

On the other hand,

$$\sigma_3 L_n^{xxx}\sigma_3 = U_n L_n^{xxx} U_n$$

where $U_n = \exp i\pi\, \Psi_n\Psi_n^*$ is an involutive ($U^2 = $ Identity) antomorphism of the representation. Therefore

$$\tau_N^{NS}(x) = \frac{1}{(is)^N}\, U\tau_N^{xxx}\left(\frac{-\lambda}{\kappa}\right)U$$

where

$$U = \overset{N}{\underset{n=1}{\pi}}\, U_n^{n-1}$$

Thus a series of local integrals for the NLS model can be written down : $I_k^{(NS)} = U\, I_k^{(xxx)} U$. For the latter the procedure outlined earlier in the lecture applies. Note that the choice of L matrix, the starting point, is dictated by two factors : i) It gives as

continuum, classical limit the usual Zakharov-Shabat linear operator
ii) It has a natural interpretation in terms of the rational YB
algebra. One can check that this is the 'right' choice by taking
the quantum continuum limit. See reference below for details.

Main reference :
[TTF] V.O. Tarasov, L.A. Takhtadzhyan, L.D. Faddeev, Theor. Math.
Fiz. <u>57</u> (1983) 163-181.

Lecture 3 : Classification of FPRs

This lecture, and the next, will be devoted to a (partial) classification of the Fundamental Poisson Brackets Relations (FPR):

$$\{L(x,\lambda) \overset{,}{\otimes} , L(y,\mu)\} = \{\tau(\lambda-\mu), L(x,\lambda) \otimes 1 + 1 \otimes L(y,\mu)\} \delta(x-y)$$

Now the dependence on the space-variable x is very simple in the above equation (we say it is 'ultralocal' because, for example, no derivatives of $\delta(x-y)$ occur), and it is clear that it suffices to classify

$$\{L(\lambda) \overset{,}{\otimes} L(\mu)\} = \{\tau(\lambda-\mu), L(\lambda) \otimes 1 + 1 \otimes L(\mu)\}$$

1. By a Poisson structure on a manifold M we mean a bilinear map $C^{\infty}(M) \times C^{\infty}(M) \to C^{\infty}(M)$, which we denote by $(f,g) \to \{f,g\}$, satisfying $\{f,g\} = -\{g,f\}$ and $\{f,\{g,h\}\} + \{g,\{h,f\}\} + \{h,\{f,g\}\} = 0$ and $\{f,gh\} = \{f,g\}h + g\{f,h\}$. (By $C^{\infty}(M)$ is meant the space of infinitely differentiable functions on M). One familiar example in the usual Poisson bracket defined between classical observables, i.e., functions on phase-space. Another important class of Poisson manifolds is the following: Let g be any Lie algebra, g^* its dual. Then there is a natural Poisson structure on g^*, given by

$$<[df(\hat{x}), dg(\hat{x})],\hat{x}> = \{f,g\} (\hat{x}) \text{ where } f,g \in C^{\infty}(g^*), \hat{x} \in g^*$$

and $[,]$ is the Lie bracket in g.

If we choose a basis $\{X_a\}$ for g, and let $\{\hat{\xi}_a\}$ denote the dual basis $(<X_a,\hat{\xi}_b> = \delta_{ab})$ the above bracket can be written

$$\{f,g\} (\hat{x}) = \sum C^c_{ab} \frac{\partial f}{\partial x_a} \frac{\partial g}{\partial \hat{x}_b} \hat{\xi}_c$$

where C^c_{ab} are the structure constants of g with respect to the basis $\{X_a\}$ so that $[X_a,X_b] = C^c_{ab} X_c$ and $\hat{x}_a$ are the co-ordinates of $\hat{x} = \sum \hat{x}_a\hat{\xi}_a$.

Now the above Poisson structure is degenerate, i.e., there exist nonconstant functions which have zero Poisson bracket with everything. But consider now a co-adjoint orbit (= orbit of the Lie group corresponding to g in g^* = integral manifold for all Hamiltonian vector fields = a sub-manifold of g^* where all 'bad' functions are constant). It is easy to define a restriction of the Poisson structure to such an orbit, and to see that the restricted structure to such an orbit, is nondegenerate. In the case when g = su(2), for example, the co-adjoint orbits are spheres with respect to the

Cartan-Killing inner product. This is a general fact - the centre of
the Poisson algebra is generated by the Casimir operators, and
equating them to constants yields nondegenerate Poisson structures.

Note that the co-ordinate functions $\hat{x}_a$ satisfy $\{\hat{x}_a, \hat{x}_b\} = C^c_{ab}\, \hat{x}_c$
and that this continues to hold (by definition) for the functions
restricted to co-adjoint orbits.

2. Fix now a finite-dimensional Lie algebra g and let $C(\lambda)$ denote the
space of (say, smooth, vanishing sufficiently fast at ∞) functions of
λ $(-\infty < \lambda < \infty)$ with values in g. This gives us an infinite-dimensional
Lie algebra - one defines for $x, y \in C(\lambda)$

$$[x,y](\lambda) = [x(\lambda), y(\lambda)]$$

Consider now the decomposition of $L^2(\mathbb{R})$ into orthogonal subspaces: any
function $f \in L^2(\mathbb{R})$ can be written uniquely as a sum $f = f^+ + f^-$, where
the Fourier transform of f^+ has support on $(0, \infty)$ and that of f^- in
$(-\infty, 0)$. The projections $f \to f^\pm$ can be written

$$f^\pm(\lambda) = \frac{1}{2\pi i} \int \frac{1}{\lambda - \mu \pm io}\, f(\mu)\, d\mu$$

Note f^+ can be analytically continued into the upper-half-plane and f^-
into the lower. This decomposition of $L^2(\mathbb{R})$ induces a decomposition
$C = C_+ \oplus C_-$. Now it is easy to verify that C_+ and C_- are subalgebras
(but not ideals). However we now define a new Lie algebra structure
on C denoting it by C_0 by requiring

$$[x,y]_o = [x^+, y^+] + [x^-, y^-]$$

Consider now the space $C^* = C^*_o$ of functions $\xi(\lambda)$ with values in g^*.
We can think of C^* as the dual of C or C_o as follows:

$$\langle x, \xi \rangle = \int \langle x(\lambda), \xi(\lambda) \rangle\, d\lambda$$

We compute the co-Poisson structure on C^* considered as the dual of C_o.
We work formally, and let $\hat{\xi}_{a,\lambda}$, defined by $\hat{\xi}_{a,\lambda}(\mu) = \delta(\lambda - \mu)\hat{\xi}_a$ be a
basis for C^*, so that we can write, for any $\xi \in C^*$

$$\xi(\mu) = \xi_a(\mu)\hat{\xi}_a = \int \xi_a(\lambda)\hat{\xi}_{a,\lambda}(\mu)\, d\lambda$$

This $\xi_a(\mu)$ can be regarded as the co-ordinates of ξ. It is clear that

$$\{\xi_a(\lambda),\ \xi_b(\mu)\}_c = \delta(\lambda - \mu) C^c_{ab} \xi_c(\lambda)$$

The decomposition of C induces a decomposition of $C^*; C^* = C^*_+ + C^*_-$ where

C_+^* consists of elements annihilated by C_- (and hence analytically continuable into the _lower_ half-plane) and similarly for C_-^*. We have then

$$\xi_a^+(\lambda) = \frac{-1}{2\pi i} \int \frac{1}{\lambda-\mu+io} \, \xi_a(\mu) d\mu \ \epsilon \ C_-^*$$

$$\xi_a^-(\lambda) = \frac{1}{2\pi i} \int \frac{1}{\lambda-\mu-io} \, \xi_a(\mu) d\mu \ \epsilon \ C_+^*$$

We have then

$$\{\xi_a^+(\lambda), \ \xi_b^+(\mu)\}_C = \frac{1}{(2\pi i)^2} \iint \frac{1}{\lambda-\rho+io} \frac{1}{\mu-\delta+io}$$

$$\{\xi_a(\rho), \ \xi_b(\delta)\} d\rho d\delta$$

$$= \frac{1}{(2\pi i)^2} \int \frac{1}{(\lambda-\rho+io)(\mu-\rho+io)} C_{ab}^c \, \xi_c(\rho) \, d\rho$$

$$= \frac{C_{ab}^c}{\lambda-\mu} \frac{1}{2\pi i} \int (\frac{1}{\lambda-\rho+io} - \frac{1}{\lambda-\rho+io}) \, \xi_c(\rho) \, d\rho$$

$$= \frac{C_{ab}^c}{2\pi i(\lambda-\mu)} \, (\xi_c^+(\lambda) - \xi_c^+(\mu))$$

Similarly,

$$\{\xi_a^-(\lambda), \xi_b^-(\lambda)\}_C = \frac{C_{ab}^c}{2\pi i(\lambda-\mu)} \, (\xi_c^-(\lambda)-\xi_c^-(\mu))$$

Thus we see that

$$\{\xi_a(\lambda), \xi_b(\mu)\} = \{\xi_a^+(\lambda), \ \xi_b^-(\mu)\} + \{\xi_a^-(\lambda), \ \xi_b^-(\mu)\}$$

$$= \frac{C_{ab}^c}{2\pi i(\lambda-\mu)} \, (\xi_c(\lambda) - \xi_c(\mu))$$

A rescaling gets rid of the 2 i factor in the denominator, and we can write

$$\{\xi_a(\lambda), \ \xi_b(\mu)\} = \frac{C_{ab}^c}{\lambda-\mu} \, (\xi_c(\lambda) - \xi_c(\mu))$$

3. We will see that the above equation can be cast in the form of an FPR. Suppose there exist matrices A^a (acting on a vector space V) and π (acting on $V \otimes V$) satisfying

$$-[\pi, \ A^a \otimes I] = [\pi, I \otimes A^a] = C_{bc}^a \, A^b \otimes A^c$$

Define $L(\lambda) = \xi_a(\lambda) A^a$. Then

$$\{L(\lambda) \otimes L(\mu)\} = \sum_{ab} \{\xi_a(\lambda), \xi_b(\mu)\} \ A^a \otimes A^b$$

$$= \frac{C^c_{ab}}{\lambda - \mu} \ A^a \otimes A^b \ (\xi_c(\lambda) - \xi_c(\mu))$$

$$= \frac{1}{\lambda - \mu} \ [\pi, \ I \otimes L(\lambda) + L(\mu) \otimes I]$$

which is as desired.

If the Lie algebra g is semi simple one can take $A^a = K^{ab} X_b$, $r = K^{ab} X_a \otimes X_b$ where K^{ab} is the inverse of the Cartan matrix $K_{ab} = (X_a, X_b)$ and (,) denotes the Killing inner product. (In the above expressions for A^a and r the X_a are regarded as matrices in any representation of the Lie algebra). For example the relation $[\pi, \ I \otimes A^a] = C^a_{bc} A^b \otimes A^c <=> C^e_{at} K_{pe} = C^e_{pa} K_{te} <=> (X_p, [X_a, X_t]) = (X_t, [X_p, X_a]) <=> ([X_a, X_p], X_t) + (X_p, [X_a, X_t]) = 0$ which is the statement of the g-invariance of the inner product.

4. We will now see that there are finite-dimensional co-orbits in C_0^*. Fix λ_0, with Im $\lambda_0 < 0$ for definiteness, and consider the set θ_{λ_0} of elements ξ_1 of C_0^* of the form $\xi_1(\lambda) = \frac{\eta}{\lambda - c}$, $\eta \in g^*$. We claim that the set of θ_{λ_0} is invariant under C_0. Note $\xi_1 \in C_-^*$. Let $x \in C_-$ and compute ad $x(\xi_1)$. This is defined by

$$<Y, \ ad \ x(\xi_1)> = <[x, y], \ \xi_1> = \int \frac{<[x(\lambda), y(\lambda)]>}{\lambda - \lambda_0} \ d\lambda$$

$$= 2\pi i \ <[x(\lambda_0), \ y(\lambda_0)], \eta> \ \text{by the residue formula}$$

$$= 2\pi i \ <y(c), \ ad \ x(\lambda_0)\eta>$$

$$= \int \frac{<y(\lambda), \ ad \ x(\lambda_0)\eta>}{\lambda - \lambda_0} \ d\lambda$$

$$= <y, \xi_1> \ \text{where} \ \xi_1(\lambda) = \frac{ad \ x(\lambda_0)\eta}{\lambda - \lambda_0}$$

Thus θ_{λ_0} is a finite-dimensional submanifold left invariant by the coadjoint action of C_0. Of course the restricted Poisson structure is not yet nondegenerate, but that can be achieved in an obvious way. For example for g = su(2) one considers the submanifold of θ_{λ_0} defined by $|\eta|^2$ = constant. Note that the only finite-dimensional C-orbit of C^* is the trivial one $\xi = 0$ unless g has a nontrivial centre.

<u>Lecture 4</u> : <u>The Fundamental Poisson Brackets (contd.)</u>
<u>The classical lattice case.</u>

1. In the last lecture we saw that a large class of solutions of the
FPR

$$\{L(\lambda) \otimes L(\mu)\} = [r(\lambda-\mu), L(\lambda) \otimes 1 + 1 \otimes L(\mu)\}$$

are associated with the current algebra C_o. The $L(\lambda)$ had matrix
entries which were functions on finite-dimensional sympletic manifolds
corresponding to the case of finite number of fields. If the current
algebra C_o is defined 'over' a semisimple Lie algebra g and $\{X_a\}$ is
a basis for g, the r matrix is given by $r(\lambda) = \frac{K^{ab}}{\lambda} X_a \otimes X_b$ where the X_a
are now thought of as matrices in any representation. On the single-
pole orbits L is given by

$$L(\lambda) = \frac{S_a A^a}{\lambda-\lambda_o}$$

where $A^a = K^{ab} X_b$ (and there has been a change of notation $\xi_a \to S_a$).
The S_a can be regarded as dynamical variables, i.e., functions on the
orbit, and satisfy (check) $\{S_a, S_b\} = C^c_{ab} S_c$. In this way we return to
the basic example with the matrix $L(x, \lambda)$ having the form $\frac{S(x)}{\lambda}$ where
$S(x)$ is defined on the co-orbit of a.

We can generalise to multipole orbits and find solutions of the
form $L(\lambda) = \sum_\alpha \sum_k \frac{S^{\alpha,k}}{\lambda-C_\alpha}$ (C_α constants) where $S^{\alpha,k} = S^{\alpha,k}_a A^a$ and one can
check that

$$\{S^{\alpha,k}_a, S^{\beta,\ell}_b\} = \delta_{\alpha\beta} C^c_{ab} S^{d,k+\ell-1}_c$$

Example : $L = \frac{S}{\lambda-C} + \frac{T}{(\lambda-C)^2}$, C a constant

$$\{S_a, S_b\} = C^c_{ab} S_c$$

$$\{S_a, T_b\} = C^c_{ab} T_c$$

$$\{T_a, T_b\} = 0$$

If $g = SU(2)$ this gives the Lie algebra of the group $E(3)$ of Euclidean
motions on 3 dimensions.

2. We now turn to FPR with trigonometric L and r matrices. We obtain
these by averaging the rational ones over a discrete group of
translations. Formally, we define (for $\omega \neq 0$),

173

$$\tilde{L}(\lambda) = \sum_{n=-\infty}^{\infty} U^n L(\lambda+n\omega) U^{-n}$$

where U is an automorphism of the space on which the L matrix acts, chosen such that $[U \otimes U, r] = 0$. We can then check

$$\tilde{L}(\lambda) \otimes \tilde{L}(\mu) = [\tilde{r}(\lambda-\mu), \tilde{L}(\lambda) \otimes I + I \otimes \tilde{L}(\mu)]$$

where

$$\tilde{L}(\lambda) = \sum_n (U^n \otimes I) r(\lambda+n\omega)(U^{-n} \otimes I)$$

Note that both r and L are now periodic with period . The elliptic case is obtained by averaging over a lattice in $\mathbb{C}^2$. One needs a second operator U' (ad-) commuting with the first, and this exists only when g = SU(n).

Example : g = SU(2). We can take $U_1 = \sigma_3$, $U_2 = \sigma_1$. Consider L,r defined in the fundamental representation. Note that for x ε g we have $\sigma_1\sigma_2 \times \sigma_2\sigma_1 = -\sigma_3 \times \sigma_3 = \sigma_2\sigma_1 \times \sigma_1\sigma_2$. The averaging procedure gives

$$L(\lambda,K) = \frac{1}{sn(\lambda,K)} [S_1\sigma_1 + S_2\sigma_2 dn(\lambda,K) + S_3\sigma_3 cn(\lambda,K)]$$

This is the L operator for the classical analogue of the XYZ model and was introduced by Sklyanin 5 years ago as a guess. As K → 0 we recover a trigonometric L operator, corresponding to a classical XXY model. One can also obtain L operators of the classical Liouville and sine-Gordon models.

3. We can retrace the path followed in lecture 1 and construct classical and quantum lattice operators L_n via a suitable modification of the co-adjoint picture. We first consider the classical lattice case. We need solutions of

$$\{L(\lambda) \otimes L(\mu)\} = \{r(\lambda-\mu), L(\lambda) \otimes L(\mu)\}$$

The right-hand side is quadratic in L, and this is a new algebraic structure which has been investigated in by Sklyanin, and hence called a 'Sklyanin algebra'.

The r matrix has to satisfy the Jacobi Identity. One considers r matrices obtained via the earlier considerations and seeks all L matrices with finite-dimensional phase-spaces satisfying the above algebra. Now the choice of the representation of the Lie algebra

174

(which entered into the definition of r and L) is relevant and there will in general be restrictions on the orbits for the algebra to be valid. In the case when $g = su(n)$, and the representation is fundamental, π is a permutation (upto addition of a constant) and we can take

$$L(\lambda) = I + \frac{i \, S_a^n \, A^a}{\lambda - \lambda_o}$$

where $\{S_a^n, S_b^n\} = C_{ab}^c S_c$, and C_{ab}^c are the SU(n) structure constants. The orthogonal and symplectic case are more difficult. For example if $g = o(n)$ and we work with the spin representation a single pole L matrix exists only on certain orbits. If one allows a second pole all orbits are allowed.

It is possible to obtain trigonometric and elliptic L matrices as well; one uses a procedure of multiplicative averaging in contrast to the additive one used in the continuum case.

References :

1) N.Yu. Reshetikhin and L.D. Faddeev, Teor. Mat. Fiz., _56_ (1983) 323-343.
2) M.A. Semenov Tyan-Shanskii, Funct. Anal. Appl. _17_, 259, 1984.
3) E.K. Sklyanin, Funct. Anal. Appl. _16_, 263 (1982); ibid _17_, 273 (1983).

<u>INTRODUCTION TO ALGEBRAIC BETHE ANSATZ</u>

L.A. Takhtajan

Steklov Mathematical Institute
Leningrad Branch, Fontanka 27
191011 Leningrad USSR

<u>Lecture I. Monodromy matrix and quantum R-matrix.</u>

Consider $M \times N$ rectangular lattice on two-dimensional torus and place arrows on every bond. Spin variable $\sigma = \pm 1$ can be used to denote arrows. Assign to each combination of arrows at every vertex a positive number ε_j - energy of combination. The total energy of a given configuration of arrows on the lattice is

$$E = \sum_{j=1}^{16} N_j \, \varepsilon_j \ , \tag{1.1}$$

where N_j is the number of vertices with j-type combination of arrows.

The <u>partition function</u>

$$Z = \sum e^{-\beta E} \ , \tag{1.2}$$

(all configurations)

where β is proportional to the inverse temperature, describes thermodynamical properties of the model. The expression (1.2) can be simplified in the following way. Consider the <u>Boltzmann weights</u> $v_j = e^{-\beta \varepsilon_j}$ and organize them into a matrix $R_\alpha^{\alpha'}(\gamma, \gamma')$ (see Fig. 1)

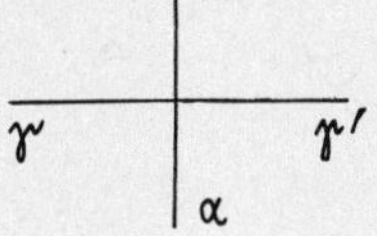

Figure 1

where for the arrows pointing up and to the right we have variables
$+1$ and -1 for those pointing down and to the left. Introducing
multiindices $\{\alpha\} = \{\alpha_1,\ldots,\alpha_N\}$, $\{\alpha'\} = \{\alpha'_1,\ldots,\alpha'_N\}$ define
the matrix T with matrix elements

$$T_{\{\alpha\},\{\alpha'\}} =$$

$$= \sum_{\gamma_1}\ldots\sum_{\gamma_N} R^{\alpha'_1}_{\alpha_1}(\gamma_1,\gamma_2)\, R^{\alpha'_2}_{\alpha_2}(\gamma_2,\gamma_3)\ldots R^{\alpha'_N}_{\alpha_N}(\gamma_N,\gamma_1)\,; \tag{1.3}$$

it can be graphically represented as follows

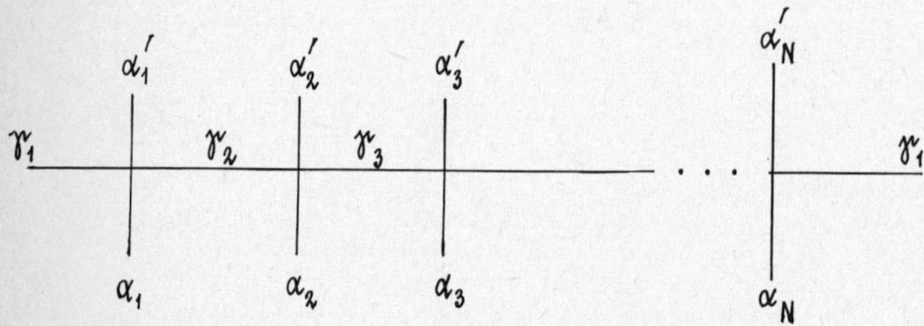

Figure 2

Matrix T is of order $2^N \times 2^N$ and is called <u>transfer-matrix</u>.
Now we can rewrite (1.2) in the form

$$Z = \mathrm{Sp}\, T^M\,, \tag{1.4}$$

where Sp denotes the matrix trace. To prove this arrange summation in (1.2) first along the horizontal directions and then along
the vertical ones. Thus we reduce the problem of evaluating the partition function to the problem of diagonalization of the corresponding transfer-matrix. What was said above are well-known facts about
the connection between $\mathcal{D}$-dimensional classical statistical mechanics and $\mathcal{D}-1$ -dimensional quantum mechanics.

The real problem is now to diagonalize the transfer-matrix. The
idea, which seems quite natural , is to embed it into the family of
commuting transfer-matrices and to try to diagonalize the whole

family simultaneously. Suppose we have two sets v_j and v_j' of Boltzmann weights and corresponding transfer-matrices T and T'. Under what conditions do we have

$$[\, T, T'\,] = 0 \,.$$
(1.5)

Can we answer this question in terms of Boltzmann weights only?

Things will be simplified if we introduce a new object - 2×2 matrix $\mathcal{T}$ with matrix elements - operators in $\mathbb{C}^{2N}$. Graphically it looks as follows

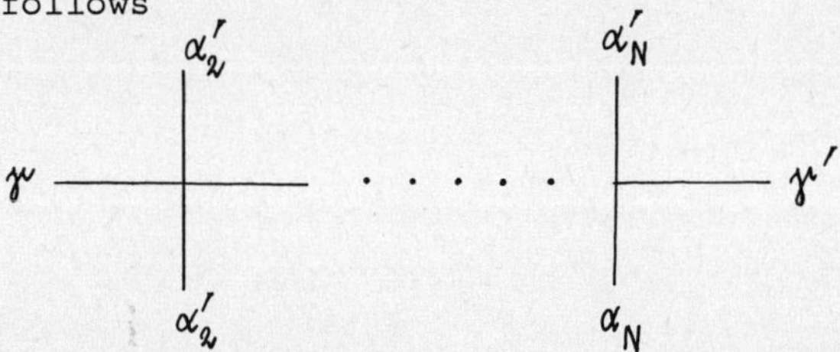

Figure 3

Now the variables γ and γ' are free and there is no summation contrary to Fig.2. It is clear that

$$T = \mathrm{tr}\, \mathcal{T},$$
(1.6)

where the trace is taken in two-dimensional space. More formally: introduce the space

$$\mathcal{S}_N = \prod_{n=1}^{N} \otimes \, h_n \,, \quad h_n \approx \mathbb{C}^2$$
(1.7)

in which T acts, write down

$$R_{\alpha'}^{\alpha'}(\gamma, \gamma') = \sum_{i,j=1}^{4} w_{ij}\, \sigma_{\gamma'\gamma}^{i}\, \sigma_{\alpha\alpha'}^{j} \,,$$
(1.8)

where $\sigma^{i}_{\alpha\alpha'}$ are the matrix elements of Pauli matrices σ^{i}, $i = 1, \ldots, 4$, $\sigma^{4} = I$, and define

$$L_n = \sum_{i,j=1}^{4} w_{ij}\, \sigma^{i} \otimes \sigma^{j}_{n} \ . \tag{1.9}$$

Here σ^{j}_{n} are matrices in $\mathfrak{H}_N$; they act non-trivially only in the space h_n , where they coincide with σ^{j} . So

$$L_n = \begin{pmatrix} \hat{\alpha}_n & \hat{\beta}_n \\ \hat{\gamma}_n & \hat{\delta}_n \end{pmatrix} \ , \tag{1.10}$$

where $\hat{\alpha}_n, \hat{\beta}_n, \hat{\gamma}_n, \hat{\delta}_n$ are operators in $\mathfrak{H}_N$ which act non-trivially only in h_n . This operator-valued matrix L_n is called the <u>local L-operator</u>. We have now

$$T = L_N \cdots L_1 = \prod_{n=1}^{N} L_n \ , \tag{1.11}$$

so

$$T = \begin{pmatrix} A & B \\ C & \mathcal{D} \end{pmatrix} \ , \qquad T = A + \mathcal{D} \ . \tag{1.12}$$

Operator-valued matrix T is called <u>monodromy matrix</u>.

In terms of monodromy matrix the sufficient (not necessary!) condition for (1.5) to be true is that

$$R(\mathcal{T}\otimes\mathcal{T}') = (\mathcal{T}'\otimes\mathcal{T})R\ , \tag{1.13}$$

where R is 4×4 matrix over complex numbers, and the tensor product $\mathcal{T}\otimes\mathcal{T}'$ is 4×4 operator-valued matrix with the following 2×2 blocks

$$\mathcal{T}\otimes\mathcal{T}' = \begin{pmatrix} A\mathcal{T}' & B\mathcal{T}' \\ C\mathcal{T}' & \mathcal{D}\mathcal{T}' \end{pmatrix}\ . \tag{1.14}$$

Indeed, rewrite (1.13)

$$\mathcal{T}\otimes\mathcal{T}' = R^{-1}(\mathcal{T}'\otimes\mathcal{T})R \tag{1.15}$$

and apply the trace in $\mathbb{C}^4$; then (1.5) follows.

It seems that to deal with (1.13) is even more difficult than with (1.5). However the main advantage of (1.13) is that it is sufficient to verify it only <u>locally</u>.

Namely, <u>if for all n we have</u>

$$R(L_n\otimes L_n') = (L_n'\otimes L_n)R\ , \tag{1.16}$$

<u>then (1.13) follows.</u>

The graphical proof is quite clear (see Fig.4)

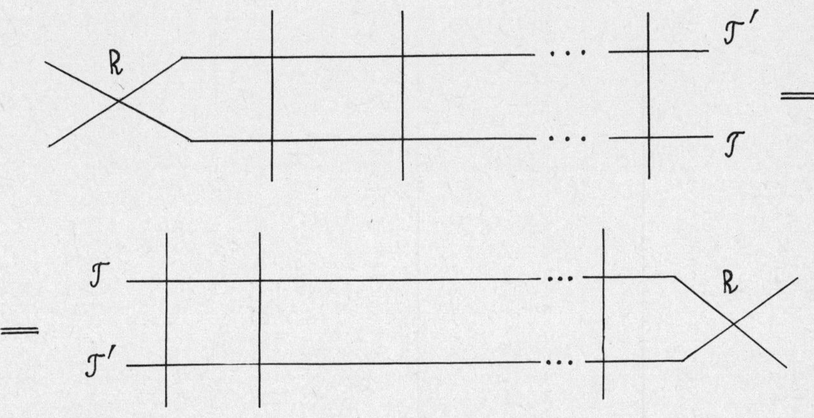

Figure 4

- the "railroad proof". The formal proof uses the formula

$$(A \otimes C)(B \otimes D) = AB \otimes CD \tag{1.17}$$

for C-number matrices which is also valid in our case because the matrix elements of L_n and L_m commute when $n \neq m$.

The matrix R is called the <u>quantum R-matrix</u> (though its matrix elements are C-numbers).

So far we have considered the general <u>16-vertex model</u>; unfortunately it has no R-matrix. However in the case of <u>8-vertex model</u> without external field R-matrix exists. In this model we have only four different combinations (see Fig.5)

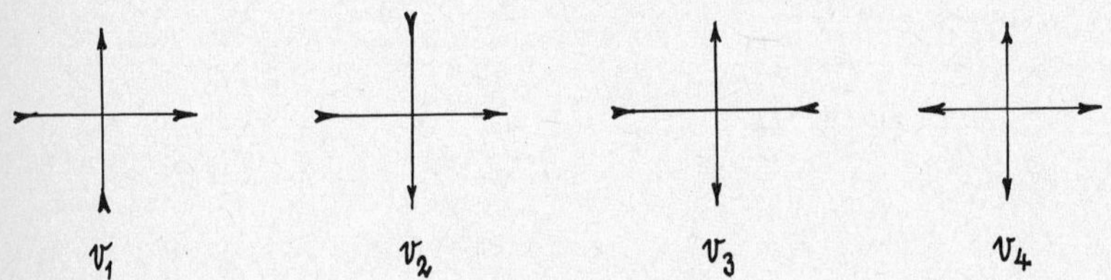

Figure 5

due to the fact that the model is invariant under the reversion of all the arrows. In the case $v_4 = 0$ we have so-called "ice rule", which corresponds to the <u>6-vertex model</u>.

Now write down

$$L_n = \sum_{j=1}^{4} w_j \, \sigma^j \otimes \sigma_n^j \ , \qquad L_n' = \sum_{j=1}^{4} w_j' \, \sigma^j \otimes \sigma_n^j \ , \tag{1.18}$$

where

$$w_1 = \frac{v_3 + v_4}{2} \ , \quad w_2 = \frac{v_3 - v_4}{2} \ ,$$

$$w_3 = \frac{v_1 - v_2}{2} \ , \quad w_4 = \frac{v_1 + v_2}{2} \tag{1.19}$$

and we have similar expressions for w_j' in terms of v_j' , and try to find R-matrix in the form

$$R = P \sum_{j=1}^{4} w_j'' \, \sigma^j \otimes \sigma^j \ . \tag{1.20}$$

Here P is the permutation matrix in $\mathbb{C}^4$:

$$P(\, e \otimes f \,) = f \otimes e \tag{1.21}$$

for all e and f in $\mathbb{C}^2$. Inserting (1.18) and (1.20) into (1.16) we get 6 independent equations. Considering coefficients w_j'' as unknown we have the following solvability conditions:

quotients $\dfrac{w_j^2 - w_\kappa^2}{w_\ell^2 - w_n^2}$ should be invariant under the change $w \longrightarrow w'$. From this one obtains 3 equations which define an elliptic curve. Its explicit parametrization is given by

$$w_3 + w_4 = \rho \, sn\,(\lambda + \eta, k)\,,$$

$$w_3 - w_4 = \rho \, sn\,(\lambda - \eta, k)\,,$$

$$w_1 + w_2 = \rho \, sn\,(2\eta, k)\,,$$

$$\tag{1.22}$$

$$w_1 - w_2 = \rho\, k\, sn(2\eta, k)\, sn(\lambda - \eta, k)\, sn\,(\lambda + \eta, k)\ .$$

Here $sn(u, k)$ denotes an elliptic sine of modulus k . Coefficients w_j' and w_j'' are obtained from (1.22) by the replacement $\lambda \longrightarrow \mu$ and $\lambda \longrightarrow \lambda - \mu + \eta$ correspondingly. Denoting $L_n = L_n(\lambda)$, $L_n' = L_n(\mu)$ we can rewrite (1.16) in the form

$$R\,(\lambda - \mu)(\, L_n(\lambda) \otimes L_n(\mu)) = (\, L_n(\mu) \otimes L_n(\lambda)) \, R\,(\lambda - \mu)\,, \tag{1.23}$$

where

$$R(\lambda) = PL(\lambda + \eta), \qquad L(\lambda) = \sum_{j=1}^{4} w_j(\lambda)\, \sigma^j \otimes \sigma^j \,. \tag{1.24}$$

Relation (1.23) is extremely important for the further exposition. The following remarks are in order.

1. "Horizontal" indices γ, γ' "belong" to <u>auxiliary space</u> $V = \mathbb{C}^2$ and "vertical" indices α, α' belong to the <u>quantum space</u> $h \,(\approx \mathbb{C}^2)$; R-matrix acts in $V \otimes V$, L_n - in $V \otimes h_n$, $\mathcal{T}$ - in $V \otimes \mathfrak{H}_N$ and T - in $\mathfrak{H}_N$.

2. Generally V and h can have any dimensions, e.g. $\dim h = \infty$. The last case occurs for field-theoretical models on the one-dimensional lattice. Algebraic scheme described above works for arbitrary V and h .

3. Formula (1.24) makes sense only when $V \approx h$. Denoting

$$S(\lambda) = PR(\lambda) \qquad (= L(\lambda + \eta)) \tag{1.25}$$

one can rewrite (1.23) in the form

$$S_{12}(\lambda - \mu)\, S_{13}(\lambda)\, S_{23}(\mu) \;=\; S_{23}(\mu)\, S_{13}(\lambda)\, S_{12}(\lambda - \mu)\,, \tag{1.26}$$

where $S_{12}(\lambda)$ is the matrix in $V \otimes V \otimes V$, which acts trivially in the third space V in the tensor product and coincides with $S(\lambda)$ in the product of the first two spaces; similarly for S_{13} and S_{23} . Relation (1.26) is called <u>factorization equation</u> or <u>triangle equation</u>. General relation (1.23) is called <u>Yang-Baxter equation</u>.

4. In the context of the scattering theory $S(\lambda)$ plays the role of two-particle S-matrix. Relation (1.26) shows that the scattering theory is <u>factorizable</u>.

5. In statistical mechanics relation (1.26) means that (see Fig. 6)

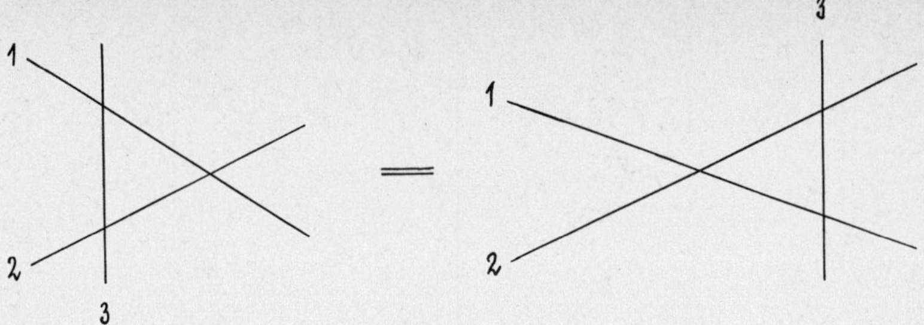

Figure 6

in the sense of partition functions. This explains the second name
of the relation (1.26).

   6. Equation (1.26) for $S(\lambda) = PR(\lambda)$ gives a sufficient
condition for possible R-matrices. Namely, consider
$L_n(\lambda) \otimes L_n(\mu) \otimes L_n(\nu)$ and rewrite the associativity rule of
the tensor product using (1.23). One obtains

$$\left[RHS\ (1.26) - LHS(1.26) \right] L_n \otimes L_n \otimes L_n = 0 . \tag{1.27}$$

   7. One can also consider the formal semigroup with generators
$A_j(\lambda),\quad j = 1,\ldots, m$ and relations

$$A_i(\lambda) A_j(\mu) = S^{i\ell}_{jk} (\lambda - \mu) A_k(\mu) A_\ell(\lambda) . \tag{1.28}$$

Then equation (1.26) ensures the associativity law.

Lecture II . Algebraic Bethe ansatz in " abc -form".

   Thus we have imbedded the transfer-matrix T into the commuting family $T(\lambda)$. Now we shall give the procedure for diagonalization of operators $T(\lambda)$. For simplicity we shall consider
the 6-vertex case, where $w_1 = w_2$. In formulas (1.22) one should
set $k \longrightarrow 0$ (and take $\rho = 1$), so elliptic functions go into
trigonometric ones. We have

$$L_n(\lambda) = \begin{pmatrix} w_4(\lambda) + w_3(\lambda)\,\sigma_n^3 & \dfrac{\sin 2\eta}{2}\,\sigma_n^- \\[2ex] \dfrac{\sin 2\eta}{2}\,\sigma_n^+ & w_4(\lambda) - w_3(\lambda)\,\sigma_n^3 \end{pmatrix}, \qquad (2.1)$$

where

$$(w_4 + w_3)(\lambda) = \sin(\lambda + \eta), \quad (w_4 - w_3)(\lambda) = \sin(\lambda - \eta) \qquad (2.2)$$

and

$$R(\lambda) = \begin{pmatrix} 1 & 0 & 0 & 0 \\ 0 & b(\lambda) & c(\lambda) & 0 \\ 0 & c(\lambda) & b(\lambda) & 0 \\ 0 & 0 & 0 & 1 \end{pmatrix}, \qquad (2.3)$$

where

$$b(\lambda) = \frac{\sin 2\eta}{\sin(\lambda + 2\eta)}, \qquad c(\lambda) = \frac{\sin \lambda}{\sin(\lambda + 2\eta)} \qquad (2.4)$$

— the "abc-form" of R-matrix with $a = 1$.

Now observe that in h_n there exists vector ω_n such that matrix $L_n(\lambda)$ acting on ω_n by all its matrix elements becomes triangular. Namely, for $\omega_n = \begin{pmatrix} 1 \\ 0 \end{pmatrix}$ we have

$$L_n(\lambda)\,\omega_n = \begin{pmatrix} \alpha(\lambda)\,\omega_n & * \\[2ex] 0 & \delta(\lambda)\,\omega_n \end{pmatrix}, \qquad (2.5)$$

where

$$\alpha(\lambda) = \sin(\lambda + \eta), \qquad \delta(\lambda) = \sin(\lambda - \eta).$$

(2.6)

The vector ω_n having these properties is called the <u>local va-cuum</u>. The vector

$$\Omega = \prod_{n=1}^{N} \otimes \, \omega_n \in \mathfrak{H}_N$$

(2.7)

plays the same role with respect to the monodromy matrix $\mathcal{T}(\lambda)$

$$\mathcal{T}(\lambda)\,\Omega = \begin{pmatrix} \alpha^N(\lambda)\,\Omega & * \\ 0 & \delta^N(\lambda)\,\Omega \end{pmatrix}.$$

(2.8)

This follows from formula (2.5), the definition of $\mathcal{T}(\lambda)$

$$\mathcal{T}(\lambda) = \prod_{n=1}^{N} L_n(\lambda) = \begin{pmatrix} A(\lambda) & B(\lambda) \\ C(\lambda) & \mathcal{D}(\lambda) \end{pmatrix}$$

(2.9)

and the multiplication rule for triangular matrices.

Thus we have

$$A(\lambda)\Omega = \alpha^N(\lambda)\Omega, \qquad \mathcal{D}(\lambda)\Omega = \delta^N(\lambda)\Omega,$$

(2.10)

so Ω is the eigenvector of $T(\lambda)$ with the eigenvalue $\alpha^N(\lambda) + \delta^N(\lambda)$. The vector Ω having these properties is called the <u>reference state</u>.

In order to construct other eigenvectors of $T(\lambda)$ one should make use of the fundamental relation with R-matrix

$$R(\lambda-\mu)(T(\lambda)\otimes T(\mu)) = (T(\mu)\otimes T(\lambda))\,R(\lambda-\mu)\,, \qquad (2.11)$$

which was introduced in the previous lecture. In (2.11) we have 16 commutation relations between operators A, B, C and D . Three of them will be of great use

$$\left[\,B(\lambda),\,B(\mu)\,\right] = 0\,, \qquad (2.12)$$

$$A(\lambda)B(\mu) = \frac{1}{c(\mu-\lambda)}\,B(\mu)A(\lambda) - \frac{b(\mu-\lambda)}{c(\mu-\lambda)}\,B(\lambda)A(\mu)\,, \qquad (2.13)$$

$$D(\lambda)B(\mu) = \frac{1}{c(\lambda-\mu)}\,B(\mu)D(\lambda) - \frac{b(\lambda-\mu)}{c(\lambda-\mu)}\,B(\lambda)D(\mu)\,. \qquad (2.14)$$

These formulas suggest that $B(\lambda)$ can be interpreted as a kind of "creation operator", so one can try to obtain eigenvectors of $T(\lambda) = A(\lambda) + D(\lambda)$ by successive application of operators $B(\lambda)$ to the reference state Ω .

Namely, suppose that the relations (2.10), (2.12) - (2.14) with odd function $\dfrac{b(\lambda)}{c(\lambda)}$ are satisfied. <u>Then the vector</u>

$$\Phi(\{\lambda_j\}) = \prod_{j=1}^{\ell} B(\lambda_j)\,\Omega \qquad (2.15)$$

is the eigenvector of $T(\lambda)$

$$T(\lambda)\,\Phi = \Lambda(\lambda)\,\Phi \;, \qquad\qquad (2.16)$$

where

$$\Lambda(\lambda;\lambda_1,\ldots,\lambda_\ell) = \alpha^N(\lambda)\prod_{j=1}^{\ell}\frac{1}{c(\lambda_j-\lambda)} +$$

$$+\,\delta^N(\lambda)\prod_{j=1}^{\ell}\frac{1}{c(\lambda-\lambda_j)} \qquad\qquad (2.17)$$

if the pairwise distinct numbers $\{\lambda_j\}$ satisfy the following system of equations

$$\frac{\alpha^N(\lambda_j)}{\delta^N(\lambda_j)} = \prod_{\substack{k=1\\k\neq j}}^{\ell}\frac{c(\lambda_k-\lambda_j)}{c(\lambda_j-\lambda_k)}\;,\qquad j=1,\ldots,\ell\;. \qquad (2.18)$$

For the proof consider the action of operator $A(\lambda)$ on vector Φ - the expression $A(\lambda)B(\lambda_1)\ldots B(\lambda_\ell)\Omega$. Using commutation relation (2.14) we can carry $A(\lambda)$ through all $B(\lambda_j)$ towards the vector Ω and apply (2.10). At first glance one gets 2^ℓ terms according to (2.14). However they can be combined into $\ell+1$ terms using the following observation. Commuting $A(\lambda)$ with $B(\mu)$ by the formula (2.14) we have two terms: in the first term A and B preserve their arguments λ and μ and in the second term their arguments are exchanged. So moving A through all B's towards Ω we will get $\ell+1$ terms: in the first term all arguments λ_j in B's are present; in the second term λ_1 disappears from the arguments of B's and so on. Thus we have the expression

$$A(\lambda)\,\Phi = M(\lambda,\{\lambda_j\})\,\Phi +$$

$$+ \sum_{k=1}^{n} M_k\,(\lambda,\{\lambda_j\})\,\Phi_k \quad, \tag{2.19}$$

where

$$\Phi_k(\lambda,\{\lambda_j\}) = \prod_{\substack{j=1 \\ j\neq k}}^{\ell} B(\lambda_j)\,B(\lambda)\,\Omega \;. \tag{2.20}$$

The term Φ in the RHS of (2.19) can be obtained only when the first terms in (2.14) are used. So

$$M = \alpha^N(\lambda)\,\prod_{j=1}^{\ell} \frac{1}{c(\lambda_j-\lambda)} \quad. \tag{2.21}$$

In order to obtain the term Φ_1 the only way is to use the second term in (2.14) when commuting $A(\lambda)$ with $B(\lambda_1)$ and to use the first term in (2.14) when commuting $A(\lambda_1)$ with remaining B's. Otherwise the argument λ_1 reappears in B's. Thus, we get

$$M_1 = -\alpha^N(\lambda_1)\,\frac{b(\lambda_1-\lambda)}{c(\lambda_1-\lambda)}\,\prod_{j=2}^{\ell} \frac{1}{c(\lambda_j-\lambda_1)} \quad. \tag{2.22}$$

In order to obtain expressions for coefficients M_k, $k \geqslant 2$, one should in principle combine more terms. However, the following trick works: use commutativity of B's (we have already used it) and put $B(\lambda_k)$ in (2.15) into first place and then repeat the reasoning presented above. One immediately obtains

$$M_k = -\alpha^N(\lambda_k)\,\frac{b(\lambda_k-\lambda)}{c(\lambda_k-\lambda)}\,\prod_{\substack{j=1\\j\neq k}}^{\ell}\frac{1}{c(\lambda_j-\lambda_k)}\ . \tag{2.23}$$

The same considerations can be applied to $\mathcal{D}(\lambda)\,\Phi$ and we get

$$\mathcal{D}(\lambda)\,\Phi = N(\lambda,\{\lambda_j\})\,\Phi\ +$$

$$+\sum_{k=1}^{\ell} N_k(\lambda,\{\lambda_j\})\,\Phi_k\ , \tag{2.24}$$

where

$$N = \delta^N(\lambda)\,\prod_{j=1}^{\ell}\frac{1}{c(\lambda-\lambda_j)} \tag{2.25}$$

and

$$N_k = -\delta^N(\lambda_k)\,\frac{b(\lambda-\lambda_k)}{c(\lambda-\lambda_k)}\,\prod_{\substack{j=1\\j\neq k}}^{\ell}\frac{1}{c(\lambda_k-\lambda_j)}\ . \tag{2.26}$$

Now, if the relations

$$M_k(\lambda,\{\lambda_j\}) + N_k(\lambda,\{\lambda_j\}) = 0\ , \tag{2.27}$$

$\kappa = 1, \dots, \ell$, are valid for all λ, then the vector $\Phi(\{\lambda_j\})$ is really the eigenvector of the family $T(\lambda)$ with the eigenvalue $\Lambda(\lambda, \{\lambda_j\})$ given by (2.17). The oddness of the function $\dfrac{b(\lambda)}{c(\lambda)}$ makes the relations (2.27) equivalent to the system (2.18). This completes the proof of our main statement. This procedure for diagonalization of commuting family of transfer-matrices is called <u>algebraic Bethe ansatz</u> (in the simplest abc-form).

<u>Lecture III. Comments on the algebraic Bethe ansatz. Quantum Hamiltonians on the lattice.</u>

1. Algebraic scheme presented in the previous lecture is quite general: it works in the case $V = \mathbb{C}^2$ and arbitrary h if the conditions (2.10), (2.12) - (2.14) are satisfied.

2. Traditionally the expression (2.17) for the eigenvalue $\Lambda(\lambda)$ and the system of equations (2.18) for the 6-vertex model are derived by <u>coordinate Bethe ansatz</u>. This method was introduced by Hans Bethe in 1931 for the XXX-Heisenberg model of spins $\dfrac{1}{2}$. We shall see later that algebraic Bethe ansatz gives already known results for this model and for one-dimensional Bose-gas. It is also applicable to the field-theoretical models such as the Sine-Gordon model. It should be noted that besides simple and transparent derivation algebraic Bethe ansatz also gives a compact form (2.15) for the eigen-vectors $\Phi(\{\lambda_j\})$ - so-called <u>Bethe states</u>.

3. Algebraic scheme also works in inhomogeneous case when $L_n(\lambda)$ is replaced by $L_n(\lambda - p_n)$. Really, equation (1.23) is valid when $L_n(\lambda) \longrightarrow L_n(\lambda - p_n)$, $L_n(\mu) \longrightarrow L_n(\mu - p_n)$, so that for

$$T(\lambda, \{p_k\}) = \prod_{n=1}^{N} L_n(\lambda - p_n) \tag{3.1}$$

relation (2.11) remains. The only change in the algebraic Bethe an-satz is the replacement $a^N(\lambda) \longrightarrow \prod_{n=1}^{N} a(\lambda - p_n)$, $d^N(\lambda) \longrightarrow \prod_{n=1}^{N} d(\lambda - p_n)$ in (2.17) - (2.18). This situation occurs for the secondary Bethe ansatz for the Hubbard model (see Sutherland's lectures).

4. When $\dim V > 2$ in the case of a generalized " abc-condition" we get <u>nested algebraic Bethe ansatz</u>. Typical example is the one-dimensional Bose-gas with inner degrees of freedom. In this case creation operators - B's aquire additional indices.

5. For the 8-vertex model it is possible to formulate the <u>generalized algebraic Bethe ansatz</u>. However, the technical details are more involved and we shall not present them here.

6. There is an interesting interpretation of the system (2.18). Namely in general R-natrix is normalized so that $R(\lambda)|_{\lambda=0} = I$, i.e. $b(0) = 1$, $c(0) = 0$. Suppose the point $\lambda = 0$ is the simple zero of $c(\lambda)$. Then the eigenvalue $\Lambda(\lambda, \{\lambda_j\})$ of operator $T(\lambda)$ will have simple poles at $\lambda = \lambda_j$, $j = 1, \ldots \ell$. But transfer-matrix $T(\lambda)$ "does not know " about these λ_j . So the residues at these poles should vanish. This condition is equivalent to the system (2.18). This observation helps in more complicate cases when only some apriori information about eigenvalues of $T(\lambda)$ is known.

7. Our exposition in previous lectures was purely algebraic: we did not bother about the Hermitian property of $T(\lambda)$ and the corresponding Boltzmann weights could even be complex. Below we shall give some really physical examples. Parameters λ_j will play the role of quasimomenta or rapidities. Reference state Ω could be the <u>physical vacuum</u> or could be not: accordingly we shall distingush <u>ferromagnetic</u> or <u>antiferromagnetic</u> cases.

8. It should be emphasized that for the validity of algebraic Bethe ansatz one needs to be sure that the Bethe state in (2.15) is not zero. There exists a more complicated algebraic scheme for evaluating its norm. There could be certain general restrictions on the number ℓ of B's in (2.15): e.g. for the XXX-Heisenberg model of spins s we have $\ell \leqslant sN$ (see lecture IV).

Now let us consider quantum Hamiltonians diagonalizable by the algebraic Bethe ansatz. The transfer-matrix $T(\lambda)$ is surely the generating function for them. The problem is how to construct the family of <u>local</u> Hamiltonians (i.e. those describing interaction of nearest neighbours on the lattice). The form (2.17) for the eigenvalue $\Lambda(\lambda, \{\lambda_j\})$ suggests an answer. Consider the value λ_0 where $\delta(\lambda_0) = 0$ (for 6-vertex model $\lambda_0 = \eta$). Then the operator

192

$\log T(\lambda_0)$ and also all $\dfrac{d^{\kappa}}{d\lambda^{\kappa}} \log T(\lambda)\Big|_{\lambda=\lambda_0}$ will have additive spectrum. So it is plausible that these operators constitute the family of local Hamiltonians.

For simplicity we shall verify this for the special so-called XXX-case. It is obtained from the 6-vertex case by rescaling $\lambda \longrightarrow \lambda\Delta$, $\eta \longrightarrow \frac{i}{2}\Delta$ and letting $\Delta \longrightarrow 0$. We get

$$L_n(\lambda) = \lambda + \frac{i}{2} \sum_{a=1}^{3} \sigma^a \otimes \sigma_n^a = \begin{pmatrix} \lambda + \frac{i}{2}\sigma_n^3 & \frac{i}{2}\sigma_n^- \\ \frac{i}{2}\sigma_n^+ & \lambda - \frac{i}{2}\sigma_n^3 \end{pmatrix}, \tag{3.2}$$

so $a(\lambda) = \lambda + \frac{i}{2}$, $d(\lambda) = \lambda - \frac{i}{2}$. For the entries $b(\lambda)$ and $c(\lambda)$ in the R-matrix we have $b(\lambda) = \dfrac{i}{\lambda+i}$, $c(\lambda) = \dfrac{\lambda}{\lambda+i}$, so

$$R(\lambda) = \frac{\lambda P + i}{\lambda + i} \tag{3.3}$$

the simplest (rational) R-matrix.

In this case $\lambda_0 = \frac{i}{2}$ and due to

$$L_n\left(\tfrac{i}{2}\right) = \frac{i}{2}(I + \vec{\sigma} \otimes \vec{\sigma}_n) = i P_{0n} \tag{3.4}$$

(here the space $V = \mathbb{C}^2$ is identified with h_0) we have

$$T\left(\tfrac{i}{2}\right) = i^N tr_0(P_{0N} \cdots P_{01}) = i^N U_N^{-1}, \tag{3.5}$$

where $\mathcal{U}_N$ is the operator of a cyclic shift on the lattice

$$\mathcal{U}_N \vec{\sigma}_n \mathcal{U}_N^{-1} = \vec{\sigma}_{n+1} \quad , \qquad N+1 \equiv 1 \ . \tag{3.6}$$

For the first derivative we get

$$\frac{d}{d\lambda} \log T(\lambda) \Big|_{\lambda=\frac{i}{2}} = i^{-N} \mathcal{U}_N \frac{dT(\lambda)}{d\lambda} \Big|_{\lambda=\frac{i}{2}} =$$

$$= \frac{1}{i} \mathcal{U}_N \sum_{n=1}^{N} tr_0 (P_{0N} \cdots \underset{n}{\underbrace{I}} \cdots P_{01}) \ . \tag{3.7}$$

As

$$\mathcal{U}_N \, tr_0 (P_{0N} \cdots \underset{n}{\underbrace{I}} \cdots P_{01}) = P_{n-1,n} =$$

$$= \frac{I + \vec{\sigma}_{n-1} \cdot \vec{\sigma}_n}{2} \tag{3.8}$$

we have

$$i \frac{d}{d\lambda} \log T(\lambda) \Big|_{\lambda=\frac{i}{2}} = \sum_{n=1}^{N} \frac{\vec{\sigma}_{n-1} \cdot \vec{\sigma}_n + I}{2} \ , \tag{3.9}$$

where $\vec{\sigma}_{N+1} = \vec{\sigma}_1$. The periodic boundary conditions reflect the fact that $T(\lambda)$ is the trace of $\mathcal{T}(\lambda)$ (see also the remark below).

In the expression in the RHS of (3.9) we recognize the well-known Heisenberg Hamiltonian of isotropic spin chain of spins $\frac{1}{2}$:

$$H = -\frac{J}{4} \sum_{n=1}^{N} \left(\vec{\sigma}_n \cdot \vec{\sigma}_{n+1} - I \right).$$ (3.10)

The spectrum of H is particle-like and one-particle energy is obtained from (2.17), (3.9) – (3.10)

$$h_j = -\frac{J}{2} \frac{1}{\lambda_j^2 + \frac{1}{4}}.$$ (3.11)

For the eigenvalues P_j of U_N (quasimomenta) we have from (2.17), (3.5)

$$e^{iP_j} = \frac{\lambda_j - \frac{i}{2}}{\lambda_j + \frac{i}{2}},$$ (3.12)

thus obtaining familiar dispersion law

$$h(\lambda) = -\frac{J}{2} \frac{dP}{d\lambda} = -J(1 - \cos P).$$ (3.13)

Higher derivatives of $\log T(\lambda)$ at $\lambda = \frac{i}{2}$ will give local Hamiltonians with more neighbours inreracting.

The fact that $V \approx h$ and the formula (3.4) were crucial for our derivation. Local L-operators with such properties are called <u>fundamental</u>.

The way of obtaining local Hamiltonians resembles the procedure of deriving the local integrals of motion in the <u>inverse scattering method</u> for integration of nonlinear differential equations of evolution type. This analogy is very deep: the algebraic Bethe ansatz method is just the <u>quantum inverse scattering method</u> for solving quantum-mechanical problems. To the local L-operator one can attach auxiliary linear problem

$$F_{n+1} = L_n(\lambda) F_n$$ (3.14)

(the quantum analog of the auxiliary linear problem for discrete
nonlinear equations) so that $\mathcal{T}(\lambda)$ is really the monodromy matrix
of problem (3.14). The trace of $\mathcal{T}(\lambda)$ gives rise to periodic
boundary conditions.

Consider now the generalization of the Heisenberg model for
higher spins: S_n^a are now the spin operators of spin S acting
in $h_n = \mathbb{C}^{2S+1}{}_n$. Naive generalization of (3.10)

$$H = \sum_n \vec{S}_n \cdot \vec{S}_{n+1} \tag{3.15}$$

is non-integrable for $S > \frac{1}{2}$. To make it integrable one should
add higher order terms of $\vec{S}_n \cdot \vec{S}_{n+1}$. A remarkable thing is that
algebraic formalism of the quantum inverse scattering method presents
this generalization. It is very difficult to obtain it using the
coordinate Bethe ansatz.

The group-theoretical foundation for this derivation will be
given in Faddeev's lectures. Here we present the results. Local L-
operator has the form

$$L_n(\lambda) = \begin{pmatrix} \lambda + i S_n^3 & i S_n^- \\ i S_n^+ & \lambda - i S_n^3 \end{pmatrix} \tag{3.16}$$

and satisfies fundamental relation (1.23) with the same R-matrix
as in the case $S = \frac{1}{2}$. Really to verify (1.23) only the commutation relations between S_n^a are needed. Local vacuum $\omega_n \in h_n$
is the highest weight of representation of $SU(2)$ of spin S :

$$S_n^+ \omega_n = 0, \qquad S_n^3 \omega_n = s \omega_n , \tag{3.17}$$

so

$$\alpha(\lambda) = \lambda + is , \qquad \delta(\lambda) = \lambda - is . \tag{3.18}$$

Thus the whole machinery of lecture II works.

The statistical model, associated with $L_n(\lambda)$, is rather unnatural: we have different objects on horizontal ($\sigma = \pm 1$) and vertical ($\sigma = 0, \ldots, 2s$) bonds. However, it is possible to construct fundamental L-operator with $V = h = \mathbb{C}^{2s+1}$. With its help one can obtain local Hamiltonian of the form

$$H = \sum_{n=1}^{N} f_{2s}(\vec{S}_n \cdot \vec{S}_{n+1}), \qquad (3.19)$$

where $f_{2s}(x)$ is a certain polynomial of degree $2s$, e.g. $f_1(x) = x$, $f_2(x) = x - x^2$. For the general formula for $f_{2s}(x)$ see Faddeev's lectures.

Hamiltonian (3.19) is an integrable generalization of (3.10) for higher spins. Its one-particle eigenvalue has the form $h_j = \dfrac{s}{\lambda_j^2 + s^2}$.

We see that though the Hamiltonian (3.19) looks rather complicated, its eigenvalues are very simple.

Thus with each irreducible representation of $SU(2)$ we have connected an integrable spin model. Another way to go to higher spins is to consider the models connected with fundamental representation of $SU(N)$. The well-known example is the Sutherland's model

$$H = \sum_{n=1}^{N} P_{n,n+1}, \qquad (3.20)$$

where $P_{n,n+1}$ is the permutation operator in $h_n \otimes h_{n+1} \approx \mathbb{C}^N \otimes \mathbb{C}^N$. It can also be represented as a polynomial of $\vec{S}_n \cdot \vec{S}_{n+1}$ of degree $N - 1 = 2s$. In this case we have $N - 1$-nested Bethe ansatz according to Dynkin diagram for $SU(N)$.

It is instructive to compare these polynomials. For the case $s = 1$ ($N = 3$) we have $x + x^2$ for (3.19) and $x^2 - x$ for (3.20). Of course they turn into each other under the transformation $\vec{S}_n \longrightarrow (-1)^n \vec{S}_n$. But for odd n this is not a canonical transformation. Thus the models (3.19) and (3.20) are really different: the symmetry group of the first model is $SU(2)$, whereas for the second model the symmetry group is $SU(N)$.

Lecture IV. Properties of the Bethe states for the XXX-Heisenberg model of spins S . Other examples of integrable models.

Consider the Bethe states for the XXX-Heisenberg model of spins S

$$\Phi(\{\lambda_j\}) = \prod_{j=1}^{\ell} B(\lambda_j)\,\Omega. \tag{4.1}$$

We shall prove that

$$S_+ \Phi = 0, \qquad S_3 \Phi = (sN - \ell)\,\Phi , \tag{4.2}$$

where $S_a = \sum_{n=1}^{N} S_n^a$ are the generators of $SU(2)$ action in $\mathcal{H}_N = \mathbb{C}^{(2s+1)\,N}$ and $S_\pm = S_1 \pm i S_2$.

This means that the Bethe states are the highest weights with respect to the action of $SU(2)$. So these states generate the multiplets $S_-^m \Phi$, $m = 0,1,\dots 2sN - 2\ell$; the corresponding eigenvalues are $2sN - 2\ell + 1$ - fold degenerated.

For the reference state Ω (4.2) (where $\ell = 0$) immediately follows from (3.17). Thus the reference state is $2sN + 1$ -fold degenerated. In order to prove (4.2) in general case we shall determine the action of $SU(2)$ on the matrix elements of the monodromy matrix $\mathcal{T}(\lambda)$, i.e. evaluate the matrix

$$\left[S_a, \mathcal{T}(\lambda)\right]_Q \overset{def}{=} \begin{pmatrix} \left[S_a, A(\lambda)\right] , & \left[S_a, B(\lambda)\right] \\[2em] \left[S_a, C(\lambda)\right] , & \left[S_a, \mathcal{D}(\lambda)\right] \end{pmatrix} . \tag{4.3}$$

Let us begin with the local L-operator

$$L_n(\lambda) = \lambda I + i \sum_a \sigma^a \otimes S_n^a .$$

(4.4)

We have

$$\left[S_a , L_n(\lambda) \right]_Q = \left[S_n^a , L_n(\lambda) \right]_Q =$$

$$= - \sum_{b,c} \varepsilon^{abc} \sigma^b \otimes S_n^c = -\frac{1}{2} \left[\sigma^a , L_n(\lambda) \right]_{aux} ,$$

(4.5)

where in the last line we have used the commutation relations $\left[\sigma^a , \sigma^b \right] = 2i\varepsilon^{abc} \sigma^c$ for Pauli matrices and by $\left[\, , \, \right]_{aux}$ we have denoted the usual commutator in the auxiliary space $V(=\mathbb{C}^2)$. From this we obtain

$$\left[S_a , T(\lambda) \right]_Q = \left[S_a , \prod_{n=1}^N L_n(\lambda) \right]_Q =$$

$$= \sum_{n=1}^N L_N(\lambda) \ldots \left[S_a , L_n(\lambda) \right]_Q \ldots L_1(\lambda) =$$

$$= -\frac{1}{2} \sum_{n=1}^N L_N(\lambda) \ldots \left[\sigma^a , L_n(\lambda) \right]_{aux} \ldots L_1(\lambda) =$$

$$= -\frac{1}{2} \left[\sigma^a , T(\lambda) \right] .$$

(4.6)

In particular we get

$$[S_\alpha , T(\lambda)] = 0 \tag{4.7}$$

(which is surely obvious) and

$$[S_3 , B(\lambda)] = -B(\lambda) , \tag{4.8}$$

$$[S_+ , B(\lambda)] = A(\lambda) - \mathcal{D}(\lambda) . \tag{4.9}$$

The second formula in (4.2) now follows from (4.8) and is valid for all λ_j . However the first formula in (4.2) is true only for the Bethe states, i.e. when $\{\lambda_j\}$ satisfy the system of equations from algebraic Bethe ansatz.

Really, we have

$$S_+ \Phi = \sum_{j=1}^{\ell} B(\lambda_1) \ldots B(\lambda_{j-1})(A(\lambda_j) - \mathcal{D}(\lambda_j)) \cdot$$

$$\cdot B(\lambda_{j+1}) \ldots B(\lambda_\ell)\, \Omega \tag{4.10}$$

and as in lecture II let us "move" $A(\lambda_j) - \mathcal{D}(\lambda_j)$ through B's toward Ω using the commutation relations (2.13) - (2.14). Again there will be many terms but we can arrange them as follows

$$S_+ \Phi = \sum_{j=1}^{\ell} Q_j(\{\lambda_j\})\, B(\lambda_1) \ldots B(\lambda_{j-1}) B(\lambda_{j+1}) \ldots B(\lambda_\ell)\, \Omega \tag{4.11}$$

with yet unknown coefficients Q_j . To calculate them we shall use the same trick as in lecture II : Q_1 is easily calculated (one should use only the first terms in (2.13) - (2.14)) and Q_j is obtained by the replacement $\lambda_1 \longleftrightarrow \lambda_j$. So as a result we get

$$Q_j = (\lambda_j + is)^N \prod_{\substack{\kappa=1 \\ \kappa \neq j}}^{\ell} \frac{1}{c(\lambda_\kappa - \lambda_j)} - (\lambda_j - is)^N \prod_{\substack{\kappa=1 \\ \kappa \neq j}}^{\ell} \frac{1}{c(\lambda_j - \lambda_\kappa)} \quad .$$

$$(4.12)$$

Now the requirement that $Q_j = 0$ for all j is equivalent to the system of equations of algebraic Bethe ansatz for the XXX-model. This completes the proof.

Thus we have shown that the Bethe states for the XXX-model are not complete in $\mathfrak{H}_N$ but are the highest weights. In the next lecture we shall present a plausible arguments to show that the system of multiplets associated with Bethe states is complete in $\mathfrak{H}_N$.

Also notice that for the highest weights $S_3 \geqslant 0$ so that we have a restriction on the number ℓ of B's in the Bethe state

$$\ell \leqslant sN \, . \tag{4.13}$$

Now we shall present two examples of integrable field-theoretical models.

1. <u>Nonlinear Schrödinger model</u>.

The Hamiltonian of the model looks like

$$H = \int (\Psi_x^+ \Psi_x + \varkappa \Psi^+ \Psi^+ \Psi \Psi) \, dx \, , \tag{4.14}$$

where $\Psi^+(x)$ and $\Psi(x)$ are usual creation-annihilation operators

$$\left[\Psi(x), \Psi^+(y) \right] = \delta(x-y) . \tag{4.15}$$

At the classical level this model is completely integrable through the inverse scattering method. The corresponding auxiliary linear problem has the form

$$L(\lambda) F = 0 , \tag{4.16}$$

where

$$L(\lambda) = \frac{d}{dx} - \frac{\lambda \sigma_3}{2i} - \sqrt{\varkappa} \begin{pmatrix} 0 & -i\overline{\Psi}(x) \\ i\Psi(x) & 0 \end{pmatrix} . \tag{4.17}$$

In the quantum case the problem (4.16) is still valid if one uses the normal operator ordering.

On the lattice the corresponding local L-operator has the form

$$L_n(\lambda) = \begin{pmatrix} 1 - \dfrac{i\lambda\Delta}{2} + \dfrac{\varkappa\Delta\Psi_n^+\Psi_n}{2} & -i\sqrt{\varkappa\Delta}\,\Psi_n^+\rho_n \\[2ex] i\sqrt{\varkappa\Delta}\,\rho_n\Psi_n & 1 + \dfrac{i\lambda\Delta}{2} + \dfrac{\varkappa\Delta\Psi_n^+\Psi_n}{2} \end{pmatrix} , \tag{4.18}$$

where

$$\left[\Psi_n, \Psi_m^+ \right] = \delta_{nm} , \qquad \rho_n = \sqrt{1 + \frac{\varkappa\Delta}{4}\Psi_n^+\Psi_n} . \tag{4.19}$$

Here $V = \mathbb{C}^2$ and $h = L^2(\mathbb{R}^1)$ — one-particle quantum space. The R-matrix has the form (2.3) where

$$b(\lambda) = \frac{i\varkappa}{i\varkappa - \lambda} , \qquad c(\lambda) = \frac{\lambda}{\lambda - i\varkappa} . \tag{4.20}$$

The local vacuum ω_n is the usual Fock vacuum

$$\Psi_n \omega_n = 0, \tag{4.21}$$

so

$$\alpha(\lambda) = 1 - \frac{i\lambda\Delta}{2} , \quad \delta(\lambda) = 1 + \frac{i\lambda\Delta}{2} . \tag{4.22}$$

In the continuous limit $\Delta \longrightarrow 0, \quad L = N\Delta$

$$\left(\frac{1 - \frac{i\lambda\Delta}{2}}{1 + \frac{i\lambda\Delta}{2}} \right)^N \longrightarrow e^{-i\lambda L} \tag{4.23}$$

and the system of equations (2.18) goes to the well-known system of equations of coordinate Bethe ansatz for the one-dimensional Bose-gas with δ-function interaction.

Matrix elements of local L-operator (4.18) are the generators of irreducible representation of Lie algebra $su(2)$ in Hilbert space $\mathcal{H}$. More precisely, operators

$$S_n^1 = \frac{i}{\sqrt{\varkappa\Delta}} \left(\Psi_n^+ \varphi_n + \varphi_n \Psi_n \right) ,$$

$$S_n^2 = \frac{1}{\sqrt{\varkappa\Delta}} \left(\varphi_n \Psi_n - \Psi_n^+ \varphi_n \right) , \tag{4.24}$$

$$S_n^3 = -\frac{2}{\varkappa\Delta} \left(1 + \frac{\varkappa\Delta}{2} \Psi_n^+ \Psi_n \right)$$

in the case $\varkappa < 0$ are the generators of $su(2)$ representation in $\mathcal{H}$ of spin $S = -\frac{2}{\varkappa\Delta}$. When S is a half-integer the usual representation of $su(2)$ in $\mathbb{C}^{2s+1}$ splits from (4.24).

We have

$$ L_n^{NS}(\lambda) = \frac{1}{is}\, \sigma_3\, L_n^{XXX}\left(-\frac{\lambda}{\varkappa}\right) , \tag{4.25} $$

so the quantum nonlinear Schrödinger model on the lattice and XXX-Heisenberg model of arbitrary spins coincide. In the continuous and semi-classical limits $\Delta \to 0$, $\hbar \to 0$ the auxiliary linear problem with local L-operator (4.18) goes into the auxiliary linear problem (4.16) - (4.17).

2. Sine-Gordon model.

This model is described by the formal Hamiltonian

$$ H = \int \left(\frac{1}{2}\pi^2 + \frac{1}{2}\varphi_x^2 + \frac{m^2}{\beta^2}(1 - \cos\beta\varphi) \right) dx \tag{4.26} $$

and commutation relations

$$ \left[\pi(x), \varphi(y) \right] = i\delta(x - y) . \tag{4.27} $$

Here we have two problems: the problem of ultraviolet divergences and that of operator ordering.

The first problem is avoided by going to the lattice. Introduce operators p_n, q_n such that

$$ \left[p_n, q_m \right] = i\delta_{nm} \tag{4.28} $$

and set

$$ \pi_n = e^{\frac{i\beta p_n}{4}} , \qquad \chi_n^{\pm} = e^{\pm\frac{i\beta q_n}{2}} . \tag{4.29} $$

Then the local L-operator for the Sine-Gordon model is

$$
L_n(\lambda) = \begin{pmatrix}
\pi_n^{-\frac{1}{2}}(1 + 2r\cos\beta q_n)\,\pi_n^{-\frac{1}{2}} & \sqrt{r}\,(e^{-\lambda}\chi_n^- - e^{\lambda}\chi_n^+) \\[2em]
\sqrt{r}\,(e^{-\lambda}\chi_n^+ - e^{\lambda}\chi_n^-) & \pi_n^{\frac{1}{2}}(1 + 2r\cos\beta q_n)\,\pi_n^{\frac{1}{2}}
\end{pmatrix},
\tag{4.30}
$$

where $r = \left(\dfrac{m\Delta}{4}\right)^2$. The R-matrix has the form (2.3) as above where

$$
b(\lambda) = \frac{i\sin\gamma}{\sin(\lambda + i\gamma)}, \quad c(\lambda) = \frac{\sin\lambda}{\sin(\lambda + i\gamma)}, \quad \gamma = \frac{\beta^2}{8}.
\tag{4.31}
$$

In this case there is no local vacuum for individual L-operator $L_n(\lambda)$. But it exists for the product $L_{n+1}(\lambda)\,L_n(\lambda)$ of two successive L-operators. Thus the algebraic scheme from lecture II works and we can diagonalize the commuting family $T(\lambda) = \mathrm{tr}\,\mathcal{T}(\lambda)$ where

$$
\mathcal{T}(\lambda) = \overset{N}{\underset{n=1}{\prod}}\, L_n(\lambda).
\tag{4.32}
$$

The corresponding λ_j play the role of rapidities. In this way the problem of operator ordering is solved. In order to obtain physical results one should remove the cutt-offs $L = N\Delta$ and Δ: perform the <u>thermodynamical limit</u> $L \to \infty$, $\Delta \to 0$. In this way one gets the exact mass-spectrum and S-matrix for the Sine-Gordon model. In the next lecture we shall illustrate this procedure by more simple example.

Lecture V. Thermodynamical limit for the XXX-Heisenberg model.

For simplicity we shall consider the Heisenberg Hamiltonian of spins $\frac{1}{2}$

$$H = \frac{J}{4} \sum_{n=1}^{N} (\vec{\sigma}_n \cdot \vec{\sigma}_{n+1} - I) , \quad \vec{\sigma}_{N+1} = \vec{\sigma}_1 . \tag{5.1}$$

1. The ferromagnetic case $J \leqslant 0$

We have

$$H = -J \sum_{n=1}^{N} (\vec{\sigma}_n \cdot \vec{\sigma}_{n+1} - I)^2 \geqslant 0 \tag{5.2}$$

so the reference state Ω is the physical vacuum - the state with zero energy.

The one-particle state is given by

$$\Phi(\lambda) = B(\lambda)\Omega , \tag{5.3}$$

where

$$\left(\frac{\lambda - \frac{i}{2}}{\lambda + \frac{i}{2}} \right)^N = e^{-ipN} = 1 . \tag{5.4}$$

As $N \longrightarrow \infty$ this equation is simplified and means that momentum p varies over the interval $[0, 2\pi)$. The energy of this state is given by (3.11)

$$h(p) = -J(1 - \cos p) . \tag{5.5}$$

The corresponding particles are called magnons.

The ℓ-magnon state is

$$\Phi(\lambda_1, \ldots, \lambda_\ell) = B(\lambda_1) \ldots B(\lambda_\ell)\Omega \tag{5.6}$$

where

$$\left(\frac{\lambda_j - \frac{i}{2}}{\lambda_j + \frac{i}{2}}\right)^N = e^{-ip_j N} = \prod_{\substack{k=1 \\ k \neq j}}^{\ell} \frac{\lambda_j - \lambda_k - i}{\lambda_j - \lambda_k + i} \qquad (5.7)$$

and λ_j are real. As $N \longrightarrow \infty$ and ℓ is fixed this system simply means that momenta $p_1, \ldots, p_\ell$ independently vary over the interval $[0, 2\pi)$. The corresponding energy is given by

$$h(p_1, \ldots, p_\ell) = \sum_{j=1}^{\ell} h(p_j) . \qquad (5.8)$$

So we really have the ℓ-particle state.

However magnons can also form <u>bound states</u>. They correspond to complex solution of system (5.7). As $N \longrightarrow \infty$ and ℓ is fixed they split into complexes called <u>strings</u>. The string of <u>lenght $2M + 1$</u> is the set of $2M + 1$ numbers

$$\lambda_m = x + im, \quad m = -M, \ldots, M \qquad (5.9)$$

where M is a half-integer. The real number x is called the <u>center</u> of a string.

We shall illustrate this by the example of 2-string. Suppose that $\lambda_1 = x_1 + iy_1$, $\lambda_2 = x_2 + iy_2$, $y_1 > 0$ satisfy the system (5.7) for $\ell = 2$. Considering modulus of the first equation in (5.7) as $N \longrightarrow \infty$ we see that $x_1 = x_2$, $y_1 - y_2 = 1$ up to the order $O(e^{-c_0 N})$. Multiplying equations in (5.7) and using these formulas we get $y_1 = \frac{1}{2}$ so $\lambda_1 = x + \frac{i}{2}$, $\lambda_2 = x - \frac{i}{2}$.

The L-string with center x describes the L-magnon's bound state with momentum

$$P_L = P\left(\frac{x}{L}\right) = \frac{1}{i} \log \frac{x + \frac{iL}{2}}{x - \frac{iL}{2}} \qquad (5.10)$$

and energy

$$h_L = -\frac{J}{L}\left(1 - \cos P_L\right).$$

(5.11)

The states presented above exhaust all the excitations of the Heisenberg model in the ferromagnetic phase.

Dynamical picture described above is developed in the Hilbert space $\mathfrak{H}_F$ - separable subspace of the "large" space $\mathfrak{H}_\infty =$
$$= \prod_{n=1}^{\infty} \otimes\, \mathbb{C}^2$$
which is incomplete tensor product in von Neumann sense of the spaces h_n adjusted to the vector $\Omega_\infty = \prod_{n=1}^{\infty} \otimes\, \omega_n$.

The space $\mathfrak{H}_F$ can be represented as incomplete tensor product of Fock spaces for all the particles in the model - L-strings, $L = 1, 2, \ldots$. The physical vacuum and the many-particle states in $\mathfrak{H}_F$ are non-degenerated because the action of the whole $SU(2)$ in $\mathfrak{H}_F$ reduces just to the action of operator $S_3 =$
$$= \lim_{N \to \infty}\left(S_3 - \frac{N}{2}\right) -$$ the number of particles operator.

2. <u>Antiferromagnetic case $J \geqslant 0$</u> .

In this case $H \leqslant 0$. The physical vacuum - ground state should be the state with minimal energy and so differs from Ω . All other physical states are characterized by having finite positive energy relative to the ground state. They are defined in terms of "Dirac Sea". In order to present this construction we shall first count the number of Bethe states.

To do this one should determine at given $\ell \leqslant \frac{N}{2}$ the number of solutions of the system of equations

$$\left(\frac{\lambda_j - \frac{i}{2}}{\lambda_j + \frac{i}{2}}\right)^N = \prod_{\substack{k=1 \\ k \neq j}}^{n} \frac{\lambda_j - \lambda_k - i}{\lambda_j - \lambda_k + i}$$

(5.12)

for which the Bethe states do not vanish. This is a difficult problem. We shall simplify it by introducing certain plausible though heuristic arguments.

a) Suppose that the <u>string hypothesis is true</u>, i.e. all solutions of (5.12) consist of strings. Then multiply the LHS of (5.12) along the given $2M+1$-string and in the RHS group together the factors

$$\frac{\lambda_j - \lambda_K - i}{\lambda_j - \lambda_K + i}$$

into the strings. We obtain

$$V_0^N\left(\frac{\lambda_{j,M}}{M+\frac{1}{2}}\right) = \prod_{M'} \prod_{\substack{k=1 \\ (K,M')\neq(j,M)}}^{\nu_{M'}} V_{M,M'}\,(\lambda_{j,M} - \lambda_{j,M'})\quad,$$

$$\tag{5.13}$$

where ν_M in the number of $2M+1$-strings with centers $\lambda_{j,M}$, $j = 1,\ldots,\nu_M$ in the Bethe state so that

$$\ell = \sum_M (2M+1)\,\nu_M \tag{5.14}$$

and

$$V_0(\lambda) = \frac{\lambda - i}{\lambda + i}\,,\quad V_{M,M'}(\lambda) = \prod_{m=-M}^{M} \prod_{m'=-M'}^{M'} V_0(\lambda + i(m+m'))\,. \tag{5.15}$$

b) Take the logatithm of (5.15) using

$$\frac{1}{i}\log V_0(\lambda) = \pi + 2\,\mathrm{arctg}\,\lambda\,, \tag{5.16}$$

where $-\frac{\pi}{2} < \text{arctg }\lambda < \frac{\pi}{2}$ is the principal branch of arctg . As the result we arrive at the following system of equations

$$2N \,\text{arctg}\, \frac{\lambda_{j,M}}{M+\frac{1}{2}} = 2\pi Q_{j,M} + \sum_{M'} \sum_{k=1}^{\nu_{M'}} \Phi_{M,M'}(\lambda_{j,M} - \lambda_{k,M'}) , \qquad (5.17)$$

where

$$\Phi_{M,M'}(\lambda) = 2 \sum_{L=|M-M'|}^{M+M'} \left(\text{arctg}\,\frac{\lambda}{L} + \text{arctg}\,\frac{\lambda}{L+1} \right) \qquad (5.18)$$

(for the case $M = M'$ the term $\text{arctg}\,\frac{\lambda}{0}$ is omitted in (5.18)) and $Q_{j,M}$ are half-integers arising from (5.16) and the winding number of the RHS of (5.13).

Numbers $Q_{j,M}$ parametrize possible solutions of system (5.18). We assume next that to each set of admissible $Q_{j,M}$

$$-Q_M^{max} \leq Q_{1,M} < Q_{2,M} < \ldots < Q_{\nu_M,M} \leq Q_M^{max}$$

there corresponds a unique solution of (5.17). The numbers Q_M^{max} are determined by the following principle: if $Q = Q_M^{max} + 2M + 1$ then corresponding λ_M – the center of $2M+1$ -string – goes to ∞. Thus one obtains

$$Q_M^{max} = \frac{N}{2} - \frac{1}{2\pi} \sum_{M'} \Phi_{M,M'}(\infty)\, \nu_{M'} - \frac{1}{2} . \qquad (5.19)$$

Assuming all this we have for ν_M distinct values of $Q_{j,M}$

$$P_M = 2 Q_M^{max} + 1$$

vacancies. So the number $Z(N,\ell)$ of Bethe states with given ℓ is

$$Z(N,\ell) = \sum_{\ell = \sum_M (2M+1)\nu_M} \prod_M C_{P_M}^{\nu_M} . \qquad (5.20)$$

It can be shown that

$$Z(N,\ell) = C_N^\ell - C_N^{\ell-1} . \qquad (5.21)$$

Now taking into accound the degeneracy of Bethe states for the total number N of vectors in their multiplets we have

$$N = \sum_{\ell=0}^{\frac{N}{2}} (N - 2\ell + 1)(C_N^\ell - C_N^{\ell-1}) = 2^N . \qquad (5.22)$$

The latter number in this line is just the dimension of the space $\mathfrak{H}_N$. Thus we have convinced ourselves of the completeness of the Bethe multiplets.

Next we shall describe the physical vacuum and the low-lying excitations above it.

The physical vacuum Ω_{phys} is the state where $\ell = \nu = \frac{N}{2}$, $\nu_M = 0$ for $M > 0$ and exists only for even N which is assumed. In this state the number ℓ of real λ_j is equal to the number $P_0 = \frac{N}{2}$ of vacancies. So it is uniquely determined and has the total spin zero, i.e. is not degenerate.

In order to be sure that Ω_{phys} is really a ground state we must show that all excitations above it have positive relative energy. Due to the particle-like structure of the spectrum we shall demonstrate this only for elementary excitations. There are only two possible ways to construct them.

$\underline{\text{T}}.$ $\quad \ell = \frac{N}{2} - 1 , \quad \nu_0 = \frac{N}{2} - 1, \quad \nu_M = 0$ $\qquad$ for $M > 0 .$

Here $P_0 = \frac{N}{2} + 1$ so <u>two</u> vacancies for $Q_{j,0}$ are free. The total spin of this state is equal to 1 .

$\underline{\text{S}}.$ $\quad \ell = \frac{N}{2} , \quad \nu_0 = \frac{N}{2} - 2 , \quad \nu_{\frac{1}{2}} = 1, \quad \nu_M = 0$ $\qquad$ for $M > 0 .$

We have $P_0 = \frac{N}{2}$ so again there are <u>two</u> free vacancies. For the only one present 2-string we have $P_{\frac{1}{2}} = 1$ so its center is uniquely determined.

Now let us calculate the energy and momentum of these states as $N \longrightarrow \infty$. We shall assume that as $N \longrightarrow \infty$ numbers $Q_{j,0}/N$ fill the interval $\left[-\frac{1}{4} , \frac{1}{4} \right]$ and $\lambda_j \rightarrow \lambda(x) , \lambda(\pm \frac{1}{4}) = \pm \infty .$

For the ground state we have $\dfrac{Q_{j,0}}{N} \longrightarrow x , \quad -\frac{1}{4} \leqslant x \leqslant \frac{1}{4}$ and

$$\frac{Q_{j,0}}{N} \longrightarrow x + \frac{1}{N} \left(\theta(x - x_1) + \theta(x - x_2) \right)$$

for the states T and S. Here $x_i = \lim\limits_{N \to \infty} \dfrac{Q_i^{(h)}}{N}, \quad i = 1, 2 ,$, where $Q_1^{(h)} < Q_2^{(h)}$ are the two <u>gaps-holes</u> in the sequence of $Q_{j,0}$.

For the ground state as $N \longrightarrow \infty$ equations (5.17) go into

$$\operatorname{arctg} \lambda(x) = \pi x + \int_{-\frac{1}{4}}^{\frac{1}{4}} \operatorname{arctg}(\lambda(x) - \lambda(y)) \, dy .$$

$$(5.23)$$

The function

$$\rho(\lambda) = \frac{1}{\left. \dfrac{d\lambda}{dx} \right|_{x = x(\lambda)}}$$

$$(5.24)$$

plays the role of density of numbers λ_j in the interval $[\lambda, \lambda + d\lambda]$ so that

$$\frac{1}{N} \sum_j f(\lambda_j) \longrightarrow \int_{-\infty}^{\infty} f(\lambda)\, \rho(\lambda)\, d\lambda \tag{5.25}$$

for the smooth function $f(\lambda)$. Remember that when $x = \pm\frac{1}{4}$ the corresponding λ's are equal to $\pm\infty$. Differentiating (5.23) with respect to x we obtain for $\rho(\lambda)$ the linear integral equation

$$\pi\rho(\lambda) + \int_{-\infty}^{\infty} \frac{\rho(\mu)}{1+(\lambda-\mu)^2}\, d\mu = \frac{2}{1+4\lambda^2} \ . \tag{5.26}$$

This equation is easily solved by the Fourier transform and we get

$$\rho(\lambda) = \frac{1}{2\,\mathrm{ch}\,\pi\lambda} \ . \tag{5.27}$$

Thus for the ground state energy and momentum we have

$$E = -\sum_j h(\lambda_j) = -N \int_{-\infty}^{\infty} h(\lambda)\, \rho(\lambda)\, d\lambda = -N \log 2 \tag{5.28}$$

and

$$P = \sum p(\lambda_j) = N \int_{-\infty}^{\infty} p(\lambda)\, \rho(\lambda)\, d\lambda = \frac{N\pi}{2} \ (\mathrm{mod}\ 2\pi) \ . \tag{5.29}$$

Here $h(\lambda)$ is the old magnon's energy with $J = -1$.

For the state T the system (5.17) reduces to the equation

$$\pi \, \rho_T(\lambda) + \int_{-\infty}^{\infty} \frac{\rho_T(\mu)}{1 + (\lambda - \mu)^2} \, d\mu = \frac{2}{1 + 4\lambda^2} -$$

$$- \frac{\pi}{N} \left(\delta(\lambda - \lambda_1) + \delta(\lambda - \lambda_2) \right) \tag{5.30}$$

so

$$\rho_T(\lambda) = \rho(\lambda) + \frac{1}{N} \left(\sigma(\lambda - \lambda_1) + \sigma(\lambda - \lambda_2) \right) , \tag{5.31}$$

where $\sigma(\lambda)$ satisfies the equation

$$\pi \, \sigma(\lambda) + \int_{-\infty}^{\infty} \frac{\sigma(\mu)}{1 + (\lambda - \mu)^2} \, d\mu = -\pi \delta(\lambda) . \tag{5.32}$$

Here $\lambda_i = \lambda(x_i)$, $i = 1, 2$ are parameters of the holes. For the relative energy and momentum we have

$$\Delta E = - \sum_j \left(h(\lambda_j^{(T)}) - h(\lambda_j^{(vac)}) \right) =$$

$$= \varepsilon(\lambda_1) + \varepsilon(\lambda_2) , \tag{5.33}$$

$$\Delta P = \sum_j \left(p(\lambda_j^{(T)}) - p(\lambda_j^{(vac)}) \right) = k(\lambda_1) + k(\lambda_2) \tag{5.34}$$

where

$$\varepsilon(\lambda) = -\int_{-\infty}^{\infty} h(\mu)\,\sigma(\lambda-\mu)\,d\mu = \frac{\pi}{2\,\mathrm{ch}\,\pi\lambda} \tag{5.35}$$

and

$$k(\lambda) = \int_{-\infty}^{\infty} p(\mu)\,\sigma(\lambda-\mu)\,d\mu = \mathrm{arctg}\,\mathrm{sh}\,\pi\lambda - \frac{\pi}{2}\,. \tag{5.36}$$

The total spin of this state is equal to 1 .

For the state T there will be contribution to (5.31) coming from the 2-string. Its center is defined from (5.17) and is equal to $\dfrac{\lambda_1+\lambda_2}{2}$. There is a simple general fact that the strings have zero relative energy and momentum. So the energy and momentum of the state S are given by (5.33) - (5.36). The only difference is that the total spin of the state S is zero.

Thus the states T and S are two-particle triplet and singlet states (this explains the notations) of the spin $\frac{1}{2}$ particle with energy and momentum $\varepsilon(\lambda)$ and $k(\lambda)$ connected by the dispersion law

$$\varepsilon(k) = \frac{\pi}{2}\,\sin k, \qquad 0 \leqslant k \leqslant \pi\,. \tag{5.37}$$

This excitation - a spin wave - is rather a kink than a particle because in all physical states there is an even number of them (remember that N is even). This can be seen from the formula for the vacancies for real λ's.

Now we can describe the dynamical picture of the Heisenberg model in the antiferromagnetic phase. The physical vacuum is an "unperturbed" Dirac Sea of negative energy particles and the physical states are obtained by making holes and embedding strings into the Sea.

The holes-kinks have spin $\frac{1}{2}$, energy and momentum ε and k and go in pairs. The strings are "non-material": they have zero energy and momentum and their role is to distinguish the states by total spin. There is no bound states of kinks.

This picture can be generalized to the case of Heisenberg model of spins S . The physical vacuum will be the Dirac Sea of $2S$ - strings and the only elementary excitation is the kink of spin $\frac{1}{2}$.

The kinks again go in pairs and have the same energy and momentum as above. However the S-matrix, which can be calculated exactly, depends on the spin S .

At this point it is natural to finish this introduction into the algebraic Bethe ansatz. I hope that the various items which were presented will help to feel the flavour of this method. I think that it would be meaningless to present the technical details (which could be very involved) or to discuss subtle points (which could be very sophisticated). So I had concentrated your attention on the algebraic foundation leaving aside all rigorous analytic details. Also I have not said anything either about purely mathematical and very intriguing concepts arising from the development of the quantum inverse scattering method, or about its physical applications. The latter is clear: find and try to solve new models, or use them in the perturbation theory for the non-integrable cases. Interesting mathematical results you will find in Faddeev's lectures.

Finally here in the list of main references and comments mostly of historical nature.

References

I. Fundamental relation (1.16) was introduced by Baxter

1. R.J.Baxter. Partition-function of the eight-vertex lattice model.
   - Ann.Phys. (N.Y.), 1972, v.70, p.323-337,
   where the solution of the 8-vertex model was given. The Yang-Baxter relation is a far going generalization of the Onsager's famous star-triangle relation. Parametrization (1.22) of Boltzmann weights was extremely important for Baxter's approach.

   The role of equation (1.23) in statistical mechanics was manifested by Baxter

2. R.J.Baxter. Solvable eight-vertex model on arbitrary planar

lattice.- Phyl.Trans. Royal Soc. London, 1978, v.289A, N 1359, p.315-346 and in scattering theory - by Zamolodchikov

3. A.B.Zamolodchikov. Factorized S-matrices and Lattice Statistical Systems. - Sov.Sci.Rev., 1980, Phys.Rev., v.2.

Baxter's solution of the triangle equation - the R-matrix (1.24) was generalized to the higher matrix dimensions by Belavin

4. A.A.Belavin. Dynamical symmetry of integrable quantum systems.- Nucl.Phys. B, 1981, v.B180, N 2, p.189-200.

The literature, devoted to the solutions of the Yang-Baxter equation, is rather extensive; for the introduction the survey paper

4. P.P.Kulish, E.K.Sklyanin. On the solutions of Yang-Baxter equation. - LOMI Proceedings, 1980, v.95, p.129-160 (in Russian) is recommended.

II. The monodromy matrix was used in the full strength in the papers

5. E.K.Sklyanin, L.A.Takhtajan, L.D.Faddeev. Quantum inverse scattering method. I. - Teor. i matem. fisika, 1979, v.40, N 3, p.194-220 (in Russian).

6. L.A.Takhtajan, L.D.Faddeev. Quantum inverse scattering method and the XYZ-Heisenberg model.- Uspechi Mat. Nauk, 1979, v.34, N 5, p.13-63 (in Russian).

In [5] the quantum inverse scattering method was proposed and exact solution of the Sine-Gordon model was given. In particular, [5] contains the formulation of algebraic Bethe ansatz in abc-form. In [6] the generalized algebraic Bethe ansatz for the 8-vertex model (and connected with it XYZ-Hamiltonian) was presented. In coordinate form it was given in the papers

7. R.J.Baxter. Eight-vertex model in lattice statistics and one-dimensional anisotropic Heisenberg chain. I.-III. - Ann.Phys. (N.Y.), 1973, v.76, p.1-24, 25-47, 48-71.

III. The coordinate Bethe ansatz method was formulated by Bethe

8. H.Bethe. Zur Theorie der Metalle. I. Eigenwerte und Eigenfunktionen Atomkette. - Zeit.Phys., 1931, B.71, N 3-4, p.205-226.

It was generalized for the XXZ-Heisenberg model by Yang and Yang

9. C.N.Yang, C.P.Yang. One-dimensional chain of anisotropic spin-spin interactions. I. - II. - Phys.Rev., 1966, v.150, p.321-

327, 327-339.

Papers [9] contain all rigorous proofs.

References to the coordinate Bethe ansatz for the one-dimensional Bose-gas see in Korepin's lectures. Nested coordinate Bethe ansatz was introduced by Yang

10. C.N.Yang. Some exact results for the many-body problem in one dimension with repulsive delta-function interaction. - Phys. Rev.Lett., 1967, v.19, p.1312-1315.

Sutherland's $SU(N)$ model was introduced and solved in the paper

11. B.Sutherland. Model for a multicomponent quantum system.- Phys. Rev.B, 1975, v.12, N 9, p.3795-3805.

It was imbedded into the framework of quantum inverse scattering method in

12. P.P.Kulish, N.Yu.Reshetikhin. Generalized Heisenberg ferromagnet and the Gross-Neveu model.- JETP , 1981, v.80, N 1, p.214-228 (in Russian).

General " abc -condition" and the nested Bethe ansatz in algebraic form were given in

13. L.A.Takhtajan. Quantum inverse scattering method and algebraic matrix Bethe ansatz. - LOMI Proceedings, 1980, v.101,p.158-183 (in Russian).

Comment N 6 in lecture II belongs to S.V.Manakov (private communication). The requirement that $\{\lambda_j\}$ in Bethe states should be pairwise distinct is not really a restriction. This was proved in the paper

14. A.G. Isergin, V.E.Korepin. Pauli principle for one dimensional bosons and the algebraic Bethe ansatz. - LMP, 1982, v.6, N 4, p.283-288.

Norms of Bethe states in the formalism of quantum inverse scattering method were calculated by Korepin

15. V.E.Korepin. Calculation of norms of Bethe wave functions.- CMP, 1982, v.86, N 3, p.381-418.

IV. General scheme for obtaining local quantum Hamiltonians on a lattice can be found in

16. V.O.Tarasov, L.A.Takhtajan, L.D.Faddeev. Local Hamiltonians for quantum integrable models on the lattice. - Teoret. i matem. fisika, 1983, v.57, N 2, p.163-181.

Integrable generalization of isotropic Heisenberg model of spins $\frac{1}{2}$ to higher spins was given by Kulish, Reshetikhin and

Sklyanin

17. P.P.Kulish, N.Yu.Reshetikhin, E.K.Sklyanin. Yang-Baxter equation
    and representation theory. I. - LMP, 1981, v.5, N 5, p.393-403.

Its solution and thermodynamical properties are given, correspondingly, in the papers

18. L.A.Takhtajan. The picture of low-lying excitations in iso-
    tropic Heisenberg chain of arbitrary spin. - Phys.Lett., 1982,
    v.87A, N 9, p.479-482.
19. H.M.Babujian. Exact solution of the one-dimensional isotropic
    Heisenberg chain with arbitrary spin S . - Phys. Lett., 1982,
    v.90A, N 9, p.479-482.

Properties of Bethe states for the XXX-Heisenberg model
were established in the papers

20. L.D.Faddeev, L.A.Takhtajan. What is the spin of a spin wave?-
    Phys.Lett., 1981, v.85A, N 6-7, p.375-377.
21. L.A.Takhtajan, L.D.Faddeev. Spectrum and scattering of excita-
    tions in one-dimensional isotropical magnet. - LOMI Proceedings,
    1981, v.109, p.134-179 (in Russian)   and [18] .

V. Complete solution of the nonlinear Schrödinger model by the
quantum inverse scattering method was given by Sklyanin

22. E.K.Sklyanin. Quantum variant of the inverse scattering method.
    - LOMI Proceedings, 1980, v.95, p.55-128 (in Russian).

The local L-operator (4.18) was introduced in the paper

23. A.G.Isergin, V.E.Korepin. Lattice model, connected with nonli-
    near Schrödinger equation. - Doklady Acad.Nauk USSR, 1981,
    v.259, N 1, p.76-79.

About the connection between this model and XXX-Heisenberg
model of arbitrary spins see [16] . Quantum Sine-Gordon model was
solved by the quantum inverse scattering method in [5] . Its
lattice version was developed by Isergin and Korepin

24. A.G.Isergin, V.E.Korepin. The lattice quantum Sine-Gordon
    equation . - LMP, 1981, v.5, p.199-205 (see also their paper
    in Nucl.Phys.),

where the local L-operator (4.30) was presented.

VI. Exposition in lecture V follows the paper [21], where an
extensive list of references was given. Here we notice only that
assumptions a) and b) were actually used by Bethe in [8] .
The dynamical picture for the case of spins S was presented in
 [18] . In this case the proof of an analog of completeness relation

(5.21) reguires more complicated combinatorics. This proof was
given by Kirillov

25. A.N.Kirillov. Combinatoric identities and the completeness
    of Bethe multiplets for the Heisenberg model. - LOMI Proceedings,
    1983, v.131, p.88-105 (in Russian).

The counting of Bethe states is very important when one has
antiferromagnetic case. Thus in $[20 - 21]$ it was shown that a spin
wave-kink has spin $\frac{1}{2}$ rather then 1 which was generally assumed
to be true.

It should be emphasized that the string hypothesis is literally
true only when $N \longrightarrow \infty$ and ℓ is fixed. However in the antiferromagnetic case we have $\ell \approx \frac{N}{2}$, so the string hypothesis
for such ℓ needs verification. In the paper

26. C.Destry , J.H.Lowenstein. Analysis of the Bethe ansatz equa-
    tions of the chiral-invariant Gross-Neveu model.- Nucl.Phys.
    v.B205, N 3, p.369-385

in was shown that in the absense of external magnetic field ($H=0$)
and at zero temperature ($T = 0$) the string hypothesis literally
is not true: complex solutions of the system (5.6) do not combine
into a strings. By they do combine into a strings if $H > 0$ or
$T > 0$.

So in lecture V the following "regularization" is assumed:
introduce a small magnetic field $H \ll 1$, which permits to write
down the system (5.12), and then set $H \longrightarrow 0$.

VI. It is impossible to give here all necessary references to
such rapidly developing subject as quantum inverse scattering
method. For mathematically oriented audience I should mention
the good pedagogical course

27. L.D.Faddeev. Integrable models in $1 + 1$ dimensions. - Les
    Houches Lectures 1982, Elsevier Sci.Publ., 1984

and the lecture

28. L.A.Takhtajan. Integrable models in classical and quantum
    field theory.- Proc.Inter.Congress of Mathem., Warsaw 1983.

For those who would like to follow the current papers on
quantum inverse scattering method I shall point to the main
journals where they are published: LOMI Proceedings, Teoret. i
matem. fisika, JETP (these are Soviet journals) and Phys.Lett,
Nucl.Phys., CMP, LMP, J.Phys. A and C, Physika D, Phys. Rev.D.
The time interval starts from 1978 - 1979 .

QUANTUM INVERSE SCATTERING METHOD AND CORRELATION FUNCTIONS

N.M.Bogoliubov, A.G.Izergin, V.E.Korepin

Leningrad Department of V.A.Steklov
Mathematical Institute of the Academy of Science,
Fontanka 27, Leningrad 191011, USSR

CONTENTS

Introduction
Part I. Coordinate Bethe Ansatz. Calculation of observable quantities
  1. Coordinate Bethe Ansatz
  2. Periodical boundary conditions
  3. Thermodynamical limit at zero temperature
  4. Excitations at zero temperature
  5. Thermodynamics of the model
  6. Yangs' equation
  7. Limiting cases
  8. Excitations at nonzero temperatures
Part II. Quantum inverse scattering method
  1. Classical inverse scattering method
  2. Quantum inverse scattering method
  3. Algebraic Bethe Ansatz
  4. Classification of monodromy matrices and of L-operators
  5. Scalar products
  6. Norms of Bethe wave functions. Proof of the Gaudin hypothesis
  7. Algebraic approach to calculation of correlation functions
Part III. Correlation functions
  1. Current correlator at zero temperature
  2. Large distance asymptotics at zero temperature
  3. Current correlator at nonzero temperatures
  4. Dressing equations
  5. Temperature dependence of correlation length
Conclusion
Appendices
References

INTRODUCTION

After its appearence in 1979 $[1,2]$, the quantum inverse scattering method (QISM) has proved to be a powerful instrument for investigating completely integrable systems in quantum field theory and statistical physics. The method allows to imbed solution of these systems into a general scheme, thus revealing an approach to their classification. It also permits to solve many new models. In these lectures the consistent treatment of QISM is given, this method being applied to the one-dimensional Bose gas with a delta-function repulsive interaction between the particles (this model is described by the quantum nonlinear Schrödinger equation and is also called the nonlinear Schrödinger (NS) model). The complete theory of the one-dimensional Bose gas is presented here, from constructing eigenfunctions of the Hamiltonian up to the calculation of correlation functions for the model. We construct the ground state of the Hamiltonian (which is the Dirac sea) and describe the excitations over it. The observable characteristics of these excitations, i.e. the dressed energy, the dressed momentum and the dressed S-matrix, are calculated. The thermodynamics of the model is considered in detail. It should be noted that many results in this model had been obtained a long time before QUISM was created, using an explicit form of the coordinate wave function given by the Bethe Ansatz $[3,$ $4]$. Part I of these lectures is based on this traditional approach. One can not, however, progress enough using this method. It appears, e.g., that it is impossible to prove the Gaudin hypothesis $[5]$ about norms of the Bethe wave functions. It is also impossible to calculate correlation functions at a finite coupling constant.

The general approach to solving these problems is given by QISM , which is consistently accounted for in Part II of our lectures. We begin with the fundamentals of the method where the algebraic generalization of the coordinate Bethe Ansatz $[2, 6]$ plays a

most important rôle. We demonstrate that the further investigation
of the algebraical structure of QUISM allows to calculate the norms
of wave functions [7] and the correlation functions [8, 9] . Correlation functions for the NS model [9, 10] are calculated in
Part III. Here the formal answer for correlation functions is presented as a series. Attention is paid mainly to the large distance
asymptoties of the correlation functions. The temperature dependence of the correlation length is also calculated. It is worth mentioning that the reader who is interested in the results for the
NS model (but not in the method of obtaining them) may read Part III
immediately after Part I. We use the following enumeration of the
formulae: (5.13) means the 13-th formula of s.5. If the reference
to the formulae of other Parts is made, then (I.5.13) means the
13-th formula of s.5 of Part I. Similarly, s.II.3 means s.3 of
Part II; s.3 means section 3.

Part I.

COORDINATE BETHE ANSATZ. CALCULATION OF OBSERVABLE QUANTITIES.

The model of quantum one-dimensional Bose gas with delta-function
interaction between the particles was originally introduced by Lieb
and Liniger [4] . The traditional method of solution of this model
based on the coordinate Bethe Ansatz is considered in this Part in
detail. This method permits to construct the ground state of the
Hamiltonian and to describe excitations over it calculating such
observable characteristics as the energies, momenta and scattering
matrix of these excitations. The thermodynamics of the model is also considered here. This part is mainly based on the fundamental
papers by Lieb and Liniger [4, 11] and Yangs [12] .

1. THE COORDINATE BETHE ANSATZ.

Consider the one dimensional Bose gas which can be described by the quantum field $\psi(x,t)$ with the Hamiltonian

$$H = \int dx [\partial_x \psi^+ \partial_x \psi + c \psi^+ \psi^+ \psi \psi], \qquad (1.1)$$

where $\psi(x,t)$ is quantum operator with the equal-time canonical boson commutation relations:

$$[\psi(x), \psi^+(y)] = \delta(x-y); \quad [\psi(x), \psi(y)] = [\psi^+(x), \psi^+(y)] = 0. \qquad (1.2)$$

The equation of motion for ψ is easily obtained to be the nonlinear Schrödinger equation. That is why this model is called the NS model. The interesting thermodynamics for this model exists only in the repulsive case, i.e., if coupling constant $c > 0$. Only this case will be considered further. The Fock vacuum $|0\rangle$ defined by

$$\psi(x)|0\rangle = 0 \qquad \forall x \qquad (1.3)$$

is of importance. It will be called the pseudovacuum and it to be distinguished from the physical vacuum which is the ground state of the Hamiltonian, representing the Dirac sea. The dual pseudovacuum $\langle 0|$ is defined as $\langle 0| = |0\rangle^+$ and satisfies

$$\langle 0|\psi^+(x) = 0 \quad ; \qquad \langle 0|0\rangle = 1 \qquad (1.4)$$

Operator Q of number of particles and momentum operator P are

$$Q = \int dx \, \psi^+(x) \psi(x) ; \qquad (1.5)$$

$$P = -\frac{i}{2} \int dx \left[\psi^+(x) \partial_x \psi(x) - (\partial_x \psi^+(x)) \psi(x) \right] \qquad (1.6)$$

They are integrals of motion: $[H,P] = [H,Q] = [P,Q] = 0$. So one can look for the common eigenfunctions $|\Psi_N\rangle$ of operator H and Q :

$$|\Psi_N(\lambda_1,...,\lambda_N)\rangle = \frac{1}{\sqrt{N!}} \int d^N z \, \chi_N(z_1,...,z_N|\lambda_1,...,\lambda_N) \psi^+(z_1)...\psi^+(z_N)|0\rangle \qquad (1.7)$$

Here χ_N is a symmetrical function of all z_j . The eigenvalue equations

$$H|\Psi_N\rangle = E_N|\Psi_N\rangle ; \quad Q|\Psi_N\rangle = N|\Psi_N\rangle \qquad (1.8)$$

result in the fact that χ_N is an eigenfunction of a quantum mechanical Hamiltonian $\mathcal{H}_N$ of the one-dimensional Bose gas:

$$\mathcal{H}_N = -\sum_{j=1}^{N}(\partial^2/\partial z_j^2) + 2c\sum_{j>k=1}^{N}\delta(z_j - z_k) ; \qquad (1.9)$$

$$\mathcal{H}_N \chi_N = E_N \chi_N . \qquad (1.10)$$

So $\mathcal{H}_N$ is the Hamiltonian describing N boson particles with delta function pair interaction, the interaction being repulsive for $c > 0$. The potential in (1.9) being equal to zero almost everywhere, one can rewrite equation (1.10) in the following way. Consider the following domain in the coordinate space defined by conditions

$$z_1 < z_2 < ... < z_N . \qquad (1.11)$$

In this domain function χ_N is an eigenfunction of the free Hamiltonian $\mathcal{H}_N^{(0)}$:

$$\mathcal{H}_N^{(0)} = -\sum_{j=1}^{N}(\partial^2/\partial z_j^2) ; \qquad (1.12)$$

$$\mathcal{H}_N^{(0)} \chi_N = E_N \chi_N \quad (z_1 < z_2 < ... < z_N). \qquad (1.13)$$

The following boundary conditions must be satisfied:

$$\left(\frac{\partial}{\partial z_{j+1}} - \frac{\partial}{\partial z_j}\right)\chi_N = c\chi_N \quad (z_{j+1}=z_j+0; \; j=1,...,N). \qquad (1.14)$$

Such function in domain (1.11) can be constructed as follows. Consider the antisymmetrical eigenfunction of the Hamiltonian $\mathcal{H}_N^{(0)}$ given as the determinant of the $N \times N$ matrix $\exp\{i\lambda_j z_k\}$:

$$\det(\exp\{i\lambda_j z_k\}). \qquad (1.15)$$

The corresponding eigenvalue is $\sum \lambda_j^2$. It is easily seen that function χ_N defined in domain (1.11) as

$$\chi_N = \text{Const} \prod_{j>k}\left(\frac{\partial}{\partial z_j} - \frac{\partial}{\partial z_k} + c\right) \det(\exp\{i\lambda_j z_k\}) \qquad (1.16)$$

satisfies equations (1.13), (1.14). The determinant can be written as a sum over all the permutations P of numbers $1,...,N$:

$$\det(\exp\{i\lambda_j z_k\}) = \sum_{P}(-1)^{[P]}\exp\{i\sum_{n=1}^{N} z_n \lambda_{Pn}\} \qquad (1.17)$$

225

where $[P]$ means the parity of the permutation. Putting

$$\text{Const} = (-i)^{\frac{N(N-1)}{2}} \, \mathbb{C} \, ; \quad \mathbb{C} \equiv \left[N! \prod_{j>k} \left((\lambda_j - \lambda_k)^2 + c^2 \right) \right]^{-\frac{1}{2}}, \tag{1.18}$$

one gets in domain (1.11):

$$\chi_N = \mathbb{C} \sum_P (-1)^{[P]} \prod_{j>k} (\lambda_{P_j} - \lambda_{P_k} - ic) \exp\{ i \sum_{n=1}^{N} z_n \lambda_{P_n} \}. \tag{1.19}$$

Let us now continue χ_N by a symmetry in z_j to the whole $\mathbb{R}_N$ as (function χ_N is continuous in z_j):

$$\chi_N = \mathbb{C} \sum_P (-1)^{[P]} \prod_{j>k} (\lambda_{P_j} - \lambda_{P_k} - ic\varepsilon(z_j - z_k)) \exp\{ i \sum_{n=1}^{N} z_n \lambda_{P_n} \}, \tag{1.20}$$

where $\varepsilon(z)$ is a sign function. The symmetry of χ_N with respect to all z_j becomes obvious if one rewrites this expression in the following form:

$$\chi_N = \mathbb{C} \left(\prod_{j>k} (\lambda_j - \lambda_k) \right) \sum_P \prod_{j>k} \left[1 - \frac{ic\varepsilon(z_j - z_k)}{(\lambda_{P_j} - \lambda_{P_k})} \right] \exp\{ i \sum_{n=1}^{N} z_n \lambda_{P_n} \}. \tag{1.21}$$

One can see from (1.20) that function χ_N possesses the following antisymmetry property in λ :

$$\chi_N(z_1,...,z_N | \lambda_1,...,\lambda_j,...,\lambda_k,...,\lambda_N) = - \chi_N(z_1,...,z_N | \lambda_1,...,\lambda_k,...,\lambda_j,...,\lambda_N). \tag{1.22}$$

Hence $\chi_N = 0$ if $\lambda_j = \lambda_k$ $(j \neq k)$. This is a basis for the Pauli principle for one-dimensional interacting bosons being of great importance for constructing the ground state which is the Dirac sea. The Pauli principle is proved in paper $[30]$ in the frame of the algebraic methods of QISM. So common eigenfunctions χ_N of operators H, P and Q are constructed, the corresponding eigenvalues being equal to

$$E_N = \sum_{j=1}^{N} \lambda_j^2 \, ; \quad P_N = \sum_{j=1}^{N} \lambda_j \, ; \quad Q_N = N \tag{1.23}$$

These coordinate wave eigenfunctions were constructed originally in paper $[4]$ and studied in detail in $[13-15]$. Consider now the eigenfunctions in the whole coordinate space $\mathbb{R}_N : -\infty < z_j < +\infty$ $(j = 1, 2,..., N)$. The normalization in this case was calculated in $[16]$:

$$\int_{-\infty}^{\infty} d^N z \, \chi_N^*(z_1,\ldots,z_N | \lambda_1,\ldots,\lambda_N) \, \chi_N(z_1,\ldots,z_N | \mu_1,\ldots,\mu_N) =$$

$$= (2\pi)^N \prod_{j=1}^{N} \delta(\lambda_j - \mu_j) \, ; \quad \lambda_1 < \lambda_2 < \ldots < \lambda_N \, ; \quad \mu_1 < \mu_2 < \ldots < \mu_N . \tag{1.24}$$

In the same paper the completeness of functions χ_N is also proved:

$$\int_{-\infty}^{\infty} d^N \lambda \, \chi_N^*(z_1,\ldots,z_N | \lambda_1,\ldots,\lambda_N) \, \chi_N(y_1,\ldots,y_N | \lambda_1,\ldots,\lambda_N) =$$

$$= (2\pi)^N \prod_{j=1}^{N} \delta(z_j - y_j) \, ; \quad z_1 < z_2 < \ldots < z_N \, ; \quad y_1 < y_2 < \ldots < y_N . \tag{1.25}$$

2. PERIODICAL BOUNDARY CONDITIONS

Let us put the system considered into the periodical box of the length L . Then the wave function χ_N (1.20) should be periodical in each z_j at all the others z_k ($k \neq j$) given:

$$\chi_N(z_1,\ldots,z_j + L,\ldots,z_N | \lambda_1,\ldots,\lambda_N) = \chi_N(z_1,\ldots,z_j,\ldots,z_N | \lambda_1,\ldots,\lambda_N) . \tag{2.1}$$

These requirements result in the following system of transcendental equations (s.t.e.) for the permitted values of momenta λ_j :

$$\exp\{i\lambda_j L\} = - \prod_{k=1}^{N} [(\lambda_{jk} + ic)/(\lambda_{jk} - ic)]; \quad j = 1,\ldots,N; \quad \lambda_{jk} \equiv \lambda_j - \lambda_k . \tag{2.2}$$

This system of equations is of primary importance; furhter its main properties are discussed in detail. Let us prove first that in the repulsive case (i.e. if $c > 0$) all the solutions λ_j of (2.2) are real numbers:

$$\operatorname{Im} \lambda_j = 0 \, ; \quad j = 1,\ldots,N; \quad c > 0 \tag{2.3}$$

To do this one uses the following properties of functions entering (2.2):

$$|\exp\{i\lambda L\}| \geqslant 1 \, (\operatorname{Im} \lambda \leqslant 0); \quad |\exp\{i\lambda L\}| \leqslant 1 \, (\operatorname{Im} \lambda \geqslant 0); \tag{2.4}$$

$$|(\lambda+ic)/(\lambda-ic)| \leqslant 1 \quad (\operatorname{Im}\lambda \leqslant 0);$$

$$|(\lambda+ic)/(\lambda-ic)| \geqslant 1 \quad (\operatorname{Im}\lambda \geqslant 0). \tag{2.5}$$

Consider the set $\{\lambda_j ; j=1,2,...,N\}$ which satisfies equation (2.2). Denote λ_j with the maximal imaginary part as λ_{max} :

$$\lambda_{max} \in \{\lambda_j\} ; \quad \operatorname{Im}\lambda_{max} \geqslant \operatorname{Im}\lambda_j \quad (j=1,...,N). \tag{2.6}$$

Taking the modulus of both sides of the equation for $\lambda_j = \lambda_{max}$ in (2.2) one makes use of the estimate (2.5) for the right hand side (as $\operatorname{Im}(\lambda_{max}-\lambda_j) \geqslant 0$) and thus obtains:

$$\left|\exp\{i\lambda_{max}L\}\right| = \left| \prod_k \left(\frac{\lambda_{max}-\lambda_k+ic}{\lambda_{max}-\lambda_k-ic} \right) \right| \geqslant 1 . \tag{2.7}$$

Due to (2.4) this results in $\operatorname{Im}\lambda_j \leqslant \operatorname{Im}\lambda_{max} \leqslant 0$. Quite similarly one proves that $\operatorname{Im}\lambda_j \geqslant \operatorname{Im}\lambda_{min} \geqslant 0$. So the only possibility remaining is (2.3) and thus the statement is proved.

Let us now prove the existence of solutions for the s.t.e. (2.2) and parametrize all the solutions uniquely. To do this one puts the system into the logarithmic form. It is convenient to introduce variables φ_j :

$$\varphi_j \equiv L\lambda_j + \sum_{\substack{k=1 \\ k \neq j}}^{N} \Phi(\lambda_j - \lambda_k) ; \qquad j=1,...,N \tag{2.8}$$

where

$$\Phi(\lambda) = i \ln\left(\frac{\lambda+ic}{\lambda-ic} \right); \quad -2\pi < \Phi(\lambda) < 0 \quad (\operatorname{Im}\lambda = 0) \tag{2.9}$$

Taking the logarithm of both sides of (2.2) one has:

$$\varphi_j = 2\pi\tilde{n}_j ; \quad j=1,...,N \tag{2.10}$$

where $\{\tilde{n}_j ; j=1,...,N\}$ is an arbitrary set of N integers. Below we will use the antisymmetrical function $\theta(\lambda)$ instead of $\Phi(\lambda)$, this function increasing monotonically in λ :

$$\theta(\lambda) \equiv \Phi(\lambda) + \pi ; \qquad \theta(\lambda) = -\theta(-\lambda) ;$$

$$\theta(\lambda_1) < \theta(\lambda_2) \text{ if } \lambda_1 < \lambda_2 ; \quad \theta(\pm\infty) = \pm\pi \tag{2.11}$$

The s.t.e. (2.10) is rewritten as

$$L\lambda_j + \sum_{k=1}^{N} \theta(\lambda_j - \lambda_k) = 2\pi n_j\,; \qquad j=1,\dots,N\,, \qquad (2.12)$$

where integer or half-integer numbers n_j are defined as

$$n_j \equiv \tilde{n}_j + (N-1)/2\,. \qquad (2.13)$$

For the s.t.e. (2.12) one proves the existence and uniqueness of the solution, using Yangs' variational approach [12]. It is based on the fact that the s.t.e. (2.12) can be obtained from the variational principle, the corresponding action S being given as

$$S = \frac{L}{2}\sum_{j=1}^{N}\lambda_j^2 - 2\pi\sum_{j=1}^{N} n_j\lambda_j + \frac{1}{2}\sum_{i,j}\theta_1(\lambda_j - \lambda_i)\,, \qquad (2.14)$$

where $\theta_1(\lambda) = \int_0^\lambda \theta(\mu)\,d\mu$. Equations (2.12) are the extremum conditions for S: $\partial S/\partial\lambda_j = 0$. It appears that the action is convex, and hence the extremum is unique, this extremum being the minimum. To prove this it is sufficient to establish that the matrix $(\partial^2 S/\partial\lambda_j\partial\lambda_k)$ of second derivatives is positively definite, i.e. that $\sum(\partial^2 S/\partial\lambda_j\partial\lambda_k)\,v_j\,v_k \geqslant 0$ for any vector v_j $(j=1,\dots,N)$ with real components. One has

$$\varphi'_{j\ell} = \partial\varphi_j/\partial\lambda_\ell = \partial^2 S/\partial\lambda_j\partial\lambda_\ell =$$
$$(2.15)$$
$$= \delta_{j\ell}\Big[L + \sum_{m=1}^{N} K(\lambda_j,\lambda_m)\Big] - K(\lambda_j,\lambda_\ell)\,,$$

where

$$K(\lambda,\mu) = \Phi'(\lambda-\mu) = \theta'(\lambda-\mu) = 2c\big[(\lambda-\mu)^2 + c^2\big]^{-1} \qquad (2.16)$$

and so:

$$\sum_{j,\ell=1}^{N}\varphi'_{j\ell}\,v_j\,v_\ell = \sum_{j=1}^{N} L v_j^2 + \sum_{j>i=1}^{N} K(\lambda_j,\lambda_i)(v_j - v_i)^2 \geqslant$$
$$(2.17)$$
$$\geqslant \sum_{j=1}^{N} L v_j^2 \geqslant 0\,.$$

So the action is indeed convex, and the extremum is unique, being the minimum. At given set $\{n_j; j=1,\dots,N\}$ (2.13) solution of

(2.12) $\{\lambda_j; j=1,...,N\}$ thus exists and the solution is unique. It is also obvious that the given solution $\{\lambda_j\}$ defines the set $\{n_j\}$ uniquely. The solution of the system (2.12) possesses the following important property:

$$\lambda_j > \lambda_k \quad \text{if} \quad n_j > n_k \,; \qquad \lambda_j = \lambda_k \quad \text{if} \quad n_j = n_k\,. \qquad (2.18)$$

To prove this one subtracts the k-th equation of the system (2.12) from the j-th equation obtaining:

$$L(\lambda_j - \lambda_k) + \sum_{\ell=1}^{N} [\theta(\lambda_j - \lambda_\ell) - \theta(\lambda_k - \lambda_\ell)] = 2\pi(n_j - n_k)\,. \qquad (2.19)$$

Due to the monotonical increase of function $\theta(\lambda)$ in λ (see (2.11)) the left hand side is of the same sign as its first term and one easily comes to (2.18). In particular, if $n_j = n_k$, then also $\lambda_j = \lambda_k$ which results in the fact that the wave function becomes equal to zero (see the discussion of the Pauli principle in s.1). So only sets $\{n_j\}$ (2.13) where all the numbers n_j entering the given set are different has to be taken into account. Summing up, one comes to the conclusion that wave functions χ_N (1.20) in the periodical box (2.1) can be uniquely parametrized by sets $\{n_j\}$ (2.13) of N integers or half-integers (for N odd or even correspondingly), the numbers n_j entering the same set being all different. The corresponding solution of (2.2) is then the unique solution of (2.12), this solution existing for every set $\{n_j\}$. One has the following estimate for the solutions of the system (2.12) [16] :

$$(\lambda_j - \lambda_k) \geqslant \frac{2\pi(n_j - n_k)}{L(1 + 2c^{-1}D)} \geqslant \frac{2\pi}{L(1 + 2c^{-1}D)} \,; \quad \lambda_j > \lambda_k, \qquad (2.20)$$

where D is the particle density in the coordinate space:

$$D \equiv N/L\,. \qquad (2.21)$$

This is obtained from equation (2.19) using properties of functions θ and K following from (2.16):

$$0 < K(\lambda, \mu) < 2c^{-1}\,; \qquad (2.22)$$

$$\theta(\lambda) - \theta(\mu) = \int_{\mu}^{\lambda} K(\lambda, \nu)d\nu \leqslant 2(\lambda - \mu)c^{-1}\,; \quad \lambda > \mu \qquad (2.23)$$

The estimate (2.20) shows that momenta λ_j permitted by the s.t.e. can't be situated too near each other. This will be of importance at the thermodynamical limit restricting the upper limit of the condensation of the particles.

To conclude this paragraph we define function $\lambda(x)$ closely connected to the solutions $\{\lambda_j\}$ of the s.t.e. (2.12). This function appears to be very useful further; it is defined by the following relation:

$$L\lambda(x) + \sum_{k=1}^{N} \theta(\lambda(x) - \lambda_k) = 2\pi L x.$$ (2.24)

Introducing the action similar to (2.14),

$$S = \frac{1}{2}L\lambda^2(x) + \sum_{k=1}^{N} \theta_1(\lambda(x) - \lambda_k) - 2\pi L x \lambda(x),$$ (2.25)

one easily proves that the unique value $\lambda(x)$ does exist for every real x , and that due to the monotonicity of function $\theta(\lambda)$ (see (2.11)) function $\lambda(x)$ is a monotonically increasing function of x with the following properties:

$$\lambda(x_1) > \lambda(x_2) \text{ if } x_1 > x_2 ; \quad \lambda(x) = -\lambda(-x) ;$$
$$\lambda(x) \to 2\pi x \quad (x \to \pm\infty).$$ (2.26)

So function $\lambda(x)$ gives the one-to-one mapping of the real axis onto itself. It should be emphasized that the value of function $\lambda(x)$ at $x = n_j/L$ is just the corresponding λ_j from the solution of the s.t.e. (2.12):

$$\lambda(x = n_j/L) = \lambda_j ; \quad n_j \in \{n_j; j = 1,...,N\},$$ (2.27)

i.e. the momentum of the particle presented at the state χ_N (1.20), (2.1). Taking now number $m \notin \{n_j\}$ (m is integer for N odd and half-integer for N even) one can call the corresponding value of $\lambda(x)$,

$$\lambda_m = \lambda(x = m/L); \quad m \notin \{n_j\}; \quad m = n_j \pmod 1,$$ (2.28)

to be "the momentum" λ_m of the hole. So each number n integer or half integer ($n = ((N-1)/2)\pmod 1$) defines a vacancy; a filled vacancy ($n \in \{n_j\}$) is a particle, and a free vacancy is a hole. The total number of particles and holes gives a complete number of vacancies. The quantity $\rho_t(\lambda)$:

$$\rho_t(\lambda) \equiv dx(\lambda)/d\lambda \tag{2.29}$$

is the density of vacancies at the thermodynamical limit (it will
be shown later). Differentiating equation (2.24) with respect to
λ one has:

$$1 + L^{-1} \sum_{k=1}^{N} K(\lambda(x) - \lambda_k) = 2\pi\rho_t(\lambda(x)) \tag{2.30}$$

3. THERMODYNAMICAL LIMIT AT ZERO TEMPERATURE

We begin to describe the theormodynamical properties of the one-
dimensional Bose gas with constructing the ground state at zero
temperature $[4]$. We will use the physical arguments in this paragraph; the complete proof that the state constructed here is indeed the ground state will be given in s. 7. The grand canonical
ensemble will be considered further which corresponds to changing
the Hamiltonian H (1.1) to

$$H_h = H - hQ , \tag{3.1}$$

where Q (1.5) is the operator of the number of particles and h
is a chemical potential. We consider only the case $h > 0$ which is
physically interesting. Eigenfunctions χ_N (1.20) of the Hamiltonian H are also eigenfunctions of Hamiltonian H_h ; hence the
s.t.e. (2.2), (2.12) remains also the same. Eigenvalues of operators
H_h, P, Q are (see (1.23)):

$$E_N^{(h)} = \sum_{j=1}^{N} (\lambda_j^2 - h); \quad P_N = \sum_{j=1}^{N} \lambda_j; \quad Q_N = N. \tag{3.2}$$

The Fock vacuum $|0\rangle$ (1.3) is the eigenstate with zero eigenvalue
of the Hamiltonians H (1.1) and H_h (3.1): $H|0\rangle = H_h|0\rangle = 0$,
being the ground state for H . For H_h , however, there exist
eigenfunctions with $E_N^{(h)} < 0$, and we'll call $|0\rangle$ the pseudo-
vacuum; it is to be distinguished from the physical vacuum $|\Omega\rangle$
which is the ground state of the Hamiltonian H_h . To obtain the
ground state of Hamiltonian H_h one makes use of the s.t.e. desc-

ribed in detail in s. 2. It follows from (3.2) that particles with
small momenta λ_j give negative contributions to the energy ; hence the corresponding vacancies must be occupied in the ground state. Consider the special set $\{n_j\}$ (2.13) corresponding to such situation. One obtaines then from (2.12):

$$L\lambda_j + \sum_k \theta(\lambda_j - \lambda_k) = 2\pi j ; \quad -\frac{(N-1)}{2} \leqslant j \leqslant \frac{(N-1)}{2} ; \quad (n_j = j). \quad (3.3)$$

So all the vacancies with $-(N-1)/2 \leqslant n_j \leqslant (N-1)/2$ (i.e. with
$-q \leqslant \lambda_j \leqslant q$; $q = \lambda_{(N-1)/2}$) are occupied with particles; all
the other vacancies are free and correspond to holes. Consider now
the thermodynamical limit, supposing that

$$L \to \infty ; \quad N \to \infty ; \quad D \equiv N/L = \text{Const}. \quad (3.4)$$

The values of λ_j at this limit fill the interval $-q \leqslant \lambda_j \leqslant q$,
where

$$q = \lim \lambda_{(N-1)/2} \quad (3.5)$$

Let us denote $\rho(\lambda)$ the density of particles in the momentum space:

$$\rho(\lambda_k) = \lim L^{-1}(\lambda_{k+1} - \lambda_k)^{-1} ; \quad -q \leqslant \lambda_k \leqslant q \quad (3.6)$$

As all the vacancies inside the interval $[-q, q]$ are occupied,
one has that $\rho(\lambda)$ is just $\rho_t(\lambda)$ (2.29) in this interval:

$$\rho(\lambda) = dx/d\lambda = \rho_t(\lambda) ; \quad -q \leqslant \lambda \leqslant q, \quad (3.7)$$

the quantity $L\rho(\lambda)d\lambda$ being equal to the number of particles in
the interval $d\lambda$. Turning now to eq. (2.30) one changes also the
sum there for the integral:

$$L^{-1}\sum_k K(\lambda(x) - \lambda_k) =$$
$$= \int K(\lambda(x) - \lambda(y))dy = \int_{-q}^{q} K(\lambda(x) - \mu)\rho(\mu)d\mu, \quad (3.8)$$

obtaining the linear integral equation for $\rho(\lambda)$:

$$\rho(\lambda) - \frac{1}{2\pi}\int_{-q}^{q} K(\lambda, \mu)\rho(\mu)d\mu = \frac{1}{2\pi} ; \quad -q \leqslant \lambda \leqslant q . \quad (3.9)$$

This equation has the unique solution for any given $q > 0$ (which follows from the nondegeneracy of operator $(1-(2\pi)^{-1}\hat{K})$, see below).

It is convenient to introduce the linear operator $\hat{K}$ with positive kernel $K(\lambda, \mu)$, this operator acting on function $\rho(\lambda)$ as follows:

$$(\hat{K}\rho)(\lambda) = \int_{-q}^{q} K(\lambda, \mu)\rho(\mu)d\mu; \quad K(\lambda, \mu) = 2c\left[(\lambda-\mu)^2 + c^2\right]^{-1}. \quad (3.10)$$

Operator $(1-(2\pi)^{-1}\hat{K})$ is also of importance. This operator is nondegenerate, which follows from the estimate for its quadratic form obtained by taking the thermodynamical limit in (2.17):

$$\int_{-q}^{q} d\lambda\, v^2(\lambda) - \int_{-q}^{q} d\lambda \int_{-q}^{q} d\mu\, \frac{K(\lambda, \mu)}{2\pi} v(\lambda)v(\mu) \geqslant \int_{-q}^{q} \frac{d\lambda\, v^2(\lambda)}{2\pi \rho(\lambda)} \geqslant \quad (3.11)$$

$$\geqslant 0.$$

Here $v(\lambda)$ is a real arbitrary function. So operator $(1-(2\pi)^{-1}\hat{K})$ is indeed nondegenerate and its eigenvalues are positive, being separated from zero by a gap:

$$\left| 1 - (2\pi)^{-1}\hat{K} \right| \geqslant (2\pi \rho_{max})^{-1}, \quad (3.12)$$

where ρ_{max} is the maximum value of $\rho(\lambda)$. One obtains from (2.20) and (3.6) that

$$(2\pi)^{-1}(1 + 2Dc^{-1}) \geqslant \rho_{max} \geqslant \rho(\lambda) \geqslant (2\pi)^{-1}. \quad (3.13)$$

From the definition of $\rho(\lambda)$ one has:

$$D \equiv N/L = \int_{-q}^{q} \rho(\lambda)d\lambda. \quad (3.14)$$

The value of quantity q playing the role of the Fermi momentum was up to now arbitrary. This value is determined if one demands that the state constructed is indeed the ground state for the Hamiltonian H_h (3.1). It is shown in s.7 that this is so if one chooses q as follows. Define function $\varepsilon(\lambda)$ as the solution of the linear integral equation

$$\varepsilon(\lambda) - \frac{1}{2\pi} \int_{-q}^{q} K(\lambda, \mu)\varepsilon(\mu)d\mu = \lambda^2 - h \equiv \varepsilon_0(\lambda) \quad (3.15)$$

demanding that

$$\varepsilon(q) = \varepsilon(-q) = 0. \tag{3.16}$$

It will be shown in s.4 that $\varepsilon(\lambda)$ is just the energy of the one-particle excitation over the ground state. The properties of this function at finite temperature are studied in detail in s.7 , where the zero temperature limit is also considered. It is shown there that function $\varepsilon(\lambda)$ satisfying (3.15).and (3.16) does exist and is unique, Fermi momentum $q>0$ being thus uniquely defined by coupling constant $c>0$ and by chemical potential $h>0$. This function $\varepsilon(\lambda)$ possesses the following important properties:

$$\varepsilon(\lambda) < 0 \ \ if \ \ -q < \lambda < q ; \ \ \varepsilon(\lambda) > 0 \ \ if \ \ \lambda > q , \ \lambda < -q ;$$
$$\varepsilon(\lambda) = \varepsilon(-\lambda). \tag{3.17}$$

So the ground eigenstate $|\Omega\rangle$ -the physical vacuum for the Hamiltonian H_h (3.1) has been described above. The energy of this physical vacuum is equal to

$$E_v = \frac{\langle\Omega|H_h|\Omega\rangle}{\langle\Omega|\Omega\rangle} = L\int_{-q}^{q}(\lambda^2 - h)\rho(\lambda)d\lambda. \tag{3.18}$$

4. EXCITATIONS AT ZERO TEMPERATURE.

We will consider excitations over the physical vacuum [11] in the sector with the zero physical charge (i.e. such excitations that the number of the particles in the excited states is the same as in the ground state). These excitations are obtained by removing some number of particles with momenta $-q < \lambda_h < q$ from the vacuum distribution of particles (i.e. making holes with momenta λ_h) and by adding equal number of particles with a momenta $|\lambda_p| > q$. First construct the state of scattering of one particle with momentum $\lambda_p > q$ and of a hole with momentum $-q < \lambda_h < q$. The particle and the hole being present, the permitted values of vacuum particles momenta are changed: $\lambda_j \to \tilde{\lambda}_j$, so that the s.t.e. (3.3) for the vacuum particles is now rewritten as:

$$L\tilde{\lambda}_j + \sum_k \theta(\tilde{\lambda}_j - \tilde{\lambda}_k) + \theta(\tilde{\lambda}_j - \lambda_p) - \theta(\tilde{\lambda}_j - \lambda_h) = 2\pi j \qquad (4.1)$$

Subtracting this from (3.3) and taking into account that $\lambda_j - \tilde{\lambda}_j = \mathcal{O}(L^{-1})$; $\theta(\lambda + \Delta) - \theta(\lambda) = \mathcal{O}(\Delta)$, one obtains:

$$L(\lambda_j - \tilde{\lambda}_j) - \theta(\lambda_j - \lambda_p) + \theta(\lambda_j - \lambda_h) +$$
$$+ (\lambda_j - \tilde{\lambda}_j)\sum_k K(\lambda_j, \lambda_k) - \sum_k K(\lambda_j, \lambda_k)(\lambda_k - \tilde{\lambda}_k) = 0. \qquad (4.2)$$

Using equations (2.29), (2.30) and (3.7), one has

$$2\pi \rho(\lambda_j) L(\lambda_j - \tilde{\lambda}_j) - \theta(\lambda_j - \lambda_p) + \theta(\lambda_j - \lambda_h) -$$
$$- \sum_k K(\lambda_j, \lambda_k)(\lambda_k - \tilde{\lambda}_k)(\lambda_{k+1} - \lambda_k)^{-1}(\lambda_{k+1} - \lambda_k) = 0. \qquad (4.3)$$

Now one uses (3.6) and introduces the "shift function" F :

$$F(\lambda_j | \lambda_p, \lambda_h) \equiv (\lambda_j - \tilde{\lambda}_j)(\lambda_{j+1} - \lambda_j)^{-1}. \qquad (4.4)$$

At the thermodinamical limit one can change the sum in (4.3) for an integral obtaining the equation

$$F(\mu | \lambda_p, \lambda_h) - \int_{-q}^{q} \frac{d\nu}{2\pi} K(\mu, \nu) F(\nu | \lambda_p, \lambda_h) = \frac{\Phi(\mu - \lambda_p) - \Phi(\mu - \lambda_h)}{2\pi}, \qquad (4.5)$$

where $\Phi(\lambda) = \theta(\lambda) - \pi$ (see (2.9), (2.11)). So we are able to describe the vacuum polarization caused by a particle and a hole. It makes possible calculation of observables for the excitations over the ground states, namely, the energy, the momentum and the scattering matrix. These observable quantities are obtained by adding the contributions due to the vacuum polarization to the corresponding "bare" quantities. First we calculate the observable energy ΔE which is equal to the energy of the excited state minus the energy of the ground state:

$$\Delta E(\lambda_p, \lambda_h) = \varepsilon_0(\lambda_p) - \varepsilon_0(\lambda_h) + \sum_j [\varepsilon_0(\tilde{\lambda}_j) - \varepsilon_0(\lambda_j)] = \qquad (4.6)$$

$$= \mathcal{E}_0(\lambda_p) - \mathcal{E}_0(\lambda_h) - \int_{-q}^{q} \mathcal{E}_0'(\mu) F(\mu | \lambda_p, \lambda_h) d\mu,$$

where $\mathcal{E}_0(\lambda) = \lambda^2 - h$. Similarly one has for the observable momentum (the bare momentum is simply equal to λ):

$$\Delta P(\lambda_p, \lambda_h) = \lambda_p - \lambda_h - \int_{-q}^{q} F(\mu | \lambda_p, \lambda_h) d\mu. \tag{4.7}$$

Let us prove that

$$\Delta E(\lambda_p, \lambda_h) = \mathcal{E}(\lambda_p) - \mathcal{E}(\lambda_h), \tag{4.8}$$

where function $\mathcal{E}(\lambda)$ is defined by equations (3.15), (3.16). Differentiating (3.15) with respect to λ , one has

$$\left[(1 - (2\pi)^{-1} \hat{K}) \mathcal{E}' \right] (\lambda) = \mathcal{E}_0'(\lambda), \tag{4.9}$$

where operator $\hat{K}$ is defined in (3.10). It is convenient to define also linear integral operator $\hat{L}$ as follows:

$$(1 + \hat{L})(1 - (2\pi)^{-1} \hat{K}) = 1 ; \quad \text{or}$$
$$(2\pi)^{-1}(1 + \hat{L}) \hat{K} = \hat{L}. \tag{4.10}$$

The kernel of operator $\hat{L}$ is the symmetric function $L(\lambda, \mu)$ $(L(\lambda, \mu) = L(\mu, \lambda))$. Equation (4.9) can be now rewritten as

$$\mathcal{E}'(\lambda) - \mathcal{E}_0'(\lambda) = \int_{-q}^{q} L(\lambda, \mu) \mathcal{E}_0'(\mu) d\mu. \tag{4.11}$$

Equation (4.5) for F can be transformed as follows:

$$\left[(1 - \tfrac{1}{2\pi} \hat{K}) F \right](\mu) = \frac{1}{2\pi} \int_{\lambda_p}^{\lambda_h} \Phi'(\mu - \nu) d\nu = \frac{1}{2\pi} \int_{\lambda_p}^{\lambda_h} K(\mu, \nu) d\nu. \tag{4.12}$$

Acting by operator $(1 + \hat{L})$ on the both sides and using (4.10), one has:

$$F(\mu | \lambda_p, \lambda_h) = \int_{\lambda_p}^{\lambda_h} L(\mu, \nu) d\nu. \tag{4.13}$$

Putting this representation into (4.6) one obtains:

$$\Delta E = \varepsilon_0(\lambda_p) - \varepsilon_0(\lambda_h) - \int\limits_{\lambda_p}^{\lambda_h} d\nu \int\limits_{-q}^{q} d\mu\, L(\nu,\mu)\,\varepsilon_0'(\mu).$$

and after using (4.11):

$$\Delta E = \varepsilon_0(\lambda_p) - \varepsilon_0(\lambda_h) - \int\limits_{\lambda_p}^{\lambda_h} d\nu\,[\varepsilon'(\nu) - \varepsilon_0'(\nu)] = \varepsilon(\lambda_p) - \varepsilon(\lambda_h) \qquad (4.14)$$

which means that (4.8) is true. Let us prove now that ΔP (4.7) can be given as

$$\Delta P(\lambda_p,\lambda_h) = \lambda_p - \lambda_h + \int\limits_{-q}^{q} [\Phi(\lambda_p-\mu) - \Phi(\lambda_h-\mu)]\,\rho(\mu)\,d\mu. \qquad (4.15)$$

The vacuum particle density ρ satisfies equation (3.9) which can be rewritten as

$$2\pi\rho(\mu) - 1 = \int\limits_{-q}^{q} L(\mu,\nu)\,d\nu = \int\limits_{-q}^{q} K(\mu,\nu)\,\rho(\nu)\,d\nu \qquad (4.16)$$

Using this equation, one obtains from (4.7), (4.13):

$$\Delta P(\lambda_p,\lambda_h) = \lambda_p - \lambda_h - \int\limits_{-q}^{q} d\mu \int\limits_{\lambda_p}^{\lambda_h} d\nu\, L(\mu,\nu) =$$

$$= \lambda_p - \lambda_h - \int\limits_{\lambda_p}^{\lambda_h} d\mu \int\limits_{-q}^{q} d\nu\, K(\mu,\nu)\,\rho(\nu), \qquad (4.17)$$

which results in (4.15).

Next we consider the particle-hole S-matrix. This S-matrix being a scalar, its calculation is reduced to the calculation of the scattering phase. The scattering phase of two bare particles is equal to $\Phi(\lambda_1-\lambda_2)$ (2.9), the scattering phase for several particles being equal to the sum of the pair phases. The particle -hole scattering matrix can be written as

$$S = \exp\{i\delta(\lambda_p, \lambda_h)\} \; ; \quad \lambda_p > q \, , \; -q < \lambda_h < q . \tag{4.18}$$

To calculate δ , one considers first the complete phase Φ_p which gains the particle moving through the whole box:

$$\Phi_p = L\lambda_p + \sum_k \Phi(\lambda_p - \tilde{\lambda}_k) - \Phi(\lambda_p - \lambda_h) . \tag{4.19}$$

It is obvious that $\Phi_p = \delta(\lambda_p, \lambda_h) + \text{const}$, where const does not depend on λ_h . Thus one has differentiating (4.19) in :

$$\frac{\partial}{\partial \lambda_h} \delta(\lambda_p, \lambda_h) = \frac{\partial}{\partial \lambda_h} \Phi_p = \Phi'(\lambda_p - \lambda_h) + \frac{\partial}{\partial \lambda_h}\left[\sum_k \Phi(\lambda_p - \tilde{\lambda}_k)\right]. \tag{4.20}$$

The sum here at the thermodynamical limit can be written as (compare with (4.2), (4.3)):

$$\sum_k \Phi(\lambda_p - \tilde{\lambda}_k) = \sum_k \Phi(\lambda_p - \lambda_k) + \sum_k \Phi'(\lambda_p - \lambda_k)(\lambda_k - \tilde{\lambda}_k) =$$

$$= L \int_{-q}^{q} \Phi(\lambda_p - \mu)\rho(\mu)d\mu + \int_{-q}^{q} K(\lambda_p, \mu) F(\mu | \lambda_p, \lambda_h)d\mu . \tag{4.21}$$

So one has from (4.20):

$$\frac{\partial}{\partial \lambda_h} \delta(\lambda_p, \lambda_h) = \Phi'(\lambda_p - \lambda_h) + \int_{-q}^{q} K(\lambda_p, \mu)\frac{\partial}{\partial \lambda_h} F(\mu | \lambda_p, \lambda_h)d\mu . \tag{4.22}$$

Using now the relation obtained from (4.5):

$$\left(1 - \frac{1}{2\pi}\hat{K}\right)\left(\frac{\partial}{\partial \lambda_h} F\right) = \frac{1}{2\pi} \Phi'(\mu - \lambda_h) , \tag{4.23}$$

one can simplify (4.22) as follows:

$$\frac{\partial}{\partial \lambda_h} \delta(\lambda_p, \lambda_h) = 2\pi \frac{\partial}{\partial \lambda_h} F(\lambda_p | \lambda_p, \lambda_h) . \tag{4.24}$$

Imposing the natural boundary condition for δ :

$$\delta(\lambda_p, \lambda_h) \to 0 \quad (\lambda_p \to +\infty) \tag{4.25}$$

one can integrate (4.24) and derive then the following integral
equation for the phase itself:

$$\delta(\mu, \lambda_h) - \frac{1}{2\pi} \int_{-q}^{q} K(\mu, \nu)\, \delta(\nu, \lambda_h)\, d\nu = - \Phi(\mu - \lambda_h). \tag{4.26}$$

The quantity $\delta(\lambda_p, \lambda_h)$ is obtained by continuation in μ to
$\mu = \lambda_p$:

$$\delta(\lambda_p, \lambda_h) = \delta(\mu, \lambda_h)\big|_{\mu = \lambda_p}. \tag{4.27}$$

(The branch of the logarithm in (2.9) is supposed to be taken for
which $\Phi(+\infty) = 0$). One sees that the scattering phase for excitations over the physical vacuum can be restored from the bare phase by means of dressing equation (4.26). This equation was originally suggested in [17] for the massive Thirring model.

We have demonstrated calculation of the main physical observables of the particles, i.e. of the excitations over the ground
state $|\Omega\rangle$ of the Hamiltonian, this ground state representing
a Dirac sea. This physical quantities are obtained from the bare
ones by means of dressing equations which are here linear integral
equations. (The bare quantities, which characterize particles over
the Fock vacuum $|0\rangle$ (the pseudovacuum), such as the energy, mo-
mentum and the S-matrix (see (3.2), (2.9)) are known explicitly).

In conclusion we would like to consider quite a different
question. The sound velocity v is an important quantity in the
model. It is shown in [11] that

$$v = \frac{\partial \mathcal{E}(\lambda)}{\partial (\Delta P(\lambda))}\bigg|_{\lambda = q} = \left\{ \frac{\partial \mathcal{E}(\lambda)}{\partial \lambda} \left[\frac{\partial \Delta P(\lambda)}{\partial \lambda} \right]^{-1} \right\}\bigg|_{\lambda = q} \tag{4.28}$$

Here $\mathcal{E}(\lambda)$ is the dressed energy (3.15), (3.16); $\Delta P(\lambda)$ is the
dressed momentum and q is a Fermi momentum. Using (4.7), (4.10),
(4.16) one can see that

$$\frac{\partial \Delta P(\lambda)}{\partial \lambda}\bigg|_{\lambda = q} = 1 + \frac{1}{2\pi} \int_{-q}^{q} \overset{\circ}{F}(\lambda | q)\, d\lambda, \tag{4.29}$$

where the function $\overset{\circ}{F}$ satisfies the following integral equation:

$$\overset{\circ}{F}(\lambda|q) - \frac{1}{2\pi} \int\limits_{-q}^{q} K(\lambda,\nu)\overset{\circ}{F}(\nu|q)d\nu = K(\lambda,q). \tag{4.30}$$

So one has for the sound velocity

$$\upsilon = \left\{\frac{\partial \varepsilon(\lambda)}{\partial \lambda}\right\}\Big|_{\lambda=q} \left[1 + \frac{1}{2\pi} \int\limits_{-q}^{q} \overset{\circ}{F}(\lambda|q)d\lambda\right]^{-1}. \tag{4.31}$$

5. THERMODYNAMICS OF THE MODEL [12]

The thermodynamics of the model at finite temperature is considered here. We deal with the grand canonical ensemble (the number of particles is not fixed), so that the grand partition function Z is given as

$$Z = tr \exp\{-H_h/T\}, \tag{5.1}$$

where the Hamiltonian H_h is given by (3.1), and T is the temperature. The chemical potential h entering H_h is supposed to be positive: $h > 0$, otherwise it is an arbitrary parameter. The thermodynamical limit is defined by (3.4): $L \to \infty$, $\overline{N} \to \infty$, $D \equiv$ $\equiv \overline{N}/L = Const$, where $\overline{N}$ is a mean value of number of particles at the gas. At the thermodynamical limit vacancies, particles and holes (all these were defined at the end of s.2) have finite densities ρ_t, ρ_p and ρ_h which are defined as follows:

$$L\rho_p(\lambda)d\lambda = \text{number of particles in } [\lambda, \lambda+d\lambda]; \tag{5.2}$$

$$L\rho_h(\lambda)d\lambda = \text{number of holes in } [\lambda, \lambda+d\lambda] \quad ; \tag{5.3}$$

$$L\rho_t(\lambda)d\lambda = \text{number of vacancies in } [\lambda, \lambda+d\lambda] , \tag{5.4}$$

the number of vacancies being simply the sum of the numbers of particles and holes:

$$\rho_t(\lambda) = \rho_p(\lambda) + \rho_h(\lambda). \tag{5.5}$$

It is easily seen that $\rho_t(\lambda)$ introduced here is just the thermodynamical limit of function ρ_t defined in (2.29). At this limit the sum in equation (2.30) turns into the integral with the density $\rho_p(\lambda)$ and one has:

$$2\pi \rho_t(\lambda) = 1 + \int_{-\infty}^{\infty} K(\lambda,\mu)\rho_p(\mu)d\mu. \tag{5.6}$$

This equation is equivalent to the system of periodical boundary conditions for the wave function. It should be emphasized that we pass now from the microscopic description of the model (the set n_j (2.13) determining the wave function uniquely) to the macroscopic description in terms of densities ρ_p, ρ_h, ρ_t . The given microscopic situation described by some wave function corresponds to the unique set of densities ρ_p, ρ_h, ρ_t . The opposite is however, not true: in general, a lot of sets of microscopic variables correspond to given functions ρ_p, ρ_h, ρ_t . Indeed, there exist many possibilities to put $(L\rho_p(\lambda)d\lambda)$ particles into $(L\rho_t(\lambda)d\lambda)$ vacancies; notating the number of this possibilities $\exp\{dS\}$, one has

$$\exp\{dS\} = \frac{[L\rho_t(\lambda)d\lambda]!}{[L\rho_p(\lambda)d\lambda]! \; [L\rho_h(\lambda)d\lambda]!} . \tag{5.7}$$

This number is large at the thermodynamical limit; using the well-known formula for the factorial asymptotics, one has

$$dS = L\left[\rho_t(\lambda)\ln\rho_t(\lambda) - \rho_p(\lambda)\ln\rho_p(\lambda) - \rho_h(\lambda)\ln\rho_h(\lambda)\right]d\lambda. \tag{5.8}$$

The situation here is somewhat analogous to that in the gauge field theory if one uses the following analogy: $\exp\{dS\}$ is analogous to the volume of the gauge group, going from one to another set of the microscopic variables at given macroscopic variables corresponds to a gauge transformation, and the macroscopic variables ρ_p, ρ_h, ρ_t play the role of the gauge invariant quantities. Turn now to the partition function Z (5.1). It can be represented in the form

$$Z = \sum_{N=0}^{\infty} Z_N \, , \qquad\qquad (5.9)$$

where Z_N corresponds to the contribution of the N-particle eigenfunctions to the trace in (5.1). Eigenfunctions of the Hamiltonian H_h (3.1) can be uniquely parametrized by sets $\{n_j\}$ of integers or half-integers n_j (see (2.13)), hence one can write

$$Z_N = \frac{1}{N!} \sum_{n_1,\dots,n_N} \exp\{- \frac{E_N^{(h)}}{T}\} = \sum_{n_1<\dots<n_N} \exp\{- \frac{E_N^{(h)}}{T}\} \, , \qquad (5.10)$$

where $E_N^{(h)} = \sum(\lambda_j^2 - h)$ (see (3.2)) remains unchanged under permutations of $n_1,\dots,n_N$; the momenta λ_j here satisfy the s.t.e. (2.12). We'll consider the thermodynamics of the gas which is at rest as a whole; hence the total momentum P_N (3.2) is equal to zero:

$$P_N = \sum_{j=1}^{N} \lambda_j = 0 \qquad\qquad (5.11)$$

Summing up all the equations of the s.t.e. (2.12) and using the fact that $\theta(-\lambda) = -\theta(\lambda)$ (2.11), one has

$$\sum_{j=1}^{N} n_j = (2\pi)^{-1} L \sum_{j=1}^{N} \lambda_j = 0 \qquad\qquad (5.12)$$

Thus one obtains for Z_N (5.10) introducing variables $n_{j+1,j} = n_{j+1} - n_j$ and using (5.12):

$$Z_N = \sum_{n_{21}=1}^{\infty} \sum_{n_{32}=1}^{\infty} \cdots \sum_{n_{N,N-1}=1}^{\infty} \exp\{- \frac{E_N^{(h)}}{T}\} \qquad\qquad (5.13)$$

Now we go to the macroscopic variables from the microscopic ones. The variable $n_{j+1,j}$ can be considered as the number of vacancies for the j-th particle (i.e. only one vacancy from $n_{j+1,j} = = n_{j+1} - n_j$ is occupied). Macroscopically the ratio of the complete number of vacancies to the number of the occupied vacancies is given by the ratio of the corresponding densities, i.e., by $\rho_t(\lambda)/\rho_p(\lambda)$. So the sum in (5.13) can be changed for the functional integral $\int D(\rho_t(\lambda)/\rho_p(\lambda))$. While doing this one has to remember, however, that many sets of microscopical variables correspond to given macroscopical variables (see (5.8), (5.9)). So the accurate change is:

$$\sum \cdots \sum \rightarrow$$

$$\rightarrow \int D(\rho_t(\lambda)/\rho_p(\lambda))\,\exp\{L\int d\lambda[\rho_t\ln\rho_t-\rho_p\ln\rho_p-\rho_h\ln\rho_h]\}. \tag{5.14}$$

The partition function Z (5.1) is the sum (5.9) of Z_N over N. The main contribution to the sum is given by Z_N with N large, $N\sim\bar{N}$ ($\bar{N}/L=\mathrm{const}$; $L\rightarrow\infty$). So one obtains the functional integral representation for Z :

$$Z = \mathrm{Const}\int\Big[\prod_\lambda d(\rho_t(\lambda)/\rho_p(\lambda))\Big]\exp\{-X/T\}, \tag{5.15}$$

where

$$X = L\int d\lambda\{\rho_p(\lambda)(\lambda^2-h)-T[\rho_t\ln\rho_t-\rho_p\ln\rho_p-\rho_h\ln\rho_h]\}. \tag{5.16}$$

It appears to be convenient to use the following notations, introducing function $\varepsilon(\lambda)$:

$$\exp\{\varepsilon(\lambda)/T\} = \rho_h(\lambda)/\rho_p(\lambda). \tag{5.17}$$

As $L\rightarrow\infty$, the steepest descent method can be used to calculate the functional integral (5.15). Consider the variation of the functional X changing function $\rho_p(\lambda)$ by $\delta\rho_p(\lambda)$:

$$\delta X = L\int d\lambda\,\delta\rho_p(\lambda)\{\lambda^2-h-\varepsilon(\lambda)-$$

$$-(T/2\pi)\int K(\lambda,\mu)\,\ln[1+\exp(-\varepsilon(\mu)/T)]\,d\mu\}. \tag{5.18}$$

The variation of function $\varepsilon(\lambda)$ is:

$$\delta\varepsilon(\lambda) = (-T/\rho_p(\lambda))[1+\exp(-\varepsilon(\lambda)/T)]\cdot$$

$$\cdot\{\delta\rho_p(\lambda)-[1+\exp(\varepsilon(\lambda)/T)]^{-1}\cdot \tag{5.19}$$

$$\times (1/2\pi) \int K(\lambda,\mu)\, \delta\rho_p(\mu)\, d\mu \Big\} .$$

The second variation of X is expressed rather simply in terms of $\delta\varepsilon(\lambda)$:

$$\delta^2 X = \frac{L}{T} \int d\lambda\, [\delta\varepsilon(\lambda)]^2\, \rho_p(\lambda)\, \Big[1 + e^{-\frac{\varepsilon(\lambda)}{T}}\Big]^{-1} > 0 \qquad (5.20)$$

So the functional X is convex in the domain of its definition and the only extremum in this domain is the minimum given by the equation $\delta X = 0$, which due to (5.18) gives

$$\varepsilon(\lambda) = \lambda^2 - h -$$
$$(5.21)$$
$$- (T/2\pi) \int K(\lambda,\mu)\, \ln\big[1 + \exp\{-\varepsilon(\mu)/T\}\big]\, d\mu .$$

The value of X at the stationary point (5.21) will be denoted Y; the calculations leads to

$$Y = (-TL/2\pi) \int \ln\big[1 + \exp\{-\varepsilon(\lambda)/T\}\big]\, d\lambda . \qquad (5.22)$$

The functional integral (5.15) can be estimated by its value at the stationary point which gives for the partition function:

$$Z = \exp\{-Y/T\} . \qquad (5.23)$$

It is well-known that $Z = \exp\{L\, Pr/T\}$, where Pr denotes the pressure in the gas. One has from (5.22), (5.33):

$$Pr = (T/2\pi) \int_{-\infty}^{\infty} d\lambda\, \ln\big[1 + \exp\{-\varepsilon(\lambda)/T\}\big] . \qquad (5.24)$$

Direct calculations show that the basic thermodynamical identity is fulfilled:

$$d\,Pr = (S/L)\, dT + D\, dh . \qquad (5.25)$$

Here D is the density in the configuration space:

$$D = \int_{-\infty}^{\infty} \rho_p(\lambda)\, d\lambda = (\bar{N}/L), \tag{5.26}$$

and S is the entropy

$$\frac{S}{L} = \int_{-\infty}^{\infty} [\rho_t(\lambda)\ln\rho_t(\lambda) - \rho_p(\lambda)\ln\rho_p(\lambda) - \rho_h(\lambda)\ln\rho_h(\lambda)]\, d\lambda. \tag{5.27}$$

The free energy is given as

$$F = h\bar{N} - LP_r \tag{5.28}$$

All the densities ρ_p, ρ_h, ρ_t here are uniquely defined by the following system of equations:

$$2\pi\rho_t(\lambda) = 1 + \int_{-\infty}^{\infty} K(\lambda,\mu)\rho_p(\mu)\, d\mu; \tag{5.29}$$

$$\varepsilon(\lambda) = \lambda^2 - h - (T/2\pi)\int_{-\infty}^{\infty} K(\lambda,\mu)\ln[1 + e^{-\frac{\varepsilon(\mu)}{T}}]\, d\mu; \tag{5.30}$$

$$\rho_h(\lambda)/\rho_p(\lambda) = \exp\{\varepsilon(\lambda)/T\}; \quad \rho_t(\lambda) = \rho_p(\lambda) + \rho_h(\lambda). \tag{5.31}$$

The chemical potential $h > 0$, the temperature $T > 0$ and the coupling constant $c > 0$ are free parameters here. So the equilibrium state of the one-dimensional Bose-gas is given by the system (5.29)-(5.31). This equilibrium state is not described by the single eigenfunction of the Hamiltonian; it is not at all the pure quantum mechanical state (as for $T = 0$), but is a mixture of many eigenfunctions. The reason is that, in general, many sets of microscopic variables $\{n_j\}$ (2.13) correspond to given $\rho_p(\lambda)$, $\rho_h(\lambda)$, $\rho_t(\lambda)$, and there is the one-to-one correspondence between sets $\{n_j\}$ and eigenfunctions of the Hamiltonian. Equation (5.30) which is a nonlinear integral equation for function $\varepsilon(\lambda)$ plays the most important role. It was obtained and investigated in detail by Yangs, and we call it the Yangs' equation. Function $\varepsilon(\lambda)$ has a clear physical interpretation, giving an energy of elementary excitations over the equilibrium state. It will be shown directly in s.8. One can, however, understand this without doing much calculation. Consider the ratio $\vartheta(\lambda)$ of the number of occupied vacancies to the number of all the vacancies in the interval $[\lambda, \lambda + d\lambda]$. Using (5.17), one has:

$$\vartheta(\lambda) = \rho_p(\lambda)/\rho_t(\lambda) = \left[1 + \exp\{\varepsilon(\lambda)/T\}\right]^{-1}. \tag{5.32}$$

Comparing this with the Fermi distribution, one concludes that $\varepsilon(\lambda)$ must be the energy mentioned above. Function $\vartheta(\lambda)$ will be called the Fermi weight; it will be often used.

In conclusion we give some useful estimates which are similar to those given in s.3 (3.11), (3.12) for $T = 0$. Using (2.20) one obtains for $T > 0$ the estimate analogous to (3.11):

$$(2\pi)^{-1} \leqslant \rho_t(\lambda) \leqslant (2\pi)^{-1}(1 + 2Dc^{-1}) \tag{5.33}$$

It is also convenient to introduce the integral operator $\hat{K}_T$ acting on arbitrary function $g(\lambda)$ as follows

$$(\hat{K}_T g)(\lambda) = \int_{-\infty}^{\infty} K(\lambda,\mu)\,\vartheta(\mu)\,g(\mu)\,d\mu. \tag{5.34}$$

Function $K(\lambda,\mu)$ here is given by (2.16) as $K(\lambda,\mu) = 2c \cdot [(\lambda-\mu)^2 + c^2]^{-1}$. Operator $\hat{K}_T$ is similar to the hermitian operator K_H with the kernel $\tilde{K}(\lambda,\mu) = \sqrt{\vartheta(\lambda)}\, K(\lambda,\mu)\sqrt{\vartheta(\mu)}$. Relation (2.17) results in the following estimate for the quadratic form of this operator

$$\int_{-\infty}^{\infty} f^2(\lambda)\left[1 - (2\pi\rho_t(\lambda))^{-1}\right]d\lambda \geqslant (2\pi)^{-1} \int_{-\infty}^{\infty} d\lambda \int_{-\infty}^{\infty} d\mu \cdot \tag{5.35}$$

$$\cdot f(\lambda)f(\mu)\sqrt{\vartheta(\lambda)}\, K(\lambda,\mu)\sqrt{\vartheta(\mu)},$$

where $f(\lambda)$ is an arbitrary **real** function. Thus one has for eigenvalues k_T of this operator:

$$0 < k_T \leqslant 4\pi D(2D+c)^{-1} < 2\pi \tag{5.36}$$

The proof is similar to that of (3.11), (3.12).

6. YANGS' EQUATION

Turn now to Yangs' equation (5.30):

$$\varepsilon(\lambda) = \lambda^2 - h - \frac{T}{2\pi} \int_{-\infty}^{\infty} K(\lambda,\mu)\, \ln[1 + \exp\{-\varepsilon(\mu)/T\}]\, d\mu; \tag{6.1}$$

$$(h > 0; \quad c > 0; \quad K(\lambda,\mu) > 0).$$

Our first task is to prove the existence of the solution. To do this one constructs the following sequence of functions :

$$\varepsilon_0(\lambda) = \lambda^2 - h; \tag{6.2}$$

$$\varepsilon_{n+1}(\lambda) = \lambda^2 - h + A_n; \quad n = 0, 1, 2, \dots$$
$$A_n = (-T/2\pi) \int K(\lambda,\mu)\, \ln[1 + \exp\{-\varepsilon_n(\mu)/T\}]\, d\mu. \tag{6.3}$$

This sequence monotonically decreases in n :

$$\varepsilon_0(\lambda) > \varepsilon_1(\lambda) > \dots > \varepsilon_n(\lambda) > \varepsilon_{n+1}(\lambda) > \dots \;, \tag{6.4}$$

and is bounded below:

$$\varepsilon_n(\lambda) \geqslant \lambda^2 + x_0, \tag{6.5}$$

where x_0 is some constant. These properties proved below mean that the limit $\varepsilon(\lambda) = \lim \varepsilon_n(\lambda)$ does exist and is a solution of equation (6.1). Property (6.4) is obvious from the fact that in (6.3) $A_n < 0$ and $\delta A_n / \delta \varepsilon_n(\lambda) > 0$:

$$\delta A_n = (2\pi)^{-1} \int_{-\infty}^{\infty} K(\lambda,\mu)[1 + \exp\{\varepsilon_n(\mu)/T\}]^{-1} \delta \varepsilon_n(\mu)\, d\mu. \tag{6.6}$$

The proof of (6.5) is not so straightforward. First we establish some important properties of $\varepsilon_n(\lambda)$, namely that

$$\varepsilon_n(\lambda) = \varepsilon_n(-\lambda), \tag{6.7}$$

and

$$\varepsilon_n(\lambda_1) > \varepsilon_n(\lambda_2) \quad \text{if} \quad \lambda_1 > \lambda_2 \geqslant 0. \tag{6.8}$$

The first property means that $\varepsilon_n(\lambda)$ is an even function of λ and is quite obvious. The second property (6.8) can be proved by induction in n . Supposing that (6.8) is true for some $\varepsilon_n(\lambda)$, one establishes that it is also true for $\varepsilon_{n+1}(\lambda)$. Indeed, one has from (6.3):

$$\varepsilon'_{n+1}(\lambda) = 2\lambda + \frac{1}{2\pi} \int_{-\infty}^{\infty} K(\lambda,\mu) \frac{\varepsilon'_n(\mu)}{[1+\exp\{\varepsilon_n(\mu)/T\}]}\, d\mu =$$

$$= 2\lambda + \frac{1}{2\pi} \int_0^{\infty} [K(\lambda,\mu)-K(-\lambda,\mu)]\frac{\varepsilon'_n(\mu)}{[1+\exp\{\varepsilon_n(\mu)/T\}]}\, d\mu.$$

$$(6.9)$$

As $K(\lambda,\mu) > K(-\lambda,\mu)$ at $\lambda > 0$ and $\mu > 0$ (see (2.16)), one has that $\varepsilon'_{n+1}(\lambda) > 0$ at $\lambda > 0$ if $\varepsilon'_n(\lambda) > 0$ at $\lambda > 0$. Taking into account that $\varepsilon_0(\lambda)$ (6.2) satisfies (6.8), one completes the proof. Functions $\varepsilon_n(\lambda)$ are thus monotonically increasing on the positive semi-axis of λ . As the term A_n in (6.3) is also an even function monotonically increasing on the positive semi-axis, one has

$$\varepsilon_n(\lambda) \geqslant \lambda^2 + \varepsilon_n(0). \qquad (6.10)$$

Taking into account equations (6.3) and (6.6) one can write the following inequality

$$\varepsilon_{n+1}(0) \geqslant -h - \frac{T}{2\pi} \int_{-\infty}^{\infty} K(0,\mu)\, \ell n[1+\exp\{-\frac{\varepsilon_n(0)+\mu^2}{T}\}]d\mu. \quad (6.11)$$

Defining function $f(x)$:

$$f(x) \equiv -h - \frac{T}{2\pi} \int_{-\infty}^{\infty} K(0,\mu)\, \ell n[1+\exp\{-\frac{x+\mu^2}{T}\}]d\mu =$$

$$(6.12)$$

$$= -h + x - \frac{T}{2\pi} \int_{-\infty}^{\infty} K(0,\mu)\, \ell n[\exp\{\tfrac{x}{T}\}+\exp\{-\tfrac{\mu^2}{T}\}]d\mu,$$

one rewrites (6.11) as follows:

$$\varepsilon_{n+1}(0) \geqslant f(\varepsilon_n(0)). \qquad (6.13)$$

Function $f(x)$ increases monotonically and also $f(x) < -h$. Function $(f(x)-x)$ decreases monotonically taking values in the interval $[-\infty, \infty]$. Thus the equation

$$f(x_0) = x_0 \qquad (6.14)$$

possesses the unique solution (it is obvious also that $x_0 = $

$= f(x_0) < -h)$. One can now prove that

$$\varepsilon_n(0) \geqslant x_0 \quad \forall n \; . \tag{6.15}$$

First one notes that

$$\varepsilon_0(0) = -h > x_0 \; . \tag{6.16}$$

Then one uses induction in n . Supposing $\varepsilon_n(0) \geqslant x_0$, one has from (6.13) and from the monotonicity of function $f(x)$ that

$$\varepsilon_{n+1}(0) \geqslant f(\varepsilon_n(0)) \geqslant f(x_0) = x_0 \; . \tag{6.17}$$

So inequality (6.15) is proved. Combining now (6.10) and (6.15), one comes to (6.5), finishing the proof of the existence of solution for the Yangs' equation (6.1). Simultaneously the following important properties of this solution are in fact also established:

$$\varepsilon(-\lambda) = \varepsilon(\lambda) \; ; \tag{6.18}$$

$$\lambda^2 - h > \varepsilon(\lambda) > \lambda^2 + x_0 \; ; \tag{6.19}$$

$$d\varepsilon(\lambda)/d\lambda > 0 \quad \text{if} \quad \lambda > 0 \; ; \tag{6.20}$$

$$\varepsilon(\lambda) \to \lambda^2 + \text{const} \quad \text{at} \quad \lambda \to \pm\infty \; . \tag{6.21}$$

It follows that function $\varepsilon(\lambda)$ possesses the only zero on the positive semi-axis; denoting $q_T > 0$ the position of this zero, one obtains that the only two solutions of the equation $\varepsilon(\lambda) = 0$ are $\lambda = \pm q_T$, i.e.

$$\varepsilon(\pm q_T) = 0 . \tag{6.22}$$

At computing the asymptotics of correlation functions (which is done in Part III) one has to make the analytical continuation of function $\varepsilon(\lambda)$ to the complex plane of λ . Into some vicinity of the real axis this continuation can be made using equation (6.1):

$$\varepsilon(\lambda) = \lambda^2 - h - \frac{T}{2\pi} \int_{-\infty}^{\infty} K(\lambda,\mu) \ln[1 + \exp\{-\tfrac{\varepsilon(\mu)}{T}\}] d\mu . \tag{6.23}$$

It seems at first sight that this continuation can be made up to $\text{Im }\lambda = c$ only, due to the pole of function $K(\lambda,\mu)$ (2.16). This pole, however, is not an essential obstacle. The matter is that $\varepsilon(\lambda)$ (6.23) being analitical in some vicinity of the real axis, one can move the integration contour into the upper half-plane which permits to continue $\varepsilon(\lambda)$ for $\text{Im}\,\lambda > c$. Supposing that at this analytical continuation the logarithm in (6.23) has no singularities, i.e. supposing that function $(1 + \exp\{-\varepsilon(\lambda)/T\})$ has no zeros in the complex plane, one comes to contradiction. Indeed, then function $\varepsilon(\lambda)$ could have been continued to the whole complex plane, being an entire function with asymptotics (6.21), i.e. to a binomial which obviously is not a solution of (6.1). It means that function $(1 + \exp\{-\varepsilon(\lambda)/T\})$ must have zeros. The zeros nearest to the real axis are especially important. Due to the following properties of function $\varepsilon(\lambda)$ which are easily established from (6.23)

$$\varepsilon(\lambda) = \varepsilon(-\lambda); \qquad \varepsilon^*(\lambda) = \varepsilon(\lambda^*), \tag{6.24}$$

one sees that the zeros are situated at the vertices of the rectangular:

$$\lambda = \alpha \; ; \; \lambda = -\alpha; \; \lambda = \alpha^*; \; \lambda = -\alpha^* \tag{6.25}$$

(where $\text{Im}\,\alpha > 0$, $\text{Re}\,\alpha > 0$). It is of primary importance that the Fermi weight $\vartheta(\lambda) = (1 + \exp\{\varepsilon(\lambda)/T\})^{-1}$ has first order poles at these points:

$$\vartheta^{-1}(\alpha) = \vartheta^{-1}(-\alpha) = \vartheta^{-1}(\alpha^*) = \vartheta^{-1}(-\alpha^*) = 0. \tag{6.26}$$

It should be mentioned that the point $\lambda = \alpha$ (as well as other points (6.25)) is not a singular point for function $\varepsilon(\lambda)$, which can be continued into the larger domain. The singularities of $\varepsilon(\lambda)$ nearest to the real axis are situated at points $(\alpha + ic), (-\alpha - ic),$ $(\alpha^* - ic), (-\alpha^* + ic)$ where the integration contour in (6.23) is pinched between the singularities of $K(\lambda, \mu)$ and of the logarithm.

7. LIMITING CASES

Let us discuss first the zero temperature limit $T \to 0$. Considering the Yangs equation (5.30), (6.1) one has that the argument of the logarithm in the integral term there becomes rather simple, namely

$$1 + \exp\{-\varepsilon(\lambda)/T\} \to 1 \; ; \; \pm\lambda > q \, , \, \varepsilon(\lambda) > 0, \; T \to 0, \tag{7.1}$$

251

or,

$$1 + \exp\{-\varepsilon(\lambda)/T\} \to \exp\{-\varepsilon(\lambda)/T\} \to \infty ;$$

$$-q < \lambda < q , \quad \varepsilon(\lambda) < 0 , \quad T \to 0. \tag{7.2}$$

Here Fermi momentum $q = \lim q_T$ at $T \to 0$ (see 6.22). So
the Yang equation becomes linear, reproducing just equations (3.15),
(3.16). The Fermi weight $\vartheta(\lambda)$ (5.32) at $T \to 0$ is given as

$$\vartheta(\lambda) = \rho_p(\lambda)/\rho_t(\lambda) = \begin{cases} 0 & \text{if } |\lambda| > q \\ 1 & \text{if } |\lambda| < q \end{cases} \quad (T = 0). \tag{7.3}$$

Thus there are no particles with $|\lambda| > q$, all the corresponding
vacancies are free:

$$\rho_p(\lambda) = 0 ; \quad \rho_t(\lambda) = \rho_h(\lambda) \quad (|\lambda| > q). \tag{7.4}$$

On the contrary, all the vacancies with $|\lambda| < q$ are occupied:

$$\rho_h(\lambda) = 0 ; \quad \rho_t(\lambda) = \rho_p(\lambda) \quad (|\lambda| < q) \tag{7.5}$$

So the equilibrium state becomes a pure quantum mechanical state
at $T = 0$; this state being the ground state for the Hamiltonian
(3.1). The equation (5.6) for densities turns into equation (3.9).
The entropy (5.27) is equal to zero at $T = 0$: $S(T=0) = 0$.
So we have justified all the results of s.3. It should be also
mentioned that the thermodynamical quantities can be continued
nonsingularly to the point $T = 0$; in this sense there is no
phase transition.

Now consider another limiting case supposing T arbitrary,
but the coupling to be strong (i.e. $c \to \infty$). As $c \to \infty$, the
kernel $K(\lambda, \mu)$ (2.16) of the operator $\hat{K}_T$ (5.34) is $K(\lambda, \mu) =$
$= \mathcal{O}(c^{-1})$, and all the integral equations can be solved explicitly.
One obtains:

$$\varepsilon(\lambda) = \lambda^2 - h - 2c^{-1}P_r + \mathcal{O}(c^{-3}), \tag{7.6}$$

where P_r is a pressure (5.24);

$$\rho_t(\lambda) = (2\pi)^{-1}(1 + 2Dc^{-1}) + \mathcal{O}(c^{-3}); \tag{7.7}$$

$$\rho_p(\lambda) = \frac{1}{2\pi} \frac{(1 + 2Dc^{-1})}{(1 + \exp\{\varepsilon(\lambda)/T\})} ; \quad \rho_h(\lambda) = \rho_t(\lambda) - \rho_p(\lambda). \tag{7.8}$$

Here $D = N/L = \int \rho_p(\lambda) d\lambda$ (5.26). It is worth mentioning that the strong coupling limit ($c = \infty$) corresponds to free fermions and has thus clear physical meaning.

8. EXCITATIONS AT NONZERO TEMPERATURES

We have already described the thermodynamical equilibrium state in s.5. This equilibrium state at a nonzero temperature is represented as a "mixture" of different eigenfunctions of the Hamiltonian. To obtain excitations over the equilibrium state we take, however, one of these eigenfunctions and then construct excitations over this eigenfunction using the same method as for the zero temperature in s.4. It appears that the observable quantities for the excitations calculated in this way (the energy, the momentum, the scattering matrix) depend on the macroscopic characteristics of the equilibrium state only and do not depend on the special eigenfunction chosen. So the observables are "gauge invariant" in the sense discussed in s.5. Turning to the s.t.e. (2.12) it is convenient to change slightly the notations. The solutions of the s.t.e. (2.12) (i.e. the particles presented) are now denoted as λ_{n_j} , and λ_j denotes the momentum of any vacancy. So (2.12) is now rewritten as

$$L\lambda_{n_j} + \sum_{k=1}^{N} \theta(\lambda_{n_j} - \lambda_{n_k}) = 2\pi n_j \ . \tag{8.1}$$

The system of equation for vacancies is

$$L\lambda_j + \sum_{k=1}^{N} \theta(\lambda_j - \lambda_{n_k}) = 2\pi j \tag{8.2}$$

where j is any integer (as explained at the end of s.2). As in s.4 the excitation with the zero observable charge will be considered. The simplest corresponding state is obtained by making a hole with the momentum λ_h in the vacuum distribution of particles and by simultaneous adding the particle with the momentum λ_p ($|\lambda_h| < q$; $|\lambda_p| > q$, q being the Fermi momentum). Taking into account that the permitted values of the vacuum particles are slightly changed ($\lambda_j \rightarrow \tilde{\lambda}_j$) , one obtains from (8.2)

$$L\tilde{\lambda}_j + \sum_k \Phi(\tilde{\lambda}_j - \lambda_{n_k}) + \Phi(\tilde{\lambda}_j - \lambda_p) - \Phi(\tilde{\lambda}_j - \lambda_h) = 2\pi j. \quad (8.3)$$

Introducing the "shift function" F analogous to (4.4):

$$F(\lambda_j|\lambda_p, \lambda_h) = (\lambda_j - \tilde{\lambda}_j)(\lambda_{j+1} - \lambda_j)^{-1}, \quad (8.4)$$

one derives for this function the following integral equation:

$$2\pi F(\lambda|\lambda_p, \lambda_h) - \int_{-\infty}^{\infty} K(\lambda, \mu)\vartheta(\mu) F(\mu|\lambda_p, \lambda_h) d\mu = \quad (8.5)$$

$$= \Phi(\lambda - \lambda_p) - \Phi(\lambda - \lambda_h).$$

It is quite similar to the equation (4.5); the only difference is
in the Fermi weight $\vartheta(\lambda)$ (5.32):

$$\vartheta(\lambda) = \rho_p(\lambda)/\rho_h(\lambda) = [1 + \exp\{\varepsilon(\lambda)/T\}]^{-1}. \quad (8.6)$$

The dressed (observable) energy is the energy of the state considered minus the energy of the vacuum state:

$$\Delta E(\lambda_p, \lambda_h) = \lambda_p^2 - \lambda_h^2 - \int_{-\infty}^{\infty} 2\mu F(\mu|\lambda_p, \lambda_h)\vartheta(\mu) d\mu, \quad (8.7)$$

and the dressed momentum is equal to

$$\Delta P(\lambda_p, \lambda_h) = \lambda_p - \lambda_h - \int_{-\infty}^{\infty} F(\mu|\lambda_p, \lambda_h) d\mu. \quad (8.8)$$

It is not difficult to prove the following formulae similar to
(4.8), (4.15):

$$\Delta E(\lambda_p, \lambda_h) = \varepsilon(\lambda_p) - \varepsilon(\lambda_h); \quad (8.9)$$

$$\Delta P(\lambda_p, \lambda_h) = \lambda_p - \lambda_h + \int_{-\infty}^{\infty} \rho_p(\mu)[\Phi(\lambda_p - \mu) - \Phi(\lambda_h - \mu)] d\mu. \quad (8.10)$$

Function $\varepsilon(\lambda)$ in (8.9) is just the solution of the Yangs' equation (6.1). So $\varepsilon(\lambda)$ indeed can be considered as the dressed energy which was stated earlier. The scattering matrix can be also computed:

$$S(\lambda_p, \lambda_h) = \exp\{-i\delta(\lambda_p, \lambda_h)\},\qquad (8.11)$$

the scattering phase δ satisfying the following integral equation similar to (4.26)

$$\delta(\mu, \lambda_h) - \frac{1}{2\pi}\int_{-\infty}^{\infty} K(\mu,\nu)\vartheta(\nu)\delta(\nu,\lambda_h)d\nu = \Phi(\mu-\lambda_h). \qquad (8.12)$$

So indeed all the observables depend on the macroscopical variables only, i.e. are "gauge invariant".

It is interesting that the following mnemonic rule can be established. All the expressions for observable quantities at zero and nonzero temperature differ only by integration measure, so that

$$\left(\int_{-q}^{q} d\lambda\right)_{T=0} \longrightarrow \left(\int_{-\infty}^{\infty} d\lambda\,\vartheta(\lambda)\right)_{T\neq 0} \qquad (8.13)$$

Here $\vartheta(\lambda)$ is the Fermi weight (8.6).

Part II.

QUANTUM INVERSE SCATTERING METHOD

The fundamentals as well as recent developments of the quantum inverse scattering method are presented here. The general scheme of the method is accounted for using the nonlinear Schrödinger (NS) model as an example (on applications of QISM to other models see, e.g. [2, 6, 18, 19]). As was demonstrated in Part I of these lectures, the traditional approach based on the explicit form of the Hamiltonian eigenfunctions (I.1.20) permits to progress rather far constructing the thermodynamics of the model. All these results can be also reproduced by QISM. There exist, however, some important problems not solved up to now in the frame of the traditional approach which can be solved by means of QISM. The example is the calculation of correlation functions for the NS model. It is the problem which the rest of the lectures are mainly devoted to. Let us now discuss this problem briefly, taking as an example the simplest correlation function at zero temperature. This is the equal-time two-point correlator of currents:

$$\langle j(x_1) j(x_2) \rangle = \langle \Omega | j(x_1) j(x_2) | \Omega \rangle / \langle \Omega | \Omega \rangle. \tag{0.1}$$

Current $j(x)$ here is defined in terms of the quantum field operator $\psi(x)$ (I.1.2):

$$j(x) = \psi^+(x) \psi(x). \tag{0.2}$$

Operators ψ^+, ψ here are taken at equal times (furhter the time dependence of operators are not written down explicitly). The ground state $|\Omega\rangle$ of the Hamiltonian H_h (3,1) is described in detail in s.I.3. So the calculation of correlation functions is reduced to calculation of the mean values of some operators (see (0.1)), which can be obtained from the meanvalues with respect to eigenfunctions $|\Psi_N\rangle$ ((I.1.7), (I.1.20)) after taking the thermodynamical limit $|\Psi_N\rangle \to |\Omega\rangle$ ($N \to \infty, L \to \infty, D = N/L = const$) described in s.I.3. The simplest of these mean values is the mean value $\langle \Psi_N | \Psi_N \rangle$ of the unit operator, which is the norm of the eigenfunction $|\Psi_N\rangle$. This is easily written in terms of the wave function χ_N (1.20) as

$$\langle \Psi_N | \Psi_N \rangle = \int_0^L d^N z \left| \chi_N (z_1, \ldots, z_N) \right|^2. \tag{0.3}$$

The hypothetical answer for the value of the integral for any was given by Gaudin [5] still in 1972; but the proof was given only by using QISM [7] (see s.6):

$$\langle \Psi_N | \Psi_N \rangle = \det_N (\varphi'), \tag{0.4}$$

where $N \times N$ -matrix φ' is defined in (2.15),

$$\varphi'_{jk} = \partial \varphi_j / \partial \lambda_k = \partial^2 S / \partial \lambda_j \partial \lambda_k, \tag{0.5}$$

(variables φ_j are defined in (I.2.8) and the Yangs' action S in (2.14)). So even the simplest mean value (0.3) can be calculated only by means of QISM. Turn now to the mean value of the currents product introducied in (0.1). It is convenient in QISM to reduce its calculation to the calculation of the mean value of operator $Q(x_2, x_1)$ defined as follows:

$$Q(x_2, x_1) = \int\limits_{x_1}^{x_2} j(z)\, dz \,. \tag{0.6}$$

This is the operator of the number of particles in the segment $[x_1, x_2]$. One easily obtains that

$$\langle \Omega | j(x_1) j(x_2) | \Omega \rangle = -\frac{1}{2} \frac{\partial^2}{\partial x_1 \partial x_2} \langle \Omega | Q^2(x_2, x_1) | \Omega \rangle. \tag{0.7}$$

Due to the translation invariance this mean value depends on $(x_2 - x_1)$ only. Thus putting $x \equiv x_2 > x_1 \equiv 0$ one has

$$\langle \Omega | j(x) j(0) | \Omega \rangle = \frac{1}{2} \frac{\partial^2}{\partial x^2} \langle \Omega | Q_1^2(x) | \Omega \rangle, \tag{0.8}$$

where $Q_1(x)$ is the operator of the number of particles in the segment $[0, x]$:

$$Q_1(x) = \int\limits_0^x j(x)\, dx. \tag{0.9}$$

The mean value $\langle \Omega | Q_1^2(x) | \Omega \rangle$ will be calculated as follows: first one calculates the mean value $\langle \Psi_N | Q_1^2 | \Psi_N \rangle$ with respect to eigenfunction $| \Psi_N \rangle$ (I.1.7), and then takes the thermodynamical limit. Using (I.1.7) one can express $\langle \Psi_N | Q_1^2 | \Psi_N \rangle$ similarly to (0.3)

$$\langle \Psi_N | Q_1^2(x) | \Psi_N \rangle = \langle \Psi_N | Q_1(x) | \Psi_N \rangle +$$
$$+ N(N-1) \int\limits_0^x d^2 y \int\limits_0^L d^{N-2} z \, | \chi_N(y_1, y_2, z_1, \ldots, z_{N-2}) |^2 \,. \tag{0.10}$$

The mean value of operator Q_1 is easily reduced to the norm using translation invariance:

$$\langle \Psi_N | Q_1(x) | \Psi_N \rangle = (xN/L)\langle \Psi_N | \Psi_N \rangle, \qquad (0.11)$$

so that the corresponding contribution to correlator (0.1) is simply xD (here $D = \lim N/L$ (I.3.4)). To calculate the contribution of the integral term in (0.10), however, is much more difficult. It can be done only using QISM (which is done in s.7 of Part II and in Part III). It is to be mentioned that some progress in calculating field correlator $\langle \Omega | \psi^+(x) \psi(y) | \Omega \rangle$ at the strong coupling limit (for $c = \infty$) was made using the traditional approach. This correlator was investigated and represented as an infinite series in [20] . This series was summed up in paper [21] where it was established that it is equal to a Painleve function. Now we begin with the consistent account of the quantum inverse scattering method.

1. CLASSICAL INVERSE SCATTERING METHOD

Only the information concerning the classical inverse scattering method (see, e.g., [22]) which is necessary for quantization is given below. Consider the classical nonlinear Schrödinger (NS) model. The Hamiltonian of the model is (compare with (I.1.1)):

$$H = \int_0^L [\partial_x \psi^+ \partial_x \psi + c \psi^+ \psi^+ \psi \psi]\, dx, \qquad (1.1)$$

where the Poisson brackets between classical canonical fields ψ and ψ^+ is given as

$$\{\psi^+(x,t), \psi(y,t)\} = i\delta(x-y). \qquad (1.2)$$

The corresponding evolution equation is the nonlinear Schrödinger equation:

$$i\partial_t \psi = -\partial_x^2 \psi + 2c \psi^+ \psi \psi; \quad i\partial_t \psi^+ = \partial_x^2 \psi^+ - 2c \psi^+ \psi^+ \psi. \qquad (1.3)$$

The essential progress in investigation of this equation was obtained in paper [23] where the Lax representation for this equation was given, which means that equation (1.3) can be represented as a zero curvature condition. Namely, one introduces the following operators U and V :

$$U(x|\lambda) = \partial_x + (i\lambda/2)\,6_3 + Q(x)\,;\qquad\qquad(1.4)$$

$$V(t|\lambda) = -\partial_t + (i\lambda^2/2)\,6_3 + \lambda Q + i6_3(\partial_x Q + c\,\psi^+\psi)\,.\qquad(1.5)$$

Here $6_3 = \mathrm{diag}(1,-1)$ is the Pauli matrix, and 2x2-matrix Q is given as

$$Q = \begin{bmatrix} 0 & ;\ i\sqrt{c}\,\psi^+(x) \\ -i\sqrt{c}\,\psi(x); & 0 \end{bmatrix}\qquad\qquad(1.6)$$

The spectral parameter λ in (1.4), (1.5) is an arbitrary complex number. Let us require that operators U, V do commute at any λ :

$$[U(\lambda), V(\lambda)] = 0\qquad\qquad(1.7)$$

It is easily shown that this is valid if and only if the NS equation (1.3) for ψ is fulfilled. This representation for the NS equation proves to be extremely useful for obtaining and investigating all its solutions. Here the attention is, however, paid mainly to the analysis of the Hamiltonian structure which is essential for quantization.

Now we inroduce the transition matrix $T(x,y|\lambda)$ which plays an important role below:

$$U(x|\lambda)\,T(x,y|\lambda) \equiv [\partial_x + \tfrac{i\lambda}{2}\,6_3 + Q]\,T(x,y|\lambda) = 0\ ,\ x \geqslant y;\qquad(1.8)$$

$$T(y,y|\lambda) = I \equiv \mathrm{diag}(1,1).$$

The periodical boundary conditions in x are supposed to be imposed, the period being equal to L . The transition matrix for the period is called the monodromy matrix $T(\lambda)$:

$$T(\lambda) \equiv T(L,0|\lambda).\qquad\qquad(1.9)$$

Using the explicit symmetries of operator $U(x|\lambda)$ one establishes the following symmetry properties of the transition matrix (as well as of the monodromy matrix):

$$\det T(x,y|\lambda) = 1\,;\qquad\qquad(1.10)$$

$$\mathfrak{S}_1 \, T^*(x,y|\lambda)\, \mathfrak{S}_1 = T(x,y|\lambda); \quad \mathfrak{S}_1 = \begin{bmatrix} 0 & 1 \\ 1 & 0 \end{bmatrix}. \tag{1.11}$$

Here the asterisk means complex conjugation. The transition matrix for a small interval Δ ($x = y + \Delta$; $\Delta \to 0$) is easily calculated, being equal to the unit matrix I at $\Delta = 0$. This quantity will be called $\mathcal{L}$-operator:

$$T(x+\Delta, x|\lambda) = I - \{(i\lambda/2)\mathfrak{S}_3 + Q(x)\}\Delta + O(\Delta^2); \tag{1.12}$$

$$\mathcal{L}(x|\lambda) \equiv I - \{(i\lambda/2)\mathfrak{S}_3 + Q(x)\}\Delta =$$

$$= \begin{bmatrix} 1 - (i\lambda\Delta/2) \; ; & -i\sqrt{c}\,\psi^+(x)\Delta \\ i\sqrt{c}\,\psi(x)\Delta \; ; & 1 + (i\lambda\Delta/2) \end{bmatrix}. \tag{1.13}$$

Any transition matrix (and hence the monodromy matrix) is genera-ted by the $\mathcal{L}$-operator in the sense that it can be represented by a matrix product of $\mathcal{L}$-operators (1.13). To obtain such a represen-tation one divides the interval $[0, L]$ into N small equal inter-vals with length $\Delta = L/N$. The coordinate of the n-th site of the lattice thus obtained is $x_n = n\Delta$. The $\mathcal{L}$-operator at the n-th site is denoted as

$$\mathcal{L}(n|\lambda) \equiv \mathcal{L}(x_n|\lambda). \tag{1.14}$$

Using the following group property of the transition matrix

$$T(x,y|\lambda)\,T(y,z|\lambda) = T(x,z|\lambda), \quad x > y > z, \tag{1.15}$$

one obtains at $\Delta \to 0$:

$$T(x,y|\lambda) = \mathcal{L}(n|\lambda)\,\mathcal{L}(n-1|\lambda)\ldots \mathcal{L}(m|\lambda)$$

$$(n = x/\Delta; \; m = y/\Delta; \; \Delta \to 0). \tag{1.16}$$

and similarly for the monodromy matrix $T(\lambda)$ (1.9):

$$T(\lambda) = \mathcal{L}(N|\lambda)\mathcal{L}(N-1|\lambda)\ldots\mathcal{L}(1|\lambda) \quad (N = L/\Delta; \; \Delta \to 0). \tag{1.17}$$

This representation appears to be useful for quantization.

    The trace $\tau(\lambda)$ of the monodromy matrix plays a particularly important role:

$$\tau(\lambda) = \mathrm{tr}\, T(\lambda) = T_{11}(\lambda) + T_{22}(\lambda).$$ (1.18)

The reason is that the NS equation (1.3) possesses infinitely many integrals of motion (as it will be shown later) and all these can be expressed in terms of $\tau(\lambda)$. The corresponding formulae are called trace identities [37] . The most important among the integrals of motion are the Hamiltonian (1.1), the momentum P and the charge Q (compare with (I.15) and (I.1.6)):

$$P = -\frac{i}{2}\int[\psi^+\partial_x\psi - (\partial_x\psi^+)\psi]dx\,; \quad Q = \int\psi^+\psi\,dx.$$ (1.19)

Below the trace identities for this three quantities are derived. Taking $\lambda \to \infty$ one representes the transition matrix as follows:

$$T(x,y|\lambda) = U(x)\,D(x,y|\lambda)\,U^{-1}(y).$$ (1.20)

Here D is a diagonal matrix, and the matrix U supposed to be represented in the following form:

$$U(x) = I + \sum_{n=1}^{\infty}\lambda^{-n}U_n(x),$$ (1.21)

U_n being an antidiagonal matrix. The sense of representation (1.20) is that the transition matrix is diagonalized by means of a gauge transformation, the differential equation (1.8) resulting in the following equation for D :

$$[\partial_x + W(x|\lambda)]\,D(x,y|\lambda) = 0\,; \quad D(y,y|\lambda) = I,$$ (1.22)

the "potential" W being equal to

$$W(x|\lambda) = U^{-1}(x)\,\partial_x\,U(x) + (i\lambda/2)U^{-1}(x)\,\sigma_3\,U(x) +$$
$$+ U^{-1}(x)\,Q(x)\,U(x).$$ (1.23)

Matrices U_n are defined by the requirement that potential W is a diagonal matrix (up to terms λ^{-n}). One establishes by direct calculations that in (1.21)

$$U_1 = i\sigma_3 Q\,; \quad U_2 = -\partial_x Q\,; \quad U_3 = i\sigma_3(Q^3 - \partial_x^2 Q),$$ (1.24)

where Q is given in (1.6). Thus the potential W (1.24) at $\lambda \to \infty$ is

$$W = (i\lambda/2)\sigma_3 + \lambda^{-1}W_1 + \lambda^{-2}W_2 + \lambda^{-3}W_3 + \mathcal{O}(\lambda^{-4}) , \qquad (1.25)$$

where

$$W_1 = -i\sigma_3 Q^2 ; \quad W_2 = -Q\partial_x Q ; \quad W_3 = i\sigma_3(Q\partial_x^2 Q - Q^4). \quad (1.26)$$

Potential W being a diagonal matrix, the equation (1.22) is solved explicitly:

$$D(x,y|\lambda) = \exp\left\{-\int_y^x W(z|\lambda)\,dz\right\}. \qquad (1.27)$$

Take now $x = L$ and $y = 0$. Due to the periodical boundary conditions, one has in (1.20) $U(x=L) = U(y=0)$. So using (1.10) and (1.9) one comes to det $D(L,0|\lambda) = 1$. Hence

$$D(L,0|\lambda) = \exp\{\sigma_3 Z(\lambda)\} \qquad (1.28)$$

where $Z(\lambda)$ is a scalar function (and not a matrix). It is then easily obtained from (1.6), (1.26) and (1.27) that as $\lambda \to \infty$,

$$Z(\lambda) = (-i\lambda/2) + ic(\lambda^{-1}Q + \lambda^{-2}P + \lambda^{-3}H) + \mathcal{O}(\lambda^{-4}) , \quad (1.29)$$

where Q, P, H are the charge, the momentum and the Hamiltonian defined in (1.19), (1.1). Because of the periodical boundary conditions it is also valid that $\tau(\lambda) = \mathrm{tr}\, T(\lambda) = \mathrm{tr}\, D(L,0|\lambda)$. As $D_{11}(L,0|\lambda) \gg D_{22}(L,0|\lambda)$ at $\lambda \to +i\infty$ (it is easily seen from (1.28), (1.29)) one can write also:

$$\ln\left[\exp\{i\lambda L/2\}\,\tau(\lambda)\right] \to ic[\lambda^{-1}Q + \lambda^{-2}P + \lambda^{-3}H] + \mathcal{O}(\lambda^{-4})$$
$$(\lambda \to +i\infty). \qquad (1.30)$$

So the trace identities are established which express the original characteristics of the model in terms of the monodromy matrix.

Consider now another question, namely, the calculation of the Poisson brackets between the matrix elements of the transition matrix (1.8) (or the monodromy matrix (1.9)). The canonical Poisson brackets (1.2) permit to calculate the Poisson brackets between the matrix elements of the $\mathcal{L}$-operators (1.13), (1.14). It appears

that if it is possible to represent these brackets in the following
form:

$$\{\mathcal{L}(n|\lambda) \overset{\otimes}{,} \mathcal{L}(m|\mu)\} =$$
$$= [\mathcal{L}(n|\lambda) \otimes \mathcal{L}(n|\mu) , r(\lambda,\mu)] \delta_{nm} + \mathcal{O}(\Delta^2), \tag{1.31}$$

then the Poisson brackets between the matrix elements of the transition matrix are given by the same formula:

$$\{T(x,y|\lambda) \overset{\otimes}{,} T(x,y|\mu)\} = [T(x,y|\lambda) \otimes T(x,y|\mu) , r(\lambda,\mu)]. \tag{1.32}$$

Symbol $[\ ,\ ]$ at the right hand sides here denotes the matrix commutator. Let us explain also other notations. As usual, $T(\lambda) \otimes T(\mu)$
denotes the tensor product of two 2x2-matrices $T(\lambda)$ and $T(\mu)$;
it is thus 4x4-matrix. Quantity $\{T(\lambda) \overset{\otimes}{,} T(\mu)\}$ is also the 4x4-
matrix, its matrix elements being equal to the Poisson bracket of
some matrix element of $T(\lambda)$ with some matrix element of $T(\mu)$.
The labeling of the matrix elements of the matrix $\{T(\lambda) \overset{\otimes}{,} T(\mu)\}$
is the same as of matrix $(T(\lambda) \otimes T(\mu))$. Quantity $r(\lambda,\mu)$ is
the so called "classical r-matrix", which was introduced in [24] .
The r-matrix here is a 4x4-matrix which depends on λ, μ only
(but does not depend on the fields). Equation (1.32) is proved in
Appendix 1. So to have (1.32) one has only to establish (1.31)
which is much simpler and is done by direct calculation using (1.2),
(1.13), (1.14). One obtains that

$$r(\lambda,\mu) = \frac{c}{\lambda - \mu} \begin{bmatrix} 1 & 0 & 0 & 0 \\ 0 & 0 & 1 & 0 \\ 0 & 1 & 0 & 0 \\ 0 & 0 & 0 & 1 \end{bmatrix}, \tag{1.33}$$

the equation (1.31) being valid up to terms of order $\mathcal{O}(\Delta^2)$. At
the continious limit $\Delta \to 0$, however, it appears to be sufficient
to have equation (1.32). So the Poisson brackets between the matrix
elements of the monodromy matrix are calculated which fix the structure of the action-angle variables. It should be remarked that for
the functional $\tau(\lambda)$ (1.18) of the fields ψ, ψ^+ one obtains
from (1.32):

$$\{\tau(\lambda), \tau(\mu)\} = 0 \tag{1.34}$$

It is the reason that in the expansion (1.30) the coefficients at λ^{-n} are "commuting" integrals of motion. It should be also mentioned that the r-matrix replaces operator V (1.5). It is possible to prove that the r-matrix and the trace identities existing, the operator $V(\lambda)$ can be restored [24]. It is valid in the quantum case also [25, 19]. That is why operator V is not used further.

To conclude this section we would like to notice that developing the ideas mentioned above one can construct the lattice model with exactly the same structure of the angle-action variables as the continuous NS model. This lattice model is generated by the following $\mathcal{L}$-operator

$$\mathcal{L}(n|\lambda) = \begin{bmatrix} 1-(i\lambda\Delta/2)+(c\,\psi_n^+\psi_n\Delta^2/2); & -i\sqrt{c}\,\psi_n^+\rho_n\Delta \\ i\sqrt{c}\,\rho_n\psi_n\Delta; & 1+(i\lambda\Delta/2)+(c\,\psi_n^+\psi_n\Delta^2/2) \end{bmatrix}. \quad (1.35)$$

Here ψ_n, ψ_n^+ are canonical Bose variables with the following Poisson brackets:

$$\{\psi_n^+, \psi_m\} = i\Delta^{-1}\delta_{mn}; \quad \{\psi_n, \psi_m\} = \{\psi_n^+, \psi_m^+\} = 0, \quad (1.36)$$

and function $\rho_n = [1+(c\,\psi_n^+\psi_n\Delta^2/4)]^{1/2}$. This $\mathcal{L}$-operator satisfies equation (1.31) exactly (and not up to $\mathcal{O}(\Delta^2)$ as the operator (1.13)), turning into $\mathcal{L}$-operator (1.13) at $\Delta \to 0$. The lattice model discussed above was constructed in [25, 26]. In Appendix 2 the idea of construction of the local lattice Hamiltonial suggested there is also discussed.

2. QUANTUM INVERSE SCATTERING METHOD

The quantum NS Hamiltonian is given by the same formula (1.1), where the boson fields ψ, ψ^+ are now quantum operators with the canonical commutation relations:

$$[\psi(x,t), \psi^+(y,t)] = \delta(x-y). \quad (2.1)$$

The periodical boundary conditions with the period L are imposed. The momentum operator P and the number of particles operator Q ("the charge operator") are given by the same formulae (1.19). In the quantum case one acts by the analogy with the classical case considered in s.1. It is shown in paper $[27]$ that the Lax representation (1.7) is valid also in the quantum case. Considering $\mathcal{L}$-operator (1.13) (with ψ, ψ^+ being now the quantum operators (2.1)) one constructs the quantum transition matrix $T(x,y|\lambda)$ by means of the same formula (1.17) as in the classical case. Taking the limit $\Delta \to 0$, one obtains analogously to (1.8):

$$: \left[\partial_x + (i\lambda/2)\sigma_3 + Q(x) \right] T(x,y|\lambda) : = 0 . \tag{2.2}$$

It should be remarked that the transition matrix is now the 2x2-matrix, its matrix elements being functionals of quantum field operators $\psi(z)$ and $\psi^+(z)$ $(x \geqslant z \geqslant y)$. The colons in (2.2) mean the usual normal ordering (i.e. that all the operators ψ^+ must stand to the left of the operators ψ). The transition matrix possesses the following property analogous to (1.11):

$$\sigma_1 T^*(x,y|\lambda^*)\sigma_1 = T(x,y|\lambda) . \tag{2.3}$$

The asterisk over T here means the hermitian conjugation of its matrix elements (and does not transpose the matrix elements). The monodromy matrix $T(\lambda)$ is defined as

$$T(\lambda) = T(L,0|\lambda) ; \tag{2.4}$$

its matrix elements are quantum operators acting in the space where operators $\psi(x)$, $\psi^+(x)$ $(L \geqslant x \geqslant 0)$ act. The matrix trace of the monodromy matrix which is called the transfer matrix $\tau(\lambda)$ in the quantum case is especially important:

$$\tau(\lambda) \equiv \operatorname{tr} T(\lambda) \tag{2.5}$$

One can express integrals of motion in terms of the scalar operator $\tau(\lambda)$ by means of trace identities, which are obtained analogously to the classical case. Using the representation (1.20), (1.21), one obtaines that U_k and W_k are given by the same formulae (1.22), (1.24)-(1.26) also in the quantum case. The logarithm, however, must be taken carefully. One seeks for $D(L,0|\lambda)$ in the following form:

$$D(L, 0 \mid \lambda) =$$

$$= \exp\{(-i\lambda/2)\sigma_3 L\}\left[1 + a_1\lambda^{-1} + a_2\lambda^{-2} + a_3\lambda^{-3} + O(\lambda^{-4})\right]. \quad (2.6)$$

and obtains using (1.25) that

$$a_1 = -\int_0^L W_1(z)dz \; ; \qquad (2.7)$$

$$a_2 = -\int_0^L W_2(z)dz + \; : \int_0^L W_1(z)dz \int_0^z W_1(y)dy : \; ; \qquad (2.8)$$

$$a_3 = -\int_0^L W_3(z)dz + \; : \int_0^L W_2(z)dz \int_0^z W_1(y)dy : +$$

$$+ \; : \int_0^L W_1(z)dz \int_0^z W_2(y)dy : - \; : \int_0^L W_1(z)dz \int_0^z W_1(y)dy \int_0^y W_1(t)dt : \; . \qquad (2.9)$$

Due to the periodical boundary conditions $U(0) = U(L)$, and $\tau(\lambda) = \mathrm{tr}\, D(L, 0 \mid \lambda)$. As $D_{22} \ll D_{11}$ at $\lambda \to i\infty$, one has that $\tau(\lambda) = D_{11}(L, 0 \mid \lambda)$ $(\lambda \to i\infty)$. Taking the logarithm, one gets then:

$$\ln\left[\exp\{(i\lambda L/2)\}\tau(\lambda)\right] = \ln\left[1 + a_1\lambda^{-1} + a_2\lambda^{-2} + a_3\lambda^{-3} + O(\lambda^{-4})\right] =$$

$$= \lambda^{-1}b_1 + \lambda^{-2}b_2 + \lambda^{-3}b_3 + O(\lambda^{-4}), \qquad (2.10)$$

where

$$b_1 = a_1; \quad b_2 = a_2 - a_1^2/2;$$

$$b_3 = a_3 - (a_1 a_2 + a_2 a_1)/2 + a_1^3/3. \qquad (2.11)$$

The quantities a_1, a_2, a_3 are normally ordered ((2.7)-(2.9)), but the quantities b_2, b_3 are not, containing the usual products of the ordered quantities. Reducing b_2, b_3 to the normally ordered form, one obtains "quantum corrections" which results in the difference of the quantum trace identities from the classical ones:

$$b_1 = ic\,Q; \quad b_2 = ic\,P + (c^2 Q\,/2);$$
$$b_3 = ic\,H + c^2 P - (ic^3 Q/3). \tag{2.12}$$

So one has the following trace identities:

$$\ln\left[\exp\{i\lambda L/2\}\,\tau(\lambda)\right]\Big|_{\lambda \to i\infty} =$$
$$= ic\left[\lambda^{-1} Q + \lambda^{-2}\{P - (ic\,Q/2)\} + \right. \tag{2.13}$$
$$\left. + \lambda^{-3}\{H - ic\,P - (c^2 Q/3)\} + \mathcal{O}(\lambda^{-4})\right].$$

These quantum trace identities were originally obtained in paper [28] .

Our next task is to calculate the commutation relation (CR) between the matrix elements of the transition matrix. In QISM it is done by means of the R-matrix [2] . Analogously to the classical case, the following theorem can be proved. If one succeeds in representing the Poisson brackets between the matrix elements of the $\mathcal{L}$ -operator in the following form:

$$R(\lambda,\mu)\left(\mathcal{L}(n|\lambda) \otimes \mathcal{L}(n|\mu)\right) = \left(\mathcal{L}(n|\mu) \otimes \mathcal{L}(n|\lambda)\right) R(\lambda,\mu), \tag{2.14}$$

and also the matrix elements of the $\mathcal{L}$ -operators at different n commute as quantum operators (this property is usually called ultralocality):

$$\left[\mathcal{L}_{ik}(n|\lambda),\; \mathcal{L}_{pq}(m|\mu)\right] = 0 \quad (m \neq n), \tag{2.15}$$

then also

$$R(\lambda,\mu)\left(T(x,y|\lambda) \otimes T(x,y|\mu)\right) =$$
$$= \left(T(x,y|\mu) \otimes T(x,y|\lambda)\right) R(\lambda,\mu). \tag{2.16}$$

Here $(A \otimes B)$ denotes a tensor product of two 2x2-matrices A and B , as in S.1; so $R(\lambda,\mu)$ is a 4x4-matrix, its matrix elements being   C -number functions of spectral parameters λ , μ only. This theorem is proved in Appendix 3. The direct calculation using (1.13) and (2.1) shows that equation (2.14) is indeed valid in the NS model, the R-matrix being equal to

$$R(\lambda,\mu)=\begin{bmatrix} f(\mu,\lambda) & 0 & 0 & 0 \\ 0 & g(\mu,\lambda) & 1 & 0 \\ 0 & 1 & g(\mu,\lambda) & 0 \\ 0 & 0 & 0 & f(\mu,\lambda) \end{bmatrix}, \qquad (2.17)$$

where functions f and g are

$$f(\mu,\lambda) = 1 + ic(\mu-\lambda)^{-1}; \quad g(\mu,\lambda)= ic(\mu-\lambda)^{-1}. \qquad (2.18)$$

Relation (2.14) for the $\mathcal{L}$-operator (1.13) is valid up to the terms of order $\mathcal{O}(\Delta^2)$, which is sufficient to construct the transition matrix as one must put $\Delta \to 0$. The R-matrix (2.17) is just the R-matrix of the XXX Heisenberg model. It was first calculated in papers $[2, 27]$. So one has for the CR of the matrix elements of the monodromy matrix obtained by putting $x = L$, $y = 0$ in (2.16)

$$R(\lambda,\mu)(T(\lambda)\otimes T(\mu)) = (T(\mu)\otimes T(\lambda))R(\lambda,\mu). \qquad (2.19)$$

Rewriting it as $R\,T(\lambda)\otimes T(\mu)\,R^{-1} = T(\mu)\otimes T(\lambda)$ and taking the trace in the space of 4x4 matrices, one obtains for the transfer matrix $\tau(\lambda)$ (2.5):

$$[\tau(\lambda), \tau(\mu)] = \tau(\lambda)\tau(\mu) - \tau(\mu)\tau(\lambda) = 0. \qquad (2.20)$$

Due to the trace identities (2.13) one has then that also

$$[H, \tau(\lambda)] = 0. \qquad (2.21)$$

which results in existing an infinitely large number of conservation laws, the corresponding operators commuting with each other.

Now we consider the action of the monodromy matrix $T(\lambda)$ on the Fock vacuum (the pseudovacuum) $|0\rangle$ (1.3). Compute first the action of the $\mathcal{L}$-operator (1.13). As $\psi(x)|0\rangle = 0$, one obtains:

$$\mathcal{L}(\lambda)|0\rangle = \begin{bmatrix} 1 - (i\lambda\Delta/2); & -i\sqrt{c}\,\psi^+(x)\Delta \\ 0 & ; & 1 + (i\lambda\Delta/2) \end{bmatrix}|0\rangle \equiv \ell(\lambda)|0\rangle. \qquad (2.22)$$

Due to (1.17) and (2.15) the action on $|0\rangle$ of the monodromy matrix can be represented as follows:

$$T(\lambda)|0\rangle = \left(\ell(N|\lambda)\, \ell(N-1|\lambda) \ldots \ell(1|\lambda) \right)|0\rangle \qquad (2.23)$$

The right hand side being a product of the triangular matrices, the left hand side is also a triangular matrix:

$$T(\lambda)|0\rangle = \begin{bmatrix} [1-(i\lambda\Delta/2)]^N \; ; & B(\lambda) \\ 0 & ; \; [1+(i\lambda\Delta/2)]^N \end{bmatrix}|0\rangle \qquad (2.24)$$

Denoting

$$T(\lambda) = \begin{bmatrix} A(\lambda) \; ; & B(\lambda) \\ C(\lambda) \; ; & D(\lambda) \end{bmatrix}, \qquad (2.25)$$

one has thus that

$$A(\lambda)|0\rangle = a(\lambda)|0\rangle; \quad D(\lambda)|0\rangle = d(\lambda)|0\rangle; \quad C(\lambda)|0\rangle = 0, \qquad (2.26)$$

where

$$a(\lambda) = \exp\{-i\lambda L/2\}; \quad d(\lambda) = \exp\{i\lambda L/2\} \qquad (2.27)$$

(we use that $(1\pm(i\lambda\Delta/2))^N = \exp\{\pm i\lambda L/2\}$ at $\Delta\to 0$; $L = N\Delta$. The action of $T(\lambda)$ on the dual Fock vacuum $\langle 0|$ (1.4) is calculated similarly:

$$\langle 0|B(\lambda) = 0; \quad \langle 0|A(\lambda) = a(\lambda)\langle 0|; \quad \langle 0|D(\lambda) = d(\lambda)\langle 0|. \qquad (2.28)$$

These formulae will be used for construction of the algebraic Bethe Ansatz in s.3.

It should be mentioned that the lattice quantum NS model can be constructed which is determined by the $\mathcal{L}$-operator (1.35) with ψ_n, ψ_n^+ being the quantum operators with the following commutation relations:

$$[\psi_n, \psi_m^+] = \Delta^{-1}\delta_{mn}, \qquad (2.29)$$

The relation (2.4) being exact. The local quantum Hamiltonian can be constructed [25, 26] by the analogy with the corresponding classical Hamiltonians as explained in Appendix 2, these Hamiltonians being simple explicit functions of local fields.

3. ALGEBRAIC BETHE ANSATZ

We begin with writing down explicitly the commutation relations (2.19) of the matrix elements of the monodromy matrix $T(\lambda)$ (2.25). The R -matrix given in (2.17), (2.18), one obtains

$$[B(\lambda), B(\mu)] = [C(\lambda), C(\mu)] = 0 ; \tag{3.1}$$

$$A(\mu)\, B(\lambda) = f(\mu,\lambda) B(\lambda) A(\mu) + g(\lambda,\mu) B(\mu) A(\lambda) ; \tag{3.2}$$

$$D(\mu)\, B(\lambda) = f(\lambda,\mu) B(\lambda) D(\mu) + g(\mu,\lambda) B(\mu) D(\lambda); \tag{3.3}$$

$$C(\lambda)\, A(\mu) = f(\mu,\lambda) A(\mu) C(\lambda) + g(\lambda,\mu) A(\lambda) C(\mu); \tag{3.4}$$

$$C(\lambda) D(\mu) = f(\lambda,\mu) D(\mu) C(\lambda) + g(\mu,\lambda) D(\lambda) C(\mu); \tag{3.5}$$

$$[C(\lambda), B(\mu)] = g(\lambda,\mu) \{ D(\mu) A(\lambda) - D(\lambda) A(\mu) \}. \tag{3.6}$$

It was already shown (2.20) that the commutativity of the transfer matrix at different spectral parameters follows from these relations:

$$[\tau(\lambda), \tau(\mu)] = 0 ; \qquad \tau(\lambda) = A(\lambda) + D(\lambda). \tag{3.7}$$

Another important consequence of these relations is that they give an opportunity to define the determinant of the monodromy matrix also in quantum case, though the matrix elements do not commute with each other [25] :

$$\det{}_q T(\lambda) \equiv$$
$$\equiv A(\lambda - (ic/2)) D(\lambda + (ic/2)) - B(\lambda - (ic/2)) C(\lambda + (ic/2)). \tag{3.8}$$

Using relations (3.1)-(3.7) one proves that $\det{}_q T(\lambda)$ does commute with operators $A(\mu)$, $B(\mu)$, $C(\mu)$ and $D(\mu)$ at any μ and is hence a C -number. Taking then the matrix element of the determinant between the Fock vacuum $|0\rangle$ and the dual vacuum $\langle 0|$ and using relations (2.26), (2.28) one obtains:

$$\det{}_q T(\lambda) = a(\lambda-(ic/2))\,d(\lambda+(ic/2)) \tag{3.9}$$

It is to be remembered that for the NS model $a(\lambda) = \exp\{-i\lambda L/2\}$; $d(\lambda) = \exp\{i\lambda L/2\}$ (2.27). It is established in paper [25] that the quantum determinant of the monodromy matrix is equal to the product of the quantum determinants of the $\mathcal{L}$-operators which is used essentially at constructing of the local lattice Hamiltonians. The quantum determinant has also been essentially used for the quantum Gelfand-Levitan equations [29] .

Let us go now to the description of the algebraic Bethe Ansatz [2, 6] which is the extremely useful generalization of the formulae obtained by H.Bethe [3] for the coordinate wave functions (for the NS model the coordinate wave function was obtained in [4] , see s.I.1). One can use equations (3.2) and (2.26) to calculate the result of action of operator $A(\mu)$ on the state $|\{\lambda_j\}\rangle_N$ defined as

$$|\{\lambda_j\}\rangle_N = \prod_{j=1}^{N} B(\lambda_j)|0\rangle . \tag{3.10}$$

We use the notation $\{\lambda_j\}$ to denote a set $\{\lambda_j;\; j=1,2,\dots,N\}$. Due to the commutativity of operators $B(\lambda)$ (3.1) this set is supposed to be symmetrical in λ_j . The C.R. (3.2) let "move" operator A from the left to the right of some of operators B in (3.10), the first term at the right hand side of (3.2) corresponding to preserving the arguments of the operators, and the second term corresponding to the exchange of the arguments. Repeating this procedure, one can move operator A to the right of all the operators B in (3.10) and then use the relation $A(\lambda)|0\rangle = d(\lambda)|0\rangle$ (2.26). It is thus clear that the final result will be given by a linear combination of the following states:

$$A(\mu)|\{\lambda_j\}\rangle_N = \Lambda|\{\lambda_j\}\rangle_N + \sum_{n=1}^{N}\Lambda_n|\mu,\{\lambda_{j\neq n}\}\rangle_N \tag{3.11}$$

where

$$|\mu,\{\lambda_{j\neq n}\}\rangle_N = B(\mu)\prod_{\substack{j=1\\ j\neq n}}^{N} B(\lambda_j)|0\rangle . \tag{3.12}$$

Let us now calculate coefficients Λ and Λ_n. To compute Λ one has to use always the first term at the right hand side of (3.2) when "moving" $A(\mu)$ from the very left of all the operators $B(\lambda_j)$ to the very right. Consider for example the result of commutating $A(\mu)$ with the very left of the operators B, namely, with $B(\lambda_1)$:

$$A(\mu)B(\lambda_1) = f(\mu,\lambda_1)B(\lambda_1)A(\mu) + g(\lambda_1,\mu)B(\mu)A(\lambda_1). \qquad (3.13)$$

At the second term operator $B(\mu)$ has appeared to the left of operator $A(\lambda_1)$. This operator $A(\lambda_1)$ should be commuted further with the other operators $B(\lambda_j)$; $j \geqslant 2$, and finally acts on vacuum and disappears. The operator $B(\mu)$, however, remains. So the second term in (3.13) does not contribute to the coefficient Λ at the first term in (3.11), as $|\{\lambda_j\}\rangle_N$ (3.10) does not contain $B(\mu)$. So only the first term in (3.13) must be used. Then one commutes $A(\mu)$ with $B(\lambda_2)$. Repeating the arguments, one obtains finally:

$$\Lambda = a(\mu) \prod_{j=1}^{N} f(\mu,\lambda_j) \qquad (3.14)$$

Compute next Λ_1 . It is clear that now one has to take only the second term in (3.13) because the vector $|\mu,\{\lambda_{j \neq n}\}\rangle_N$ does not contain $B(\lambda_1)$. The next step is commuting $A(\lambda_1)$ with $B(\lambda_2)$:

$$A(\lambda_1)B(\lambda_2) = f(\lambda_1,\lambda_2)B(\lambda_2)A(\lambda_1) + g(\lambda_2,\lambda_1)B(\lambda_1)A(\lambda_2). \qquad (3.15)$$

Only the first term must be used here, as the second one contains $B(\lambda_1)$. Commuting $A(\lambda_1)$ with the following $B(\lambda_j)$ one must also use only the first term of (3.2), obtaining that $\Lambda_1 = a(\lambda_1) \cdot g(\lambda_1,\mu) \prod_{j \geqslant 2} f(\lambda_1,\lambda_j)$. Due to (3.1) there is a symmetry in λ_j and one restores Λ_n for an arbitrary n :

$$\Lambda_n = a(\lambda_n)g(\lambda_n,\mu) \prod_{\substack{j=1 \\ j \neq n}}^{N} f(\lambda_n,\lambda_j) \qquad (3.16)$$

The result of action of $D(\mu)$ on vector $|\{\lambda_j\}\rangle_N$ can be obtained similarly:

$$D(\mu)|\{\lambda_j\}\rangle_N = \widetilde{\Lambda}|\{\lambda_j\}\rangle_N + \sum_{n=1}^{N} \widetilde{\Lambda}_n|\mu,\{\lambda_{j\neq n}\}\rangle_N \; ; \qquad (3.17)$$

$$\widetilde{\Lambda}=d(\mu)\prod_{j=1}^{N}f(\lambda_j,\mu); \quad \widetilde{\Lambda}_n=d(\lambda_n)g(\mu,\lambda_n)\prod_{\substack{j=1\\j\neq n}}^{N}f(\lambda_j,\lambda_n). \quad (3.18)$$

Now everything is ready to obtain the formulae of the algebraic Bethe Ansatz reproducing for the NS model the formulae of the coordinate Bethe Ansatz from s.I.1. The algebraic Bethe Ansatz permits construction of eigenfunctions of the transfer matrix $\tau(\mu)=$ $=A(\mu)+D(\mu)$. These eigenfunctions are of the form (3.10), the set $\{\lambda_j\}$ introduced there being not arbitrary, but satisfying some system of equations. The result of action of operator $\tau(\mu)$ on the state $|\{\lambda_j\}\rangle_N$ is given by the sum of the right hand sides of (3.11) and (3.17). If only the first terms there were present, then $|\{\lambda_j\}\rangle_N$ were the eigenfunction of $\tau(\lambda)$. So let us demand that the sums at the right hand sides of (3.11) and (3.17) cancel, i.e. that $\Lambda_n + \widetilde{\Lambda}_n = 0$. This requirement is easily rewritten as

$$\frac{a(\lambda_n)}{d(\lambda_n)} \prod_{\substack{j=1\\j\neq n}}^{N} \frac{f(\lambda_n,\lambda_j)}{f(\lambda_j,\lambda_n)} = 1 \qquad (n=1,\dots,N) \qquad (3.19)$$

Substituting here the explicit form of functions a, d (2.27) and of f (2.18) one sees that it is just the system of the periodical boundary condition equations (I.2.2) for the coordinate wave function. So the eigenfunctions of the transfer matrix are constructed:

$$\tau(\mu)|\{\lambda_j\}\rangle_N = \Theta_N(\mu,\{\lambda_j\})|\{\lambda_j\}\rangle_N ; \qquad (3.20)$$

$$\Theta_N(\mu,\{\lambda_j\}) = a(\mu)\prod_{j=1}^{N}f(\mu,\lambda_j) + d(\mu)\prod_{j=1}^{N}f(\lambda_j,\mu). \qquad (3.21)$$

It is to be emphasized that λ_j in (3.20), (3.21) are not arbitrary but satisfy the system (3.19). Due to the trace identities (2.13) one obtains from $[\tau(\lambda),\tau(\mu)]=0$ that also $[H,\tau(\mu)]=$ $=[P,\tau(\mu)] = [Q,\tau(\mu)]=0$ (as well as $[H,P]=[H,Q]=[P,Q]=0$).

Operators H, P and Q here are the Hamiltonian (1.1), the momentum and the number of particles (1.19). So eigenfunctions of the transfer matrix are also eigenfunctions of these operators. To calculate the corresponding eigenvalues E_N, P_N, Q_N one has to use trace identities (2.13) which gives:

$$\ln\left[\exp\{i\mu L/2\}\,\theta_N(\mu,\{\lambda_j\})\right] = ic\{\mu^{-1}Q_N +$$
$$+\mu^{-2}(P_N - (ic/2)Q_N) + \tag{3.22}$$
$$+\mu^{-3}(E_N - icP_N - (c^2/3)Q_N) + \mathcal{O}(\mu^{-4})\}\quad (\mu\to i\infty);$$

$$\theta(\mu) = \exp\{-\frac{i\mu L}{2}\}\prod_{j=1}^{N}\left(\frac{\mu-\lambda_j+ic}{\mu-\lambda_j}\right) + \exp\{\frac{i\mu L}{2}\}\prod_{j=1}^{N}\left(\frac{\lambda_j-\mu+ic}{\lambda_j-\mu}\right).\tag{3.23}$$

The second term at the right hand side of (3.23) is negligibly small as $\mu\to+i\infty$, and one obtains from (3.22):

$$Q_N = N;\quad P_N = \sum_{j=1}^{N}\lambda_j;\quad E_N = \sum_{j=1}^{N}\lambda_j^2 \tag{3.24}$$

reproducing (I.1.23). It is to be remarked that the wave eigenfunction (3.14) is equal to the wave function $|\Psi_N\rangle$ (I.1.7) up to a normalization factor, which can be reproduced by comparing the normalization (0.4), (6.25) and (6.8), (6.3). Discuss now the Pauli principle. Formulae (3.11) and (3.17) are valid only if all the λ_j are different. The case where some of these are equal was considered in detail in paper [30]. It was shown there that new equations are to be added to the system (3.19) in this case, this system becoming unsolvable. So the Pauli principle could be established by means of the algebraic Bethe Ansatz. We have demonstrated that all the results of the coordinate Bethe Ansatz are reproduced, this allowing to reproduce also all the results of Part I in the frame of QISM.

Continue the account of the algebraic Bethe Ansatz. The result of action of the operator $C(\mu)$ on vector $|\{\lambda_j\}\rangle_N$ (λ_j - arbitrary) was obtained in paper [7]:

$$C(\mu)|\{\lambda_j\}\rangle_N = \sum_{n=1}^{N} M_n|\{\lambda_{j\neq n}\}\rangle_{(N-1)} +$$

$$+ \sum_{k>n=1}^{N} M_{kn} |\mu, \{\lambda_{j\neq n,k}\}\rangle_{(N-1)} \, . \tag{3.25}$$

Coefficients M here are

$$M_n = g(\mu,\lambda_n)\, a(\mu)\, d(\lambda_n) \prod_{j\neq n}^{N} f(\lambda_j,\lambda_n) f(\mu,\lambda_j) +$$
$$+ g(\lambda_n,\mu)\, a(\lambda_n)\, d(\mu) \prod_{j\neq n}^{N} f(\lambda_j,\mu) f(\lambda_n,\lambda_j) \, ; \tag{3.26}$$

$$M_{kn} = d(\lambda_k)\, a(\lambda_n)\, g(\mu,\lambda_k)\, g(\lambda_n,\mu)\, f(\lambda_n,\lambda_k) \times$$
$$\times \prod_{j\neq k,n}^{N} f(\lambda_j,\lambda_k) f(\lambda_n,\lambda_j) + d(\lambda_n)\, a(\lambda_k)\, g(\mu,\lambda_n) \times$$
$$\times g(\lambda_k,\mu)\, f(\lambda_k,\lambda_n) \prod_{j\neq k,n}^{N} f(\lambda_j,\lambda_n) f(\lambda_k,\lambda_j) \, . \tag{3.27}$$

This formulae can be obtained analogously to (3.11), they will play
an important role below. Making a concluding remark, it is to mention that dual eigenfunctions of operator $\tau(\mu)$ can be also constructed:

$$_N\langle\{\lambda_j\}| = \langle 0| \prod_{j=1}^{N} C(\lambda_j) \, . \tag{3.28}$$

By the complete analogy with the previous result one obtains:

$$_N\langle\{\lambda_j\}| \, \tau(\mu) = \theta(\mu,\{\lambda_j\}) \, _N\langle\{\lambda_j\}| \, , \tag{3.29}$$

if the system (3.19) for λ_j is valid, eigenvalue $\theta(\mu,\{\lambda_j\})$
being just the same as before (3.21). Comparing equations (3.29)
and (3.20) one comes to the conclusion that

$$\langle 0| \prod_{j=1}^{N} C(\lambda_j^C) \prod_{k=1}^{N} B(\lambda_k^B) |0\rangle = 0 \quad (\{\lambda_j^C\} \neq \{\lambda_k^B\}), \tag{3.34}$$

if $\{\lambda_j^C\}$ and $\{\lambda_k^B\}$ are different sets of solutions of the system (3.19).

4. CLASSIFICATION OF MONODROMY MATRICES AND OF L-OPERATORS

The algebraic structure of QISM considered in detail in s.2, s.3 is quite general and can be used to solve many other integrable models of quantum field theory and of statistical physics. The example which is most well-known, is the Heisenberg XXX ferromagnet [31] solved by means of QISM in [2]. The general scheme of solving of the model by means of QISM can be described as follows. First one finds the L-operator and the monodromy matrix $T(\lambda)$ using formula (1.17). The matrix elements of the monodromy matrix A, B, C, D (2.25),

$$T(\lambda) = \begin{bmatrix} A(\lambda) & ; & B(\lambda) \\ C(\lambda) & ; & D(\lambda) \end{bmatrix}, \tag{4.1}$$

are quantum operators acting in the quantum space of the system considered. Commutation relations between these operators should be represented in the form (2.19):

$$R(\lambda,\mu)(T(\lambda)\otimes T(\mu)) = (T(\mu)\otimes T(\lambda))R(\lambda,\mu). \tag{4.2}$$

The Hamiltonian of the system considered should be expressed by means of trace identities in terms of the transfer matrix $\tau(\lambda) = A(\lambda) + D(\lambda)$. As $[\tau(\lambda), \tau(\mu)] = 0$ (3.7), one obtains then as infinitely many conservation laws in the model. Finally, if the Fock vacuum $|0\rangle$ (2.26) can be found satisfying relations (the definition of the dual vacuum $\langle 0|$ is also to be made similarly to (2.28)):

$$A(\lambda)|0\rangle = a(\lambda)|0\rangle; \quad D(\lambda)|0\rangle = d(\lambda)|0\rangle; \quad C(\lambda)|0\rangle = 0, \tag{4.3}$$

then eigenfunctions and eigenvalues of the Hamiltonian can be obtained by using the algebraic Bethe Ansatz described in s.3 (operator $C(\lambda)$ playing the role of the annihilation operator, and operator $B(\lambda)$ of the creation operator). Thus the algebraic sheme of the method is quite general, the concrete models with the given R -matrix differing only by functions $a(\lambda)$ and $d(\lambda)$ in (4.3) (for the NS model they are given in (2.27) as $a(\lambda) = \exp\{-i\lambda L/2\}$; $d(\lambda) = \exp\{i\lambda L/2\}$). Now we shall prove that the algebraic structure can be realised for arbitrary given c-number functions $a(\lambda)$ and $d(\lambda)$. This appears to be of primary importance to make progress in calculating correlation functions. Formulate the statement more precisely. Consider the monodromy matrix (4.1), the commutation relations between its elements being given by relation (4.2) (for the explicit form see (3.1)-(3.6)). The R -matrix is given by (2.17). It is also supposed that the Fock vacuum $|0\rangle$ exists, the monodromy matrix elements acting on this vector as given in (4.3). Then it is possible to construct such monodromy matrix $T(\lambda)$ for any given complex functions $a(\lambda)$ and $d(\lambda)$ in (4.3). This statement was proved in paper [32] . Here we give the main points of the proof. One can realize operators $A(\lambda)$, $B(\lambda)$, $C(\lambda)$ and $D(\lambda)$ in the Fock space of the one-dimensional bosons, the operators being defined by their action on the basis vectors $|\{\lambda_j\}\rangle_N$ of this space ($|\{\lambda_j\}\rangle_N = |\lambda_1, \lambda_2, ..., \lambda_N\rangle$). The operator $B(\lambda)$ is defined to be a creation operator:

$$B(\mu) | \{\lambda_j\}\rangle_N = |\mu, \{\lambda_j\}\rangle_{(N+1)} \tag{4.4}$$

The action of the operators $A(\mu), D(\mu)$ and $C(\mu)$ on the basis vectors is defined by the formulae of the algebraic Bethe Ansatz (3.11), (3.17), (3.25). It can be now proved (using direct but very bulky calculations) that the commutation relations of the operators A, B, C, D thus obtained are given by equation (4.2) for any given complex functions $a(\lambda)$ and $d(\lambda)$. So these functions can be considered to be arbitrary functional parameters which parametrize monodromy matrices (with a given R -matrix). As it is seem from (4.2), the multiplication of all the matrix elements of $T(\lambda)$ by the same arbitrary function of λ, μ does not change the commutation relations. So the essentially different monodromy matrices are parametrized by the only arbitrary function $r(\lambda)$:

$$r(\lambda) \equiv a(\lambda)/d(\lambda). \tag{4.5}$$

It should be emphasized again that this functional arbitrariness proves to be of primary importance in calculation of norms of Bethe eigenfunctions and of correlation functions.

The classification of all possible monodromy matrices being done, one can also classify $\mathcal{L}$-operators, possessing the same properties (4.2), (4.3). It appears that the most general $\mathcal{L}$-operator for the R-matrix of the XXX-type (2.17), (2.18) can be easily obtained from the $\mathcal{L}$-operator of the lattice NS model (1.35):

$$\mathcal{L}(n|\lambda)=\begin{bmatrix} a_n^{(1)}(\lambda+ic\psi_n^+\psi_n)+a_n^{(0)}; & \psi_n^+ \\[2mm] c\rho_n\psi_n; & d_n^{(1)}(\lambda-ic\psi_n^+\psi_n)+d_n^{(0)} \end{bmatrix}. \quad (4.6)$$

Here $a_n^{(0,1)}$ and $d_n^{(0,1)}$ are four arbitrary complex parameters, ψ_n, ψ_n^+ are quantum boson operators:

$$[\psi_n, \psi_m^+]= \delta_{mn}; \quad [\psi_n, \psi_m]=[\psi_n^+, \psi_m^+]= 0, \quad (4.7)$$

and function ρ_n is defined as

$$\rho_n= i\left[a_n^{(1)}d_n^{(0)} - a_n^{(0)}d_n^{(1)}\right]+ c\, d_n^{(1)} a_n^{(1)} \psi_n^+\psi_n. \quad (4.8)$$

It can be proved [33] that the most general monodromy matrix (i.e. with any parameter $r(\lambda)$ (4.5)) can be generated by means of this $\mathcal{L}$-operator:

$$T(\lambda) = \mathcal{L}(N|\lambda)\mathcal{L}(N-1|\lambda)\ldots \mathcal{L}(2|\lambda)\mathcal{L}(1|\lambda). \quad (4.9)$$

Operator $\mathcal{L}$ (4.6) possesses the Fock vacuum $(\psi_n|0>=0)$; the vacuum eigenvalues of its diagonal elements being equal to :

$$a_n(\lambda) = a_n^{(1)}\lambda + a_n^{(0)}; \quad d_n(\lambda)=d_n^{(1)}\lambda + d_n^{(0)}. \quad (4.10)$$

Using the same arguments as in obtaining (2.24), one gets then the vacuum eigenvalues $a(\lambda)$ and $d(\lambda)$ of the diagonal elements (i.e. of operators $A(\lambda)$ and $D(\lambda)$ of the monodromy matrix (4.9)) are equal to

$$a(\lambda)= \prod_{j=1}^{N} a_j(\lambda); \quad d(\lambda) = \prod_{j=1}^{N} d_j(\lambda), \quad (4.11)$$

so that the functional parameter $r(\lambda)$ (4.5) corresponding to the monodromy matrix is:

$$r(\lambda) = \frac{a(\lambda)}{d(\lambda)} = \prod_{n=1}^{N} \frac{a_n^{(1)}\lambda + a_n^{(0)}}{d_n^{(1)}\lambda + d_n^{(0)}} \; .$$ (4.12)

All the numbers $a_n^{(0,1)}$, $d_n^{(0,1)}$ can be taken to be quite arbitrary. So the right hand side of (4.12) can be made to be equal to any given rational function of λ. Taking the limit $N \to \infty$ and choosing parameters a_n, d_n correspondingly, one can obtain any given analitical function of λ (with, e.g., any cut or any other singularity). So it is possible to generate the monodromy matrix with arbitrary $r(\lambda)$ by means of $\mathcal{L}$-operator (4.6).

5. SCALAR PRODUCTS

From now on we begin step-by-step calculation of correlation functions. It is done as the example of the current correlator of the NS model defined in (0.1). The first problem here which QISM allows to solve is the calculation of the normalization of the eigenfunctions of the transfer-matrix (and hence of the Hamiltonian). It is done in s.6, where the Gaudin hypothesis (0.4) is thus proved. In s.7 matrix elements of the operator Q_1^2 (0.9) are studied, which is necessary due to (0.8). It turns out that QISM allows to obtain results concerning these matrix elements which ensure calculating correlation functions. Finally this calculation is made in Part III. It is to be mentioned that the rest of the lectures are based mainly on papers [7-10].

Here the properties of scalar products are investigated [7] which prove to be extremely useful in obtaining the results of s.6, 7. The scalar product will be called the following quantity:

$$S_N = \langle 0 | \prod_{j=1}^{N} C(\lambda_j^C) \prod_{k=1}^{N} B(\lambda_k^B) | 0 \rangle,$$ (5.1)

which is symmetric under permutations of λ_j^C and also under permutations of all λ_k^B (separately) due to (3.1). All $2N$ momenta λ^B, λ^C here are different and arbitrary (the system (3.19) is not supposed to be fulfilled). Note that the number of operators B in (5.1) is equal to the number of operators C ; otherwise the scalar product is equal to zero. As is seen later, it will be useful to change the normalization of the creation and annihilation operators $B(\lambda)$ and $C(\lambda)$, introducing operators $\mathbb{B}(\lambda)$ and $\mathbb{C}(\lambda)$ as follows:

$$\mathbb{B}(\lambda) \equiv B(\lambda)/d(\lambda); \quad \mathbb{C}(\lambda) \equiv C(\lambda)/d(\lambda), \tag{5.2}$$

where the C-number function $d(\lambda)$ is the vacuum eigenvalue (2.26) of the operator $D(\lambda)$. The scalar product is then also redefined:

$$\mathbb{S}_N \equiv \langle 0| \prod_{j=1}^{N} \mathbb{C}(\lambda_j^C) \prod_{k=1}^{N} \mathbb{B}(\lambda_k^B)|0\rangle =$$

$$= \left(\prod_{j=1}^{N} d(\lambda_j^C) d(\lambda_j^B) \right)^{-1} S_N . \tag{5.3}$$

Scalar product can be in principle calculated by means of commutation relations (3.1)-(3.6), using also properties (2.26), (2.28). To calculate thus directly $\mathbb{S}_N$ for general N is, however, impossible because the calculations become very bulky. It is nevertheless possible to restore the properties of $\mathbb{S}_N$ by recursion in N [7]. Consider first the simplest scalar product for $N=1$. Taking the vacuum mean value of the commutation relation (3.6) and using the known results of action of the operators A, D, C on the vacuum (2.26), (2.28), one obtains

$$\mathbb{S}_1 = ic(\lambda^C - \lambda^B)^{-1} \{a(\lambda^C)d(\lambda^B) - a(\lambda^B)d(\lambda^C)\}. \tag{5.4}$$

Remind now that functions $a(\lambda), d(\lambda)$ in QISM can be considered as quite arbitrary (see s.4). So $\mathbb{S}_1$ (5.1) can be considered as a functional of these functions. This functional is, however, rather special, depending on the values of arbitrary functions at points λ^C, λ^B only. Considering this values as independent complex variables $a^{C,B} \equiv a(\lambda^{C,B}); \quad d^{C,B} \equiv d(\lambda^{C,B})$, one concludes that $\mathbb{S}_1$ is a function of 6 complex variables: $\mathbb{S}_1 = \mathbb{S}_1(\lambda^C, \lambda^B, a^C, a^B, d^C, d^B)$. It is to emphasize that variables $a^{C,B}$ and $d^{C,B}$ can be considered as independent of $\lambda^{C,B}$, because the value of an arbitrary function at a given point is quite arbitrary. So, e.g., by changing arbitrary function $a(\lambda)$ one can obtain that quantity $a^C = a(\lambda^C)$ remains unchanged at changing λ^C. The number of varialbes is, however, somewhat redundant. The matter is that due to the linearity of commutation relations with respect to each of the operators A, B, C, D only the dependence on the ratio $r(\lambda) = a(\lambda)/d(\lambda)$ (4.5) is essential (this ratio parametrizing the monodromy matrices which are really different, see s.4). Indeed, one easily obtains, that $S_1 = d^C d^B S_1$, where

$$S_1 = S_1(\lambda^C, \lambda^B; r^C, r^B) =$$
$$= ic(\lambda^C - \lambda^B)^{-1}(r^C - r^B), \tag{5.5}$$

and

$$r^C \equiv r(\lambda^C); \quad r^B \equiv r(\lambda^B); \quad r(\lambda) = a(\lambda)/d(\lambda). \tag{5.6}$$

That is why the normalization (5.2), (5.3) is convenient. It is seen from (5.5) that function S_1 has the first order pole at $\lambda^C \to \lambda^B$, the residue being equal to $ic(r^C - r^B)$. Make now the following comment which is very essential for understanding of what follows. Variables λ^C, λ^B, r^C, r^B here are considered as completely independent. At the same time at some specific model function $r(\lambda) = a(\lambda)/d(\lambda)$ is a given function of λ . In physically interesting cases it is a continuous differentiable function (e.g., $r(\lambda) = \exp\{-i\lambda L\}$ (2.27) in the NS model). In such models the residue at the pole at (5.5) vanishes, and function S_1 is, of course, smooth. But in the generic case functions $a(\lambda), d(\lambda)$ and hence $r(\lambda)$ can be chosen to be quite arbitrary. In particular, these can be discontinuous functions, such that $\lim r(\lambda^C) \neq \lim r(\lambda^B)$ at $\lambda^C \to \lambda^B$. In this case the pole in (5.5) really exists. The consideration of such discontinuous functions $r(\lambda)$ proves to be convenient to formulate recursion properties in N of scalar products S_N. We return to the continuous differentiable functions $r(\lambda)$ at the end of this section.

Consider now the scalar product S_N (5.3) at N arbitrary. It can be proved that S_N depends on the following $4N$ arguments:

$$S_N = S_N(\{\lambda^C\}_N, \{\lambda^B\}_N, \{r^C\}_N, \{r^B\}_N). \tag{5.7}$$

where the following notation is used : $\{x\}_N = \{x_j; j=1,2,...,N\}$ (so that N denotes the number of elements in the set $\{x\}_N$: $N = card\{x\}_N$). Variables r are defined as in (5.6):

$$r_j^{\,C} \equiv r(\lambda_j^{\,C}) \; ; \qquad r_j^{\,B} \equiv r(\lambda_j^{\,B}). \tag{5.8}$$

Due to (3.1), $\mathbb{S}_N$ is a symmetric function with respect to replacement of pairs $(\lambda_j^{\,C}, r_j^{\,C}) \leftrightarrow (\lambda_k^{\,C}, r_k^{\,C})$ and, similarly, with respect to $(\lambda_j^{\,B}, r_j^{\,B}) \leftrightarrow (\lambda_k^{\,B}, r_k^{\,B})$. The dependence of $\mathbb{S}_N$ on variables r is rather simple and can be extracted explicitly:

$$\mathbb{S}_N = \sum_{\text{part}} \left(\prod_{j=1}^{N} r(\lambda_j^{(pr)}) \right) \mathbb{K}_N \begin{bmatrix} \{\lambda^C\}_N \, , \; \{\lambda^B\}_N \\ \{\lambda^{(pr)}\}_N , \; \{\lambda^{(ab)}\}_N \end{bmatrix} \tag{5.9}$$

The sum here contains C_{2N}^{N} terms and is taken over all the partitions of the set $\{\lambda^C\} \cup \{\lambda^B\}$ into two disjoint subsets $\{\lambda^{(pr)}\}$ and $\{\lambda^{(ab)}\}$:

$$\{\lambda^C\}_N \cup \{\lambda^B\}_N = \{\lambda^{(pr)}\}_N \cup \{\lambda^{(ab)}\}_N ; \; \{\lambda^{(pr)}\}_N \cap \{\lambda^{(ab)}\}_N = \varnothing. \tag{5.10}$$

so that

$$\text{card}\,\{\lambda^{(pr)}\} = \text{card}\,\{\lambda^{(ab)}\} = \text{card}\,\{\lambda^C\} = \text{card}\,\{\lambda^B\} = N. \tag{5.11}$$

Coefficients $\mathbb{K}_N$ do not depend on variables r being rational functions of 2N variables λ^C, λ^B decreasing in each λ as $1/\lambda$ at $\lambda \to \infty$ and other λ's fixed. Properties of scalar products can be restored from paper $[7]$. The most important property is that the scalar product $\mathbb{S}_N$ has a first order pole at $\lambda_j^{\,C} \to \lambda_k^{\,B}$ $(j,k = 1,..,N)$ and the other varialbes given, the residue being reduced to $\mathbb{S}_{N-1}$. For example, as $\lambda_N^{\,C} \to \lambda_N^{\,B} \to \lambda_N$ one has (the general case is obvious due to the symmetry):

$$\mathbb{S}_N \left(\{\lambda^C\}_N, \{\lambda^B\}_N, \{r^C\}_N, \{r^B\}_N \right) \Big|_{\lambda_N^{\,C} \to \lambda_N^{\,B} \to \lambda_N} \longrightarrow$$

$$\longrightarrow ic(\lambda_N^{\,C} - \lambda_N^{\,B})^{-1} \prod_{j=1}^{N-1} f_{Nj}^{\,B} f_{Nj}^{\,C} \; \times \tag{5.11}$$

$$\times \; \mathbb{S}_{N-1} \left(\{\lambda_{j \neq N}^{\,C}\}_{N-1}, \{\lambda_{j \neq N}^{\,B}\}_{N-1}, \{\tilde{r}_{j \neq N}^{\,C}\}_{N-1}, \{\tilde{r}_{j \neq N}^{\,B}\}_{N-1} \right) + O(1).$$

Here $f_{N,j}^{\,B,C} \equiv f(\lambda_N, \lambda_j^{\,B,C})$; each set of the arguments of the function $\mathbb{S}_{N-1}$ contains $(N-1)$ elements, e.g., $\{\lambda_{j \neq N}\}_{N-1} =$

$=\{\lambda_j; j=1,...,N-1\}$. "Modified" variables $\tilde{r}^{B,C}$ are the values at points $\lambda_j^{B,C}$ $(j=1,...,N-1)$ of the function $\tilde{r}(\lambda)$ given as

$$\tilde{r}(\lambda) = r(\lambda)\left[f(\lambda,\lambda_N)/f(\lambda_N,\lambda)\right] \tag{5.12}$$

which means that the scalar product S_{N-1} at the right hand side of equation (5.11) should be calculated using modified vacuum eigenvalues $\tilde{a}(\lambda) = a(\lambda)f(\lambda,\lambda_N)$ and $\tilde{d}(\lambda) = d(\lambda)f(\lambda_N,\lambda)$. It is essential that coefficients K_{N-1} (given by (5.9) at $N \to N-1$) in S_{N-1} do not depend on vacuum eigenvalues and thus are not modified. The formula (5.11) is of primary importance for calculating the norms of Bethe eigenvectors and of correlation functions.

In concrete physical models variables r_j (5.8) are the values of a smooth function $r(\lambda)$ at different points. In this special case the residue in equation (5.11) becomes zero; the corresponding limit is thus finite. Consider, for example, the situation $\lambda_N^C \to \lambda_N^B \to \lambda_N$. As $r(\lambda_N^C) \to r(\lambda_N^B) \to r(\lambda_N)$ (the function $r(\lambda)$ is now smooth), it seems at first sight that the pair of variables (r_N^C, r_N^B) gives only one new variable r_N. However, that is not the case. The dependence of the scalar product on the vacuum eigenvalues at point λ_N is now represented nevertheless in terms of two variables: of the variable $r_N = r(\lambda_N)$ and of new variable z_N :

$$z_N = i\partial \ln r(\lambda)/\partial\lambda \Big|_{\lambda=\lambda_N}. \tag{5.13}$$

The matter is that due to arbitrariness of $r(\lambda)$ quantities r_N and z_N can be considered as independent complex variables. The new variable z_N appears only in the pole term in (5.11), the dependence on z_N being linear. The coefficient at z_N is given by the residue at a pole:

$$\partial\langle 0|\left(\prod_{j=1}^{N-1} C(\lambda_j^C)\right) C(\lambda_N) B(\lambda_N)\left(\prod_{k=1}^{N-1} B(\lambda_k^B)\right)|0\rangle/\partial z_N =$$

$$= cr_N\left(\prod_{j=1}^{N-1} f_{Nk}^B f_{Nk}^C\right) \times \tag{5.14}$$

$$\times S_{N-1}\left(\{\lambda_{j\neq N}^C\}_{N-1}, \{\lambda_{j\neq N}^B\}_{N-1}, \{\tilde{r}_{j\neq N}^C\}_{N-1}, \{\tilde{r}_{j\neq N}^B\}_{N-1}\right).$$

Let us consider now the scalar product at the limit $\lambda_j^C \to \lambda_j^B \to \lambda_j$ $(j=1,...,N)$, all λ_j being defferent. In this case the scalar product depends on $3N$ complex variables $\{\lambda\}_N$, $\{z\}_N$, $\{r\}_N$:

$$S_N(\{\lambda^C\}_N, \{\lambda^B\}_N, \{r^C\}_N, \{r^B\}_N) \equiv$$
$$\equiv S_N(\{\lambda\}_N, \{z\}_N, \{r\}_N) \quad (\lambda_j^C = \lambda_j^B = \lambda_j ; \; r(\lambda_j^C) = r(\lambda_j^B)), \tag{5.15}$$

where

$$z_j = i\,\partial \ln r(\lambda)/\partial\lambda\Big|_{\lambda=\lambda_j} \qquad (j=1,...,N). \tag{5.16}$$

Quantity S_N is a linear function of each z_j ; the coefficient at z_N is given as

$$\frac{\partial}{\partial z_N} S_N(\{\lambda\}_N, \{z\}_N, \{r\}_N) = c r_N \Big(\prod_{k=1}^{N-1} f_{Nk}^B f_{Nk}^C\Big) \times \tag{5.17}$$
$$\times S_{N-1}(\{\lambda_{j\neq N}\}_{N-1}, \{\tilde{z}_{j\neq N}\}_{N-1}, \{\tilde{r}_{j\neq N}\}_{N-1}).$$

Here not only r_j must be modified to $\tilde{r}_j$ according to (5.12), but also z_j to $\tilde{z}_j$:

$$\tilde{z}_j = z_j + K_{jN} ; \qquad j = 1, 2, ..., N-1 \tag{5.18}$$

where K_{jN} is given in (I.2.16) as

$$K_{jN} = K(\lambda_j - \lambda_N) = i\frac{\partial}{\partial\lambda}\ln\frac{f(\lambda,\lambda_N)}{f(\lambda_N,\lambda)}\Big|_{\lambda=\lambda_j} = \frac{2c}{(\lambda_j-\lambda_N)^2+c^2} \tag{5.19}$$

The formulae for the coefficients at $z_j \neq z_N$ can be easily obtained using the symmetry of S_N with respect to permutations $(\lambda_j, z_j, r_j) \leftrightarrow (\lambda_k, z_k, r_k)$.

6. NORMS OF BETHE WAVE FUNCTIONS. PROOF OF THE GAUDIN HYPOTHESIS

Consider some Bethe eigenfunction $|\psi_N(\{\lambda\}_N)\rangle$ of the transfer matrix $\tau(\lambda)$ (2.7):

$$|\psi_N(\{\lambda_j\}_N)\rangle = \prod_{j=1}^{N} \mathbb{B}(\lambda_j)|0\rangle,\qquad\qquad (6.1)$$

where momenta λ_j have to satisfy the s.t.e. (3.19):

$$r(\lambda_j)\prod_{\substack{k=1\\k\neq j}}^{N}[f(\lambda_j,\lambda_k)/f(\lambda_k,\lambda_j)]=1,\quad j=1,...,N.\qquad (6.2)$$

This eigenfunction differs from eigenfunction $|\{\lambda_j\}\rangle_N$ (3.10) used in s.3 only by the normalization factor which is because now operators $\mathbb{B}(\lambda)$ (5.2) are used instead of $B(\lambda)$:

$$|\psi_N(\{\lambda_j\}_N)\rangle \equiv \left(\prod_{j=1}^{N}d(\lambda_j)\right)^{-1}|\{\lambda_j\}\rangle_N\,.\qquad (6.3)$$

The dual eigenfunction $\langle\psi_N|$ is:

$$\langle\psi_N(\{\lambda_j\}_N)| = \langle 0|\prod_{j=1}^{N}\mathbb{C}(\lambda_j).\qquad\qquad (6.4)$$

The matrix element $\langle\psi_N|\psi_N\rangle$:

$$\mathcal{N}_N=\langle\psi_N(\{\lambda\}_N)|\psi_N(\{\lambda\}_N)\rangle=\langle 0|\prod_{j=1}^{N}\mathbb{C}(\lambda_j)\prod_{k=1}^{N}\mathbb{B}(\lambda_k)|0\rangle\qquad (6.5)$$

is of special interest. Due to involution (2.3) it is reduced directly to the square of the norm of the eigenfunction (it should also be noted that $\langle\psi_N(\{\lambda^c\})|\psi_N(\{\lambda^B\})\rangle=0$ if $\{\lambda^c\}\neq\{\lambda^B\}$). Quantity $\mathcal{N}_N$ can be obtained from the generic matrix element $\mathbb{S}_N$ (5.3) in two steps. The first step is to take the limit $\lambda_j^c\to\lambda_j^B\to\lambda_j$ $(j=1,2,...,N)$. One obtains thus just the quantity S_N (5.15). The second step is to impose the s.t.e. (6.2) on λ_j , which permits to express all r_j as explicit functions of variables λ_k. So $\mathcal{N}_N$ is a function of 2N complex variables λ_j and z_j :

$$\langle\psi_N(\{\lambda\}_N)|\psi_N(\{\lambda\}_N)\rangle \equiv \mathcal{N}_N(\{\lambda_j\}_N,\{z_j\}_N)=\qquad (6.6)$$

$$= S_N(\{\lambda_j\}_N,\{z_j\}_N,\{r_j=\prod_{k\neq j}[f(\lambda_k,\lambda_j)/f(\lambda_j,\lambda_k)]\}_N).$$

It is obvious that $\mathcal{N}_N$ is symmetric under the replacement of pairs $(\lambda_j, z_j) \leftrightarrow (\lambda_k, z_k)$. The dependence on z_j remains linear as for $\mathcal{S}_N$, the coefficient at z_N being obtained from (5.17)(the coefficients at $z_j \neq z_N$ are restored using the symmetry)

$$\frac{\partial \mathcal{N}_N}{\partial z_N} = c \left(\prod_{k=1}^{N-1} f_{Nk} f_{kN} \right) \mathcal{N}_{N-1} \left(\{\lambda_{j \neq N}\}_{N-1} , \{\widetilde{z}_{j \neq N}\}_{N-1} \right) , \qquad (6.7)$$

where z_j is modified to $\widetilde{z}_j$ according to the rule (5.18). It is to be mentioned that function $\mathcal{N}_{N-1}$ at the right hand side is also a square of the norm, as the set $\{\lambda_{j \neq N}\}$ satisfies the modified s.t.e:

$$\widetilde{r}_j = \prod_{k \neq j}^{N-1} \left[f(\lambda_k, \lambda_j) / f(\lambda_j, \lambda_k) \right] , \qquad j = 1, \dots, N-1,$$

with $\widetilde{r}_j$ given in (5.12). Now everything is ready to prove the following formula [7] :

$$\langle \psi_N(\{\lambda_j\}_N) | \psi_N(\{\lambda_j\}_N) \rangle = c^N \left(\prod_{\substack{j,k=1 \\ j \neq k}}^{N} f(\lambda_j, \lambda_k) \right) \det_N(\varphi') . \qquad (6.8)$$

Here φ' is the $N \times N$ -matrix which is defined in terms of variables φ_j (compare with (I.2.8) and (0.5)):

$$\varphi'_{jk} = \partial \varphi_j / \partial \lambda_k = \delta_{jk} \left[z_k + \sum_{\ell=1}^{N} K(\lambda_k, \lambda_\ell) \right] - K(\lambda_k, \lambda_j), \qquad (6.9)$$

where z_k and K are given by (5.16), (5.19). So formula (6.8) gives quantity $\langle \psi_N | \psi_N \rangle$ as an explicit function of independent variables λ_j, z_j $(j = 1, 2, \dots, N)$. Variables z_j enter only the determinant of matrix φ' (index N in $\det_N(\varphi')$ denotes the dimension of the matrix); we put by definition $\det_0(\varphi') \equiv 1$, so that

$$\langle \psi_0 | \psi_0 \rangle \equiv \langle 0 | 0 \rangle \equiv \det_0(\varphi') \equiv 1 \qquad (6.10)$$

It is convenient to introduce the following notation:

$$\| \lambda_1, \lambda_2, ..., \lambda_N \|_N = c^{-N} \left(\prod_{j,k=1}^{N} f_{j,k} \right)^{-1} \langle \psi_N(\{\lambda\}_N) | \psi_N(\{\lambda\}_N) \rangle \qquad (6.11)$$

Prove now that

$$\| \lambda_1, \lambda_2, ..., \lambda_N \|_N = \det{}_N(\varphi'), \qquad (6.12)$$

which is equivalent to (6.8).

First one studies in detail properties of the determinant of the matrix φ' (6.9). The following 5 properties prove to be of primary importance. (1) The determinant $\det_N(\varphi')$ does not change tat the permutations of pairs of variables:

$$(z_j, \lambda_j) \leftrightarrow (z_k, \lambda_k). \qquad (6.13)$$

(2) The determinant is a linear function of each z_j . (3) The coefficient at z_N is equal to:

$$\partial \det{}_N(\varphi')/\partial z_N = \det{}_{(N-1)}(\varphi'{}^{(mod)}). \qquad (6.14)$$

The $(N-1) \times (N-1)$ matrix $\varphi'{}^{(mod)}$ is obtained from the $N \times N$ matrix φ' by removing the N-th row and the N-th column; the modification in the remaining part of the matrix has to be made as in (6.7):

$$z_j \rightarrow \widetilde{z}_j = z_j + K_{jN} \ (j = 1, 2, ..., N-1). \qquad (6.15)$$

(4) If all the $z_j = 0$, then the determinant is equal to zero at $N > 0$:

$$\det{}_N(\varphi') \Big|_{z_1 = z_2 = ... = z_N = 0} = \delta_{N0} \qquad (6.16)$$

It follows from the fact that in this case matrix φ' (6.9) possesses the eigenvector with the zero eigenvalue (all the components of this vector are equal to 1).

(5) At $N = 1$ one obtains:

$$\det{}_1(\varphi') = z_1 . \qquad (6.17)$$

The proof that $\| \lambda_1, ..., \lambda_N \|_N$ possesses the same five properties is equivalent to the proof of formula (6.12), due to the following theorem.

THEOREM. To prove the equality (6.12), it is sufficient to

establish that quantity $\|\lambda_1, \lambda_2, \ldots, \lambda_N\|_N$ depends on λ_j, z_j only and possesses the following five properties:

(1) The invariance with respect to permutations of pairs:

$$(\lambda_j, z_j) \leftrightarrow (\lambda_k, z_k). \tag{6.18}$$

(2) The linearity in each z_j .

(3) The coefficient at z_N is equal to:

$$\partial\|\lambda_1, \ldots, \lambda_N\|_N / \partial z_N = \|\lambda_1, \ldots, \lambda_{N-1}\|^{(mod)}, \tag{6.19}$$

the modification meaning the change $z_j \to \tilde{z}_j$ (6.15).

(4) At λ_j given,

$$\|\lambda_1, \ldots, \lambda_N\|_N = 0 \quad \text{if} \quad z_1 = z_2 = \ldots = z_N = 0 . \tag{6.20}$$

(5) $\|\varnothing\|_0 = 1 ; \quad \|\lambda_1\|_1 = z_1$ $$\tag{6.21}$$

The proof of the theorem is done by induction in N . The base of the induction is fulfilled (compare (6.21), (6.16), (6.17)). Suppose now that equation (6.12) is valid for $N = 1, 2, \ldots, q-1$ and prove that then it is valid also for $N = q$. To do this one considers the difference Δ_q :

$$\Delta_q = \|\lambda_1, \ldots, \lambda_q\|_q - \det_q(\varphi'). \tag{6.22}$$

This is a linear function of z_q , the corresponding coefficient being equal to

$$\partial\Delta_q / \partial z_q = \|\lambda_1, \ldots, \lambda_{q-1}\|_{q-1}^{(mod)} - \det_{q-1}(\varphi'^{(mod)}) = 0, \tag{6.23}$$

which is valid due to the inductive assumption. So Δ_q does not depend on z_q ; due to the symmetry Δ_q then does not depend on $z_j \, (j = 1, \ldots, N)$ at all. But one has also due to properties (4):

$$\Delta_q = 0 \quad \text{if} \quad z_1 = \ldots = z_q = 0. \tag{6.24}$$

This means that $\Delta_q \equiv 0$ at any z_j . That concludes construction of the induction step, and the theorem is thus proved.

So the proof of equations (6.8), (6.12) is reduced to the proof of properties (1)-(5) for $\|\lambda_1, \ldots, \lambda_N\|_N$. Properties (1)-(3) were already proved at the beginning of this section (see (6.6), (6.7)).

Property (5) is easily obtained by direct calculation using (6.10) and (5.5). Property (4) was proved in $[7]$. So the equation (6.8) (which contains, in particular, the Gaudin hypothesis (0.4)) is proved. Thus one can compute the norms of the wave eigenfunctions in the periodical finite box. In the NS model the norm of the wave function χ_N (I.1.20), (0.3) is given by $\langle \Psi_N | \Psi_N \rangle = \det_N (\varphi')$ (0.4) with $N \times N$ -matrix φ' given in (0.5). The determinant is essentially simplified at the theormodinamical limit. This limit at zero termperature was described in detail in s.I.3. One can obtain that

$$\det_N (\varphi') = \Big(\prod_{j=1}^{N} 2\pi \rho(\lambda_j) \Big) \Big[\det \big(1 - \tfrac{1}{2\pi} \hat{K} \big) \Big] \big(L^N \big) \qquad (6.25)$$

Here $\rho(\lambda)$ is the momentum density of the Dirac sea (I.3.6); $\hat{K}$ is the integral linear operator defined in (I.3.10). This formula was established in paper $[9]$.

7. ALGEBRAIC APPROACH TO CALCULATION OF CORRELATION FUNCTIONS

The method of calculation of correlators $[8, 9]$ in QISM is very similar to that of the norms. It was quite unessential at calculation of the norms that the monodromy matrix could be represented as the product of $\mathcal{L}$ -operators (1.17). All this information was contained in functions $a(\lambda)$ and $d(\lambda)$ only. The monodromy matrix was considered as an indivisible algebraic object $T(\lambda)$ defined by commutation properties (4.2) and by vacuum eigenvalues $a(\lambda)$ and $d(\lambda)$ (4.3). In this sense one can assume that the whole lattice contains one site only. To compute correlation function such one-site lattice is not sufficient. One needs a lattice with at least two sites. It was shown at the beginning of this Part that calculation of the current correlator in the NS model is reduced to calculation of the mean value of the operator $Q_1^2(x)$ (0.9) with respect to eigenfunctions:

$$\langle \Psi_N | Q_1^2(x) | \Psi_N \rangle, \qquad (7.1)$$

where Q_1 is the operator of the number of particles at the segment $[0, x]$. This leads to the natural decomposition of the monodromy matrix $T(\lambda)$ (1.17) into the matrix product of two monodromy matrices $T_1(\lambda)$, $T_2(\lambda)$:

$$T(\lambda) = T_2(\lambda)\, T_1(\lambda). \tag{7.2}$$

Here $T_1(\lambda) \equiv T(x, 0 | \lambda)$ and $T_2(\lambda) = T(L, x | \lambda)$ are transition matrices (1.16) for the NS model. Commutation relations between matrix elements of these transition matrices are given by the same formula (4.2):

$$R(\lambda, \mu)\, T_i(\lambda) \otimes T_i(\mu) = T_i(\mu) \otimes T_i(\lambda)\, R(\lambda, \mu) \quad (i=1,2), \tag{7.3}$$

the matrix elements of $T_1(\lambda)$ commuting with the matrix elements of $T_2(\lambda)$. The R -matrix is given in (2.17). The vacuum state $|0\rangle$ (2.26) is also the vacuum state for $T_1(\lambda)$ and $T_2(\lambda)$, the vacuum eigenvalues of diagonal matrix elements being denoted as $a_i(\lambda)$, $d_i(\lambda)$ $(i=1,2)$. In the NS model one has:

$$a_1(\lambda) = \exp\{-i\lambda x/2\}, \quad d_1(\lambda) = \exp\{i\lambda x/2\};$$
$$a_2(\lambda) = \exp\{i\lambda(x-L)/2\}, \quad d_2(\lambda) = \exp\{i\lambda(L-x)/2\}. \tag{7.4}$$

In calculating the norms of wave functions in s.s. 4-6 it was extremely useful to go from the monodromy matrix of the NS model to the general monodromy matrix $T(\lambda)$ with arbitrary vacuum eigenvalues $a(\lambda)$ and $d(\lambda)$. Similarly, to calculate matrix elements of operator Q_1^2 (7.1) one introduces the two-site generalized model, considering $T_1(\lambda)$ and $T_2(\lambda)$ in (7.2) as monodromy matrices with arbitrary vacuum eigenvalues $a_1(\lambda), d_1(\lambda)$ and $a_2(\lambda), d_2(\lambda)$. We write T_1 and T_2 in the following form:

$$T_i(\lambda) = \begin{bmatrix} A_i(\lambda) & ; & B_i(\lambda) \\ C_i(\lambda) & ; & D_i(\lambda) \end{bmatrix} ; \qquad i = 1, 2.$$

The normalization of the creation and annihilation operators is done similarly to (5.2):

$$\mathbb{B}_i(\lambda) = B_i(\lambda)/d_i(\lambda); \quad \mathbb{C}_i(\lambda) = C_i(\lambda)/d_i(\lambda). \tag{7.5}$$

It will be also useful to introduce functions $\ell(\lambda)$ and $m(\lambda)$ similar to $r(\lambda)$ (4.5):

$$\ell(\lambda) \equiv a_1(\lambda)/d_1(\lambda); \quad m(\lambda) \equiv a_2(\lambda)/d_2(\lambda); \quad \ell(\lambda)m(\lambda) = r(\lambda), \quad (7.6)$$

in the generalized model they are arbitrary functions. The operator Q_1 is defined now as follows

$$Q_1 \prod_{j=1}^{N} B_1(\lambda_j)|0\rangle = N \prod_{j=1}^{N} B_1(\lambda_j)|0\rangle; \quad Q_1|0\rangle = 0; \quad [Q_1, B_2] = 0;$$

$$(7.7)$$

$$\langle 0| \prod_{j=1}^{N} C_1(\lambda_j) Q_1 = N \langle 0| \prod_{j=1}^{N} C(\lambda_j); \quad \langle 0| Q_1 = 0; \quad [Q_1, C_2] = 0.$$

In the case of the NS model it is just the operator Q_1 (0.9). To calculate mean value (7.1) one has to act by the operator Q_1^2 on the state $\prod_j B(\lambda_j)|0\rangle$. Now one can present $B(\lambda)$ as:

$$B(\lambda) = [A_2(\lambda)/d_2(\lambda)] B_1(\lambda) + [D_1(\lambda)/d_1(\lambda)] B_2(\lambda), \quad (7.8)$$

and separate in state $\prod_j B(\lambda)|0\rangle$ contributions of the first and of the seconds sites. For $N=1$ one obtains directly from (7.8):

$$B(\lambda)|0\rangle = m(\lambda) B_1(\lambda)|0\rangle + B_2(\lambda)|0\rangle. \quad (7.9)$$

For N arbitrary it is proved in [8] that

$$\prod_{j=1}^{N} B(\lambda_j)|0\rangle = \sum_{\{\lambda\}=\{\lambda_I\}\cup\{\lambda_{II}\}}^{n_1+n_2=N} \left(\prod_I \prod_{II} m(\lambda_I) f(\lambda_I, \lambda_{II}) \right) \times$$

$$(7.10)$$

$$\times \left(\prod_{II} B_2(\lambda_{II}) \right) \left(\prod_I B_1(\lambda_I) \right) |0\rangle.$$

Here the sum is taken over all the partitions of the set $\{\lambda\}_N$ into two disjoint subsets $\{\lambda_I\}$ and $\{\lambda_{II}\}$; $\mathrm{card}\{\lambda_I\} = n_1$, $\mathrm{card}\{\lambda_{II}\} = n_2$ and $n_1 + n_2 = N = \mathrm{card}\{\lambda\}_N$. Product $\prod_I \prod_{II}$ is an independent product over all $\lambda \in \{\lambda_I\}$ and $\lambda \in \{\lambda_{\underline{II}}\}$ and thus contains $n_1 n_2$ factors. Function $m(\lambda)$ is given in (7.6). Similar representation can be obtained also for the dual state:

$$\langle 0| \prod_{j=1}^{N} \mathbb{C}(\lambda_j) = \sum_{\{\lambda\}=\{\lambda_I\}\cup\{\lambda_{\underline{II}}\}}^{n_1+n_2=N} \langle 0| \left(\prod_I \mathbb{C}_1(\lambda_I) \right) \left(\prod_{\underline{II}} \mathbb{C}_2(\lambda_{\underline{II}}) \right) \times \tag{7.11}$$

$$\times \left(\prod_I \prod_{\underline{II}} f(\lambda_{\underline{II}}, \lambda_I) \, l(\lambda_{\underline{II}}) \right).$$

It is to be mentioned that both these formulae are valid for λ_j arbitrary. Using equations (7.7), (7.10), (7.11) one reduces matrix element $\langle 0| \prod \mathbb{C}(\lambda^C) Q_1^2 \prod \mathbb{B}(\lambda^B)|0\rangle$ to scalar products $\langle 0| \prod \mathbb{C}_i(\lambda^C) \prod \mathbb{B}_i(\lambda^B)|0\rangle$ $(i=1,2)$ investigated in detail in **s.5**.

To obtain the mean value $\langle \Psi_N| Q_1^2 |\Psi_N\rangle$ with respect to an eigenfunction $|\psi_N\rangle$ (6.1) one has to take $\{\lambda^C\}_N = \{\lambda^B\}_N \equiv \{\lambda\}_N$ and impose the s.t.e. (6.2) on λ_j. It is convenient to define the following quantities $\langle Q_1^2\rangle_N$ $(N=0,1,2,...)$:

$$\langle Q_1^2\rangle_N \equiv c^{-N} \left(\prod_{j\neq k} f_{jk} \right)^{-1} \langle \psi_N(\{\lambda\}_N)| Q_1^2 |\psi_N(\{\lambda\}_N)\rangle \tag{7.12}$$

These quantities can be investigated similarly to the norms of eigenfunctions in s.6 [8]. First of all one establishes that $\langle Q_1^2\rangle_N$ is a function of momenta $\lambda_1,...,\lambda_N$ and a functional of arbitrary functions $l(\lambda)$ and $m(\lambda)$ (7.6), this functional being rather special. Namely, $\langle Q_1^2\rangle_N$ depends only on values of function $l(\lambda)$ at points λ_j and on values of the first derivatives of functions $l(\lambda)$, $m(\lambda)$ at points $\lambda_j \in \{\lambda\}$. So it can be considered as a function of 4N independent complex variables:

$$\langle Q_1^2\rangle_N \equiv \langle Q_1^2\rangle_N (\{\lambda_j\}_N, \{x_j\}_N, \{y_j\}_N, \{l_j\}_N), \tag{7.13}$$

where

$$l_j \equiv l(\lambda_j); \quad x_j \equiv x(\lambda_j); \quad y_j \equiv y(\lambda_j), \tag{7.14}$$

$$x(\lambda) = i\partial \ln l(\lambda)/\partial\lambda; \quad y(\lambda) = i\partial \ln m(\lambda)/\partial\lambda. \tag{7.15}$$

This function $\langle Q_1^2\rangle_N$ possesses the following properties which arc quite similar to the properties of the norms listed in s.6.

(1) It is invariant under replacement of "quartets":

$$(\lambda_j, x_j, y_j, \ell_j) \longleftrightarrow (\lambda_k, x_k, y_k, \ell_k); \quad j, k = 1, \ldots, N. \tag{7.16}$$

(2) It is a linear function of x_N and y_N.

(3) The coefficients at x_N and y_N are:

$$\partial \langle Q_1^2 \rangle_N / \partial y_N =$$
$$= \langle Q_1^2 \rangle_{N-1} (\{\lambda_j\}_{N-1}, \{x_j\}_{N-1}, \{y_j + K_{jN}\}_{N-1}, \{\ell_j\}_{N-1}); \tag{7.17}$$

$$\partial \langle Q_1^2 \rangle_N / \partial x_N =$$
$$= \langle (Q_1 + 1)^2 \rangle_{N-1} (\{\lambda_j\}_{N-1}, \{x_j + K_{jN}\}_{N-1}, \{y_j\}_{N-1}, \{\tilde{\ell}_j\}_{N-1}). \tag{7.18}$$

Here

$$\tilde{\ell}_j = \ell_j \left(f(\lambda_j, \lambda_N) / f(\lambda_N, \lambda_j) \right); \quad j = 1, \ldots, N-1, \tag{7.19}$$

and

$$\langle (Q_1 + 1)^2 \rangle_N \equiv \langle Q_1^2 \rangle_N + 2 \langle Q_1 \rangle_N + \langle 1 \rangle_N. \tag{7.20}$$

(4) The value of $\langle Q_1^2 \rangle_N$ at $x_1 = y_1 = x_2 = y_2 = \ldots = x_N = y_N = 0$ is of primary importance. It is called the irreducible part I_N:

$$I_N(\{\lambda_j\}_N, \{\ell_j\}_N) \equiv$$
$$\equiv \langle Q_1^2 \rangle_N (\{\lambda_j\}_N, \{x_j = 0\}_N, \{y_j = 0\}_N, \{\ell_j\}_N). \tag{7.21}$$

The dependence of I_N on variables ℓ_j can be extracted explicitly as follows:

$$I_N(\{\lambda_j\}_N, \{\ell_j\}_N) = \sum_{\substack{0 \leqslant n \leqslant [N/2] \\ \{\lambda\} = \{\lambda_+\} \cup \{\lambda_-\} \cup \{\lambda_0\}}} \left(\prod_{(+)} \ell(\lambda_+) \right) \left(\prod_{(-)} \ell(\lambda_-) \right) \times \tag{7.22}$$

$$\times \mathcal{A}_N^n (\{\lambda_+\}_n, \{\lambda_-\}_n, \{\lambda_0\}_{N-2n}).$$

The sum here is taken over all the partitions of the set $\{\lambda\}_N$ into three disjoint subsets: $\text{card}\,\{\lambda_+\} = \text{card}\,\{\lambda_-\} = n$; $\text{card}\,\{\lambda\}_0 =$

$= N - 2n$; $0 \leqslant n \leqslant [N/2]$. Coefficients $\mathcal{A}_N^n$ are the Fourier coefficients of the irreducible part I_N . They do not depend on l_j but only on λ_j being a rational functions of λ . Fourier coefficients depend on the R-matrix only and do not depend on the concrete model. All the dependence on concerete models enters through vacuum values $l(\lambda)$ and is written explicitly. In paper [8] irreducible parts were studied in detail and the recursion procedure of their calculation was given. The irreducible parts for N small can be calculated using (4.2), (4.3) or directly in the NS model (see s.III.1). For $N = 0, 1, 2, 3$ one has:

$$I_0 = I_1 = 0 ; \tag{7.23}$$

$$I_2(\{\lambda_1,\lambda_2\},\{l_1,l_2\}) = \mathcal{A}_2^1(\lambda_1,\lambda_2)[l_1 l_2^{-1} - 1] + \mathcal{A}_2^1(\lambda_2,\lambda_1)[l_2 l_1^{-1} - 1] ;$$

$$\mathcal{A}_2^1(\lambda_1,\lambda_2) = -\frac{2}{\lambda_{12}^2}\left(\frac{\lambda_{12} + ic}{\lambda_{12} - ic}\right) ; \quad (\lambda_{jk} \equiv \lambda_j - \lambda_k) ; \tag{7.24}$$

$$I_3(\{\lambda_1,\lambda_2,\lambda_3\},\{l_1,l_2,l_3\}) = \sum_P \mathcal{A}_3^1(\lambda_{P_1},\lambda_{P_2},\lambda_{P_3})[l_{P_1} l_{P_2}^{-1} - 1] ;$$

$$\mathcal{A}_3^1(\lambda_1,\lambda_2,\lambda_3) = 8c(\lambda_{12})^{-2}(\lambda_{12} + ic)(\lambda_{12} - ic)^{-1} \times \tag{7.25}$$

$$\times[(\lambda_{32}/\lambda_{31}) + (\lambda_{31}/\lambda_{32})](\lambda_{31} + ic)^{-1}(\lambda_{23} + ic)^{-1}.$$

The sum here is taken over all the permutations of λ_1, λ_2, λ_3 . (5) The last important property of $\langle Q_1^2 \rangle_N$ is that its value for $N = 1$ is easy to calculate:

$$\langle Q_1^2 \rangle_1 = x_1. \tag{7.26}$$

The five properties listed above are established in paper [8] . These properties (especially (7.17) and (7.18)) let expressing quantities $\langle Q_1^2 \rangle_N$ in terms of irrducible parts I_n $(n \leqslant N)$ [9] . That is why irreducible parts I_N (7.22) are of primary importance. The formulae expressing $\langle Q_1^2 \rangle_N$ is terms of irreducible parts are rather bulky, but they are somewhat simplified at the thermodynamical limit; the formulae at this limit are given at s.III.1. The corresponding calculations are similar to those of s.6.

In this Part of the lectures we have thus demonstrated the main ideas of QISM using the N$ model as an example. From the results obtained for the N$ model which we can't include here due to the lack of place, the results conserning the quantum version of the Gelfand-Levitan-Marchenko equation [29, 34] should be mentioned.

Part III.

CORRELATION FUNCTIONS

Here the equal time correlation function of currents is calculated for the one-dimensional Bose gas. It is done for zero as well as for nonzero temperature. The asymptotics of the correlator at large distances is discussed. The temperature dependence of the correlation length is also calculated.

1. CURRENT CORRELATOR AT ZERO TEMPERATURE

The ground state $|\Omega\rangle$ of the Hamiltonian (I.3.1) at zero temperature was described in detail in s.I.3. Consider the equal-time current correlator (II.01) [9] :

$$\langle j(x)\,j(0)\rangle \equiv \langle\Omega|j(x)\,j(0)|\Omega\rangle/\langle\Omega|\Omega\rangle. \tag{1.1}$$

Current $j(x)$ here is $j(x) = \psi^{+}(x)\,\psi(x)$. We will use formula (II.07):

$$\langle\Omega|\,j(x)\,j(0)|\Omega\rangle = \frac{1}{2}\frac{\partial^2}{\partial x^2}\langle\Omega|\,Q_1^{2}(x)|\Omega\rangle. \tag{1.2}$$

Operator Q_1 (II.0.8) here is the operator of the number of particles in the segment $[0,x]$. As was explained at the beginning of Part II, one has first to study the mean value of operator $Q_1^{2}(x)$:

$$\langle \Psi_N | Q_1^2(x) | \Psi_N \rangle, \tag{1.3}$$

where $|\Psi_N\rangle$ (I.1.20) is an eigenvector of the Hamiltonian, and then take the thermodynamical limit $|\Psi_N\rangle \to |\Omega\rangle$. The normalization of function $|\Psi_N\rangle$ is

$$\langle \Psi_N | \Psi_N \rangle = \det{}_N(\varphi'), \tag{1.4}$$

the $N \times N$ -matrix φ' being equal to

$$\varphi'_{jk} = \delta_{jk}\left(L + \sum_{m=1}^{N} K(\lambda_m, \lambda_k)\right) - K(\lambda_j, \lambda_k) \tag{1.5}$$

(see (II.0.4), (II.0.5), (II.6.8), (I.2.8), (I.2.15)). It should be noted that the determinant is a polynomial in the length L of the box:

$$\langle \Psi_N | \Psi_N \rangle = \det{}_N(\varphi') = \sum_{k=1}^{N} L^k c_k(\{\lambda\}), \tag{1.6}$$

$c_k(\{\lambda\})$ being rational functions of momenta λ_j . The explicit expression of the matrix element (1.3) was given in (II.0.10). It is easily calculated for N small:

$$\langle 0| Q_1^2(x) |0\rangle = 0 \; ; \; \langle \Psi_1 | Q_1^2(x) | \Psi_1 \rangle = x;$$

$$\langle \Psi_2 | Q_1^2(x) | \Psi_2 \rangle = 4x^2 + 2xy + [16\,cx\,(\lambda_{12}^2 + c^2)^{-1}] +$$
$$+ 2(\lambda_{12})^{-2}(\lambda_{12} + ic)(\lambda_{12} - ic)^{-1}[1 - \exp\{-ix\lambda_{12}\}] +$$
$$+ 2(\lambda_{12})^{-2}(\lambda_{21} + ic)(\lambda_{21} - ic)^{-1}[1 - \exp\{-ix\lambda_{21}\}]. \tag{1.7}$$

Here $|\Psi_2\rangle = |\Psi_2(\{\lambda_1, \lambda_2\})\rangle$; $\lambda_{12} = \lambda_1 - \lambda_2$; $y = L - x$. One thus sees that the mean value $\langle \Psi_2 | Q_1^2 | \Psi_2 \rangle$ depends on the distance in two essentially different ways, that is, polynomially and exponentially. The exponential part (which is the sum of the last two terms) is called the irreducible part I_2 (II.7.24). Indeed, for the NS model function $l(\lambda)$ (II.7.6) entering (II.7.24) is give just as $l(\lambda) = \exp\{-i\lambda x\}$ (II.7.4). For any N the definition of the irreducible part I_N (II.7.22) can also be given directly in terms of the NS model. It is proved in papers [8, 9] that the meanvalue $\langle \Psi_N | Q_1^2 | \Psi_N \rangle$ in the NS model can be represented in the following form:

$$\langle \Psi_N | Q_1^2 | \Psi_N \rangle = \sum_{n=0}^{N} \sum_{m=0}^{N-1} x^n L^m J_{m,n}^{(N)} . \qquad (1.8)$$

Coefficients $J_{m,n}^{(N)}$ are rational functions of λ_j and of $\exp\{i x \lambda_j\}$. If one puts by definition

$$I_N(\{\lambda_j\}) \equiv J_{0,0}^{(N)} , \qquad (1.9)$$

then one reproduces just the irreducible part I_N (II.7.22) at

$$\ell(\lambda) = \exp\{-i\lambda x\}. \qquad (1.10)$$

In particular, irreducible part I_3 can be calculated in terms of the NS model similar to I_2 (1.7); one obtains thus the same result (II.7.25). Similarly, the Fourier coefficients $\mathcal{A}_N^n$ can be also introduced directly in terms of the NS model. Noting that I_N in the NS model depends on x only exponentially through $\ell(\lambda)$ (1.10), one reproduces the representation (II.7.22):

$$I_K(\{\lambda_j\}) = \sum_{\{\lambda\}_K = \{\lambda_+\}_n \cup \{\lambda_-\}_n \cup \{\lambda_0\}_{K-2n}} \exp\{-i x \sum_{j=1}^{n} (\lambda_j^+ - \lambda_j^-)\} \times \qquad (1.11)$$

$$\times \mathcal{A}_K^n(\{\lambda_+\}_n, \{\lambda_-\}_n, \{\lambda_0\}_{K-2n}) .$$

Here $\operatorname{card}\{\lambda_+\} = \operatorname{card}\{\lambda_-\} = n$, the sum has the same structure as in (II.7.22). Fourier coefficients $\mathcal{A}_K^n$ do not depend on x and are rational functions of λ_j . The properties of irreducible parts and of Fourier coeffcients can be investigated in detail, however, only by means of QISM [7]. That is the point where QISM is necessary for investigation of correlation functions. It is to be mentioned that irreducible parts as well as Fourier coefficients are small at the limit of the weak as well of the strong coupling:

$$I_K \sim c^{K-2} ; \quad \mathcal{A}_K^n \sim c^{K-2} \quad \text{if} \quad c \to 0 \quad (K \geqslant 2); \qquad (1.12)$$

$$I_K \sim c^{2-K} ; \quad \mathcal{A}_K^n \sim c^{2-K} \quad \text{if} \quad c \to \infty \quad (K \geqslant 2). \qquad (1.13)$$

Turn now to investigation of the meanvalue. Due to (1.6), (1.8) one has

$$\frac{\langle \Psi_N | Q_1^2 | \Psi_N \rangle}{\langle \Psi_N | \Psi_N \rangle} = \frac{\left(\sum_{n=0}^{N} \sum_{m=0}^{N-1} x^n L^m J_{m,n}^{(N)} \right)}{\sum_{k=1}^{N} L^k c_k(\{\lambda\})} . \tag{1.14}$$

One is interested in the thermodynamical limit of (1.14). The limit of the norm $\langle \Psi_N | \Psi_N \rangle$ is already calculated (II.6.25). The contribution of the nominator is rather similar. It was proved in [9] that the following phenomenon takes place at this limit. Polinomial contributions $(x^n L^m)$ in (1.14) can be summed up to form the exponents, these exponents dressing the "bare" exponents introducing (1.11). So the result of summing up of the polynomials is reduced to the following transformation of the bare exponents:

$$\exp\{-ix \sum_{j=1}^{n} (\lambda_j^+ - \lambda_j^-)\} \rightarrow \exp\{-ix \sum_{j=1}^{n} (\lambda_j^+ - \lambda_j^-)\} \times$$

$$\times \exp\{x \int_{-q}^{q} P_n(t, \{\lambda_n^+\}, \{\lambda^-\}_n) dt \tag{1.15}$$

(q is the Fermi momentum (I.3.5)), (I.3.16). Function P_n here is generated by some universal dressing equation. That is the nonlinear integral equation:

$$1 + 2\pi P_n(t) = S_n(t) \exp\{ \int_{-q}^{q} K(t,s) P_n(s) ds \}. \tag{1.16}$$

Here $K(t,s)$ is the standard kernel (I.2.16):

$$K(t,s) = 2c[(t-s)^2 + c^2]^{-1} \tag{1.17}$$

and $S_n(t)$ is the bare scattering matrix

$$S_n(t) = \prod_{j=1}^{n} \frac{(\lambda_j^+ - t + ic)}{(\lambda_j^+ - t - ic)} \frac{(\lambda_j^- - t - ic)}{(\lambda_j^- - t + ic)} . \tag{1.18}$$

The solution $P_n(t)$ of the equation (1.16) is complex-valued. One has to take such a solution that

$$\operatorname{Re} P_n(t) \leqslant 0 . \tag{1.19}$$

It is proved in s.4 that the solution for the system consisting of equation (1.16) and of inequality (1.19) does exist and is unique. It should be noted that equation (1.16) is obtained in the frame of algebraic Bethe Ansatz as a differential equation in variational derivatives. So after summing up the polynomials the "bare" irreducible part I_K (1.11) must be replaced by the dressed one:

$$I_K \rightarrow I_K^d = \sum_{\{\lambda\} = \{\lambda_+\} \cup \{\lambda_-\} \cup \{\lambda_0\}} \exp\{x\, p_n(\{\lambda_+\}_n, \{\lambda_-\}_n)\} \times \tag{1.20}$$

$$\times \mathcal{A}_K^n (\{\lambda_+\}_n, \{\lambda_-\}_n, \{\lambda_0\}_{K-2n}).$$

Here

$$p_n(\{\lambda_+\}_n, \{\lambda_-\}_n) = -i \sum_{j=1}^{n} (\lambda_j^+ - \lambda_j^-) + \tag{1.21}$$

$$+ \int_{-q}^{q} P_n(t, \{\lambda_+\}_n, \{\lambda_-\}_n)\, dt .$$

The Fourier coefficients remain unchanged. It is important that I_K^d is a real symmetrical function of all the λ_j (at real λ_j).

Now everything is ready to represent the current correlator at the thermodynamical limit as a series. The sum over the partitions in (1.20) is to be changed for an integral at the thermodynamical limit, the contribution of I_K^d to the mean value (1.3) being equal to Γ_K :

$$\Gamma_K(x) = \frac{1}{K!} \int\limits_{-q}^{q} \cdots \int\limits_{-q}^{q} \left\{ \prod_{j=1}^{K} \frac{\omega(\lambda_j)\, d\lambda_j}{2\pi} \right\} I_K^d(\{\lambda_j\}_K). \qquad (1.22)$$

Weight $\omega(\lambda)$ here is given as:

$$\omega(\lambda) = \exp\left\{ -\frac{1}{2\pi} \int\limits_{-q}^{q} K(\lambda,\mu)\, d\mu \right\}. \qquad (1.23)$$

It is essential that $\omega(\lambda) < 1$; namely

$$e^{-1} \leqslant \omega(\lambda) \leqslant \omega_0 < 1 ; \qquad (\mathrm{Im}\ \lambda = 0),$$

$$\omega_0 = \exp\left\{ -2cq\,\pi^{-1}(c^2 + 4q^2)^{-1} \right\} < 1 . \qquad (1.24)$$

Now it is possible to formulate the answer for the current correlator at the thermodynamical limit (see (1.1), (1.2)) (as usual, $: j(x)\, j(0): \equiv \psi^+(x)\,\psi^+(0)\,\psi(x)\,\psi(0)$)

$$\langle j(x)\, j(0) \rangle = \langle : j(x)\, j(0): \rangle + \delta(x)\langle j(0) \rangle ; \qquad (1.25)$$

$$\langle j(0) \rangle = \int\limits_{-q}^{q} \rho(\lambda)\, d\lambda = \langle \Omega | j(0) | \Omega \rangle / \langle \Omega | \Omega \rangle \qquad (1.26)$$

(for $\rho(\lambda)$ see (I.3.9), (I.3.16));

$$\langle : j(x)\, j(0): \rangle \equiv \langle j(0) \rangle^2 + \langle\!\langle j(x)\, j(0) \rangle\!\rangle . \qquad (1.27)$$

The nontrivial part of the correlator, $\langle\!\langle j(x)\, j(0) \rangle\!\rangle$, is an even function of x ; at $x > 0$ one obtains using (1.2) and (1.22):

$$\langle\!\langle j(x)\, j(0) \rangle\!\rangle = \frac{1}{2} \frac{\partial^2}{\partial x^2} \sum_{K=2}^{\infty} \Gamma_K(x) . \qquad (1.28)$$

Using explicit expressions of $\mathcal{A}_2^1$ and $\mathcal{A}_3^1$ (II.7.24), (II.7.25), (1.10) one can write the first two terms explicitly as:

$$\frac{1}{2}\frac{\partial^2}{\partial x^2}\Gamma_2(x) = -\int\limits_{-q}^{q}\frac{d^2\lambda}{(2\pi)^2}\,\omega(\lambda_1)\omega(\lambda_2)\left(\frac{\lambda_{12}+ic}{\lambda_{12}-ic}\right)\times$$

$$\times\left[\frac{p_1(\lambda_1,\lambda_2)}{\lambda_{12}}\right]^2\exp\{x\,p_1(\lambda_1,\lambda_2)\};\tag{1.29}$$

$$\frac{1}{2}\frac{\partial^2}{\partial x^2}\Gamma_3(x) = \frac{c}{2\pi^3}\oint\limits_{-q}^{q}d^3\lambda\,\omega(\lambda_1)\omega(\lambda_2)\omega(\lambda_3)\left(\frac{\lambda_{12}+ic}{\lambda_{12}-ic}\right)\times$$

$$\times\left[\frac{p_1(\lambda_1,\lambda_2)}{\lambda_{12}}\right]^2\left(\frac{\lambda_{32}}{\lambda_{31}}+\frac{\lambda_{31}}{\lambda_{32}}\right)\frac{\exp\{x\,p_1(\lambda_1,\lambda_2)\}}{(\lambda_{31}+ic)(\lambda_{23}+ic)}\,.\tag{1.30}$$

The integral in (1.30) is defined in the sense of the principal va-
lue. It should be also remarked that $p_1(\lambda_1,\lambda_2)\to 0$ at $\lambda_1\to\lambda_2$.

Let us now comment upon the structure of the series (1.28).
It has already been mentioned that at $c=\infty$ the NS model is equivalent to free fermions. One can easily extract the $(1/c)$ expansion from the series above. It can be proved [9] using (1.13)
that also $\Gamma_K(x)\sim c^{2-K}$ at $c\to\infty$. However, Γ_K is not
a monomial in c. The series (1.28) thus gives the improved version of the $(1/c)$ expansion. Already the first term of this
series gives the correct large distance asymptotics, as shown in
s.2. The k-th power of the weight $\omega(\lambda)<1$ (1.23), (1.24)
introduces the k-th term of the series (1.28). This results in
convergence of the series. The physical meaning of the expansion
(1.28) is as follows. To obtain the k-th term of the series one
extracts k particles from the Dirac sea and first calculate the
bare k-particle correlator $\langle\Psi_k|j(x)j(0)|\Psi_k\rangle$. Then one uses
dressing equation (1.16) and makes the averaging over the particles
chosen. Thus the k-th term of the series is obtained. So calculation of the correlators is similar to calculation of other observable quantities as explained in s.I.4.

2. LARGE DISTANCE ASYMPTOTICS AT ZERO TEMPERATURE

First use the formulae obtained in s.1 to study the strong coupling limit. At $c=\infty$ only the first term of the series (1.28) survives, this only term being essentially simplified. The matter is that at $c=\infty$ one has $S_n(t)\equiv 1$ in (1.16), and hence $P_n(t)=0$. So one obtains

$$\langle\!\langle j(x)j(0)\rangle\!\rangle = -\left[(\sin(qx))/(\pi x)\right]^2 ;$$

$$\langle j(0)\rangle = q/\pi \qquad (c=\infty).$$

(2.1)

Let us now study corrections in $(1/c)$. As $S_n(t)\to 1$ at $c\to\infty$, function $P_n(t)$ is small. Then the solution of equation (1.16) is given as

$$P_n(t,\{\lambda^+\}_n,\{\lambda^-\}_n)=\frac{i}{\pi c}\left(1+\frac{2q}{\pi c}\right)\sum_{j=1}^{n}(\lambda_j^+ - \lambda_j^-) -$$

$$-\frac{1}{\pi c^2}\Big[\sum_{j=1}^{n}(\lambda_j^+ - \lambda_j^-)\Big]^2 + \mathcal{O}(c^{-3}).$$

(2.2)

Using the first two terms (1.29), (1.30), one thus easily obtains the correlator up to terms of order $(1/c)^2$:

$$\langle\!\langle j(x)j(0)\rangle\!\rangle = -\left(1+\frac{2}{c}\frac{\partial}{\partial x_r}\right)\left(\frac{\sin(qx_r)}{\pi x_r}\right) - \frac{8q}{\pi c}\left(\frac{\sin qx}{\pi x}\right)^2 +$$

$$+\frac{2}{c\pi^2}\frac{\partial}{\partial x}\left[\frac{\sin qx}{\pi x}\int_{-q}^{q}\sin(\lambda x)\ln\left(\frac{q+\lambda}{q-\lambda}\right)d\lambda\right]$$

(2.3)

Here $x_r=(1+2q\pi^{-1}c^{-1})x$.

Now investigate the large distance asymptotics $(x\to\infty)$ at coupling constant c arbitrary $(0<c<\infty)$. Consider the contribution of the first term $\Gamma_2(x)$ (1.29) of the series. Integrating by parts one obtains

$$\frac{1}{2}\frac{\partial^2}{\partial x^2}\Gamma_2(x)\Bigg|_{q^{-1}\ll x\to\infty}\longrightarrow -\frac{\omega^2(q)}{2\pi^2 x^2}\,. \tag{2.4}$$

Among the correction terms the following one must be mentioned:

$$\text{Const Re}\left(x^{-2}\exp\{x\,p_1(q,-q)\}\right). \tag{2.5}$$

Due to (1.19), $\text{Re}\,p_1(q,-q)\leqslant 0$. In s.4 it is proved that $\text{Re}\,p_1(q,-q)$ <0 at $0<c<\infty$ and $q\neq 0$. Hence, the correction term (2.5) decreases exponentially at $x\to\infty$. This term also oscillates as $\text{Im}\,p_1(q,-q)\neq 0$. So the following two scales are important in the asymptotics:

$$r_1\equiv 2/q\,;\qquad r_2=-\left(\text{Re}\,p(q,-q)\right)^{-1}>0. \tag{2.6}$$

In the region $r_1\ll x\ll r_2$ the asymptotics is oscillating:

$$\langle\!\langle j(x)j(0)\rangle\!\rangle\longrightarrow -\,\alpha x^{-2}\times$$
$$\times\left[1-\theta\exp\{x\,\text{Re}\,p_1(q,-q)\}\cos\{x\,\text{Im}\,p_1(q,-q)\}\right]. \tag{2.7}$$

At $x\gg r_2$ the oscillations are absent, and

$$\langle\!\langle j(x)j(0)\rangle\!\rangle\longrightarrow -\,a/x^2. \tag{2.8}$$

Here a,θ are some constants. The analysis of other terms of the series (1.28) confirms this picture of the asymptotics given by (2.8). It should be mentioned also that $r_2\to\infty$ at $c\to\infty$:

$$r_2=-\left(\text{Re}\,p_1(q,-q)\right)^{-1}=\pi c^2/8q^3\to\infty\quad(c\to\infty) \tag{2.9}$$

This results in the fact that the asymptotics (2.1) at $c=\infty$ does contain oscillations.

3. CURRENT CORRELATOR AT NONZERO TEMPERATURES

The current correlator at $T>0$ is defined as follows:

303

$$\langle j(x) j(0) \rangle_T = \frac{tr(\exp\{-H_h/T\}\, j(x) j(0))}{tr(\exp\{-H_h/T\})} \quad , \tag{3.1}$$

the Hamiltonian H_h given in (I.3.1). The trace at the denominator was already calculated in s.I.5. The trace at the numerator can be calculated by the same method. This trace can be represented as the sum over the N-particle traces:

$$tr_N\{\exp(-H/T)\, j(x) j(0)\} =$$

$$= (N!)^{-1} \sum_{\{n_j\}} \exp\{-E_N/T\}\, \frac{\langle \Psi_N | j(x) j(0) | \Psi_N \rangle}{\langle \Psi_N | \Psi_N \rangle} \tag{3.2}$$

Here $|\Psi_N\rangle$ (I.1.20) is the eigenfunction of the Hamiltonian corresponding to the set $\{n_j\}$ (I.2.13) of integer or half integer numbers. Formula (3.2) is to be compared with (I.5.10). At the thermodynamical limit the trace can be rewritten as a functional integral (compare this with (I.5.14), (I.5.15)):

$$tr(\exp\{-H/T\}\, j(x) j(01) =$$

$$= Const \int \left\{ \prod_\lambda d\left(\frac{\rho_t(\lambda)}{\rho_p(\lambda)}\right)\right\} \exp\{-\frac{X}{T}\}\, \frac{\langle \Psi_N | j(x) j(0) | \Psi_N \rangle}{\langle \Psi_N | \Psi_N \rangle} \tag{3.3}$$

Here X is given by (I.5.16). Direct calculation of the mean values

$$\langle \Psi_N | j(x) j(0) | \Psi_N \rangle / \langle \Psi_N | \Psi_N \rangle \tag{3.4}$$

with respect to eigenfunctions of the Hamiltonian is made in paper [10] at the thermodynamical limit. These calculation done similarly to s. 1 shows that the mean value (3.4) depends on the macroscopic characteristics ρ_p, ρ_h, ρ_t of the eigenstate $|\Psi_N\rangle$, but not on the microspic ones. Hence the mean value (3.4) is a "gauge invariant" quantity (in the same sense as all the observable quantities in s.I.8). At the thermodynamical limit ($L \rightarrow \infty$, $N \rightarrow \infty$, $D = L/N = Const$) the functional integral (3.3) can be calcu-

lated by the method of steepest descent. The answer obtained for
(3.1) is then due to (3.4):

$$\langle j(x)j(0)\rangle_T = \langle \Omega_T| j(x) j(0) |\Omega_T\rangle / \langle \Omega_T|\Omega_T\rangle. \tag{3.5}$$

Here $|\Omega_T\rangle$ is any of the eigenstates of the Hamiltonian composing
the state of the thermodynamical equilibrium which was described
in detail in s.I.8. This equilibrium state is not a pure quantum
mechanical state being the mixture of many eigenstates $|\Omega_T\rangle$.
Any of these states $|\Omega_T\rangle$ can be used in (3.5), the answer not
depending on the concrete $|\Omega_T\rangle$ due to the "gauge invariance"
of the matrix element (3.4).

So the correlator (3.1) at the equilibrium state is reduced
by (3.5) to the mean value with respect to the eigenfunction $|\Omega_T\rangle$
of the Hamiltonian. This mean value can be calculated similarly to
that considered in s.1. The answer thus obtained corresponds to the
mnemonic rule (I.8.13). One has to take the zero temperature answers
and change the integration measure as follows:

$$\left(\int_{-q}^{q} d\lambda\right)_{T=0} \rightarrow \left(\int_{-\infty}^{\infty} d\lambda\,\vartheta(\lambda)\right)_{T\neq0} \tag{3.6}$$

The Fermi weight $\vartheta(\lambda)$ here is given in (I.5.32) as

$$\vartheta(\lambda) = (1 + \exp\{\varepsilon(\lambda)/T\})^{-1} \tag{3.7}$$

Let us formulate the answer for the correlator:

$$\langle j(x)j(0)\rangle_T = \langle : j(x) j(0):\rangle_T + \delta(x)\langle j\rangle_T \tag{3.8}$$
$$(:j(x) j(0): = \psi^+(x)\psi^+(0) \psi(x)\psi(0)) \, ;$$

$$\langle j\rangle_T = \frac{\langle \Omega_T| j(0)|\Omega_T\rangle}{\langle \Omega_T|\Omega_T\rangle} = \int_{-\infty}^{\infty} \rho_p(\lambda)d\lambda = D \, ; \tag{3.9}$$

$$\langle : j(x) j(0):\rangle_T \equiv \langle j\rangle_T^2 + \langle\!\langle j(x) j(0)\rangle\!\rangle_T =$$

$$= \langle \Omega_T| : j(x) j(0):|\Omega_T\rangle / \langle \Omega_T|\Omega_T\rangle. \tag{3.10}$$

The nontrivial part $\ll j(x)\,j(0)\gg$ of the correlator is an even function of x . At $x>0$ one has:

$$\ll j(x)\,j(0)\gg_T = \frac{1}{2}\frac{\partial^2}{\partial x^2}\sum_{K=2}^{\infty}\Gamma_K(x|T) \qquad (3.11)$$

Now quantity $\Gamma_K(x|T)$ is to be defined. It is done similarly to $\Gamma_K(x)$ in s.1. The "bare" irreducible part and its Fourier coefficients are defined by the old formulae (1.9) and (1.11). The dressing equation (1.16) is changed due to the mnemonic rule (3.6):

$$1 + 2\pi P_n(t) = S_n(t)\exp\Big\{\int_{-\infty}^{\infty} K(t,s)\vartheta(s)P_n(s)\,ds\Big\} \qquad (3.12)$$

The S -matrix S_n and the Kernel K here are given by the old formulae (1.17), (1.18). The same condition for the real part of P_n remains valid:

$$\mathrm{Re}\, P_n(t)\leqslant 0 \qquad (3.13)$$

Function $p_n(\{\lambda^+\}_n,\{\lambda^-\}_n)$ is now defined as

$$p_n(\{\lambda^+\}_n,\{\lambda^-\}_n)=-i\sum_{j=1}^{n}(\lambda_j^+-\lambda_j^-)+\int_{-\infty}^{\infty}dt\,\vartheta(t)P_n(t,\{\lambda^+\}_n,\{\lambda^-\}_n) \qquad (3.14)$$

Taking this into account one has the same formula (1.20) for the dressed irreducible part:

$$I_K^d = \sum_{\{\lambda\}=\{\lambda^+\}\cup\{\lambda^-\}\cup\{\lambda^0\}} \exp\{x\,p_n(\{\lambda^+\}_n,\{\lambda^-\}_n)\}\times$$

$$\times \mathcal{A}_K^n(\{\lambda^+\}_n,\{\lambda^-\}_n,\{\lambda^0\}_{K-2n}). \qquad (3.15)$$

The K-th term of the series (3.11) is given as:

$$\Gamma_K(x|T) = \frac{1}{K!}\int_{-\infty}^{\infty}\Big\{\prod_{j=1}^{K}\frac{\omega_T(\lambda_j)\vartheta(\lambda_j)\,d\lambda_j}{2\pi}\Big\}I_K^d(\{\lambda_j\}) \qquad (3.16)$$

Here

$$0 < \omega_T(\lambda) = \exp\left\{ -\frac{1}{2\pi} \int_{-\infty}^{\infty} K(\lambda,\mu)\vartheta(\mu)\,d\mu \right\} < 1 \,. \tag{3.17}$$

Write also explicitly the first term of the series (3.11):

$$\frac{1}{2}\frac{\partial^2}{\partial x^2}\Gamma_2(x|T) = -\frac{1}{4\pi^2}\int_{-\infty}^{\infty} d\lambda_1\,\omega_T(\lambda_1)\vartheta(\lambda_1)\int_{-\infty}^{\infty} d\lambda_2\,\omega_T(\lambda_2)\vartheta(\lambda_2)\times \tag{3.18}$$

$$\times\left(\frac{\lambda_1-\lambda_2+ic}{\lambda_1-\lambda_2-ic}\right)\left[\frac{p_1(\lambda_1,\lambda_2)}{\lambda_1-\lambda_2}\right]^2 \exp\{x\,p_1(\lambda_1,\lambda_2)\}\,.$$

The series (3.11) possesses the same properties as the series (1.28) for the zero temperature case. The large distance asymptotics of the correlator is calculated in s.5, where also the strong coupling limit is studied. In s.4 we investigate in detail properties of the dressing equation (3.12) which proves to be very usefull for calculating asymptotics.

4. DRESSING EQUATIONS FOR CORRELATION FUNCTIONS

Consider the equation which is a little more general than the equation (3.12):

$$1 + 2\pi P(t) = S(t)\exp\left\{ \int_{-\infty}^{\infty} K(t,s)\vartheta(s)P(s)\,ds \right\}. \tag{4.1}$$

Here $S(t)$ is some function possessing the property

$$|S(t)| \leqslant 1 \,. \tag{4.2}$$

The additional requirement for function $P(t)$ is

$$\mathrm{Re}\,P(t) \leqslant 0 \,. \tag{4.3}$$

Prove first the uniqueness of the solution. Suppose that $P_I(t)$ and $P_{II}(t)$ are two different solutions of the system (4.1) and (4.3). Subtracting the equations for them one has:

$$2\pi\left[P_{\underline{I}}(t) - P_{\underline{II}}(t)\right] =$$

$$= S(t)\left[\exp\{(\hat{K}_{T}P_{\underline{I}})(t)\} - \exp\{(\hat{K}_{T}P_{\underline{II}})(t)\}\right]. \tag{4.4}$$

Here we use the notation (I.5.34). As operator $\hat{K}_{T}$ has the positive kernel, it follows from (4.3) that

$$Re\,(\hat{K}_{T}P)(t) \leqslant 0. \tag{4.5}$$

Using the well-known inequality

$$\left|\exp\{z_{1}\} - \exp\{z_{2}\}\right| \leqslant |z_{1} - z_{2}| \quad (Re\,z_{1} \leqslant 0;\ Re\,z_{2} \leqslant 0), \tag{4.6}$$

one can estimate the modulus of the left hand side of equation (4.4):

$$2\pi|P_{\underline{I}}(t) - P_{\underline{II}}(t)| \leqslant$$

$$\leqslant \left|\exp\{(\hat{K}_{T}P_{\underline{I}})(t)\} - \exp\{(\hat{K}_{T}P_{\underline{II}})(t)\}\right| \leqslant \tag{4.7}$$

$$\leqslant |(\hat{K}_{T}(P_{\underline{I}} - P_{\underline{II}}))(t)| \leqslant (\hat{K}_{T}|P_{\underline{I}} - P_{\underline{II}}|)(t)$$

Multiplying this inequality by $\vartheta(t)|P_{\underline{I}}(t) - P_{\underline{II}}(t)|$ and integrating over t, one obtains

$$2\pi\int_{-\infty}^{\infty}v^{2}(t)dt \leqslant \int_{-\infty}^{\infty}v(t)dt\int_{-\infty}^{\infty}v(s)ds\,\sqrt{\vartheta(t)}\,K(s,t)\sqrt{\vartheta(s)}. \tag{4.8}$$

Here $v(t) = \sqrt{\vartheta(t)}\,|P_{\underline{I}}(t) - P_{\underline{II}}(t)|$. The inequality (4.8) contradicts the estimate (I.5.35) for the quadratic form of the operator $\hat{K}_{T}$. So $P_{\underline{I}}(t) = P_{\underline{II}}(t)$, the uniqueness of the solution being thus proved. It follows immediately that

$$P(t) \equiv 0 \quad if \quad S(t) = 1. \tag{4.9}$$

Prove now the existence theorem for the system (4.1), (4.3). To do this one constructs the sequence of functions $P_{j}(t)$:

$$P_{0}(t) \equiv 0;$$

$$P_{j+1}(t) = (2\pi)^{-1}\left[S(t)\exp\{(\hat{K}_{T}P_{j})(t)\} - 1\right]. \tag{4.10}$$

First one notices that it is valid that

$$Re\,P_{j} \leqslant 0. \tag{4.11}$$

That can be proved using induction in j . For $j=1$, $P_1(t) = (2\pi)^{-1} \times$ $\times [S(t)-1]$, and one has $\mathrm{Re}\, P_1(t) \leqslant 0$ due to (4.2). Then suppose that (4.11) is valid and establish then that $\mathrm{Re}\, P_{j+1} \leqslant 0$. The proof is obvious from (4.10). Indeed, as (4.11), (4.2) and the positivity of the kernel of operator $\hat{K}_T$ means that $|S(t)\exp\{(\hat{K}_T P_j)\}|$ $\leqslant 1$, one has that $\mathrm{Re}\, S(t)\exp\{(\hat{K}_T P_j)(t)\} \leqslant 1$. So (4.11) is established. The difference $|P_{j+1}(t) - P_j(t)|$ can be estimated using arguments similar to (4.6), (4.7); one gets:

$$\left|P_{j+1}(t) - P_j(t)\right| \leqslant (2\pi)^{-1}\left(\hat{K}_T |P_j - P_{j-1}|\right)(t) \tag{4.12}$$

As the eigenvalues of operator $\hat{K}_T/(2\pi)$ are less than 1 as shown in (I.5.35), one now proves the convergence of the sequence (4.10) using the contracting mapping method. So the limit of the sequence (4.10) is the solution of the system (4.1), (4.3), the existence of a solution being also proved.

Now investigate properties of the solutions $P(t)$. The important property is that

$$|P(t)| \leqslant 1/\pi \tag{4.13}$$

which follows from

$$|P(t)| = (2\pi)^{-1}\left|S(t)\exp\{(\hat{K}_T P)(t)\} - 1\right| \leqslant$$

$$\leqslant (2\pi)^{-1}\left|\exp\{(\hat{K}_T P)(t)\}\right| + (2\pi)^{-1}.$$

This property permits estimating contribution of the vacuum polarization into (3.14):

$$\int_{-\infty}^{\infty} dt\, \vartheta(t) P_n(t, \{x^+\}, \{x^-\}) \leqslant \frac{1}{\pi} \int_{-\infty}^{\infty} \vartheta(t)\, dt \leqslant 2D, \tag{4.14}$$

which means that the behavior of P_n is, roughly speaking, as this of the "bare" term:

$$P_n(\{x^+\}, \{x^-\}) \sim -i \sum_{j=1}^{n} (x_j^+ - x_j^-) \tag{4.15}$$

The next property of the solution $P(t)$ is that if $S(t)$ is not identically equal to 1, then $\mathrm{Re}\, P(t) < 0$ at any real finite t . It is also to be mentioned that if $S(t)$ is a real function, then $P(t)$ is also a real function. It means, e.g., that

$$\operatorname{Im} P(t, \alpha, \alpha^*) = 0 \qquad (\operatorname{Im} t = 0 ,\ \operatorname{Im} S(t) = 0). \tag{4.16}$$

At last, consider function $P_1(t, \lambda^+, \lambda^-)$ (3.12). For this function

$$S(t) = \left(\frac{\lambda^+ - t + ic}{\lambda^+ - t - ic} \right) \left(\frac{\lambda^- - t - ic}{\lambda^- - t + ic} \right) = S_1(t) \tag{4.17}$$

It is obvious that function $P_1(t, \lambda^+, \lambda^-)$ can be analitically continued in λ^+, λ^- . Continuing into the lower half-plane in λ^+ and into the upper half-plane in λ^- , one obtains that then

$$\left| S_1(t) \right| \leqslant 1 \qquad (\ \operatorname{Im} \lambda^+ \leqslant 0 ;\ \ \operatorname{Im} \lambda^- \geqslant 0) \tag{4.18}$$

Finally, it is to discuss the strong coupling limit. As $S_n \to 1$, $P_n \to 0$ the exponent in (4.1) can be expanded into the series. Thus one obtains for $p_1(\lambda^+, \lambda^-)$:

$$p_1(\lambda^+, \lambda^-) = -i(\lambda^+ - \lambda^-)(1 + \frac{2D}{c}) - \frac{2D}{c^2}(\lambda^+ - \lambda^-)^2 + O(c^{-3}) \tag{4.19}$$

5. TEMPERATURE DEPENDENCE OF CORRELATION LENGTH

First consider the current correlator at the strong coupling limit. As $c = \infty$, only the first term of the series (3.11) remains, this term simplifying essentially due to $P_1 \equiv 0$. So one has for the correlator:

$$\langle\!\langle j(x) j(0) \rangle\!\rangle_T = -\frac{1}{4\pi^2} \left[\int_{-\infty}^{\infty} \frac{\exp\{i\lambda x\} d\lambda}{1 + \exp\{(\lambda^2 - h)/T\}} \right]^2 ;\quad x > 0,\, c = \infty. \tag{5.1}$$

$$\langle j(x) \rangle_T = \langle j(0) \rangle_T = \frac{1}{2\pi} \int_{-\infty}^{\infty} \frac{d\lambda}{1 + \exp\{(\lambda^2 - h)/T\}} ;\quad x > 0;\, c = \infty. \tag{5.2}$$

So the correlator is given essentially by the square of the Fourier transform of the Fermi weight. To obtain the asymptotics at $x \to \infty$ one shifts the integration contour in the integral introducing (5.1) into the upper half-plane. The nearest singularity of the integrand is the first order pole of the Fermi weight at

$$\lambda = \alpha = \sqrt{h + i\pi T} \quad ; \quad \mathrm{Im}\,\alpha > 0 \qquad (c = \infty) \qquad (5.3)$$

So the asymptotics at $x \to \infty$ is:

$$\langle\!\langle j(x)\, j(0)\rangle\!\rangle \big|_{x \to \infty} = -\frac{T^2}{2|\alpha|^2}\, \exp\{-2x\,\mathrm{Im}\,\alpha\} \times$$

$$(5.4)$$

$$\times \left[1 - \frac{\alpha^*}{2\alpha}\, \exp\{2ix\,\mathrm{Re}\,\alpha\} - \frac{\alpha}{2\alpha^*}\, \exp\{-2ix\,\mathrm{Re}\,\alpha\}\right] (c = \infty).$$

Consider now the large distance asymptotics of the correlator at the coupling constant c and the temperature T arbitrary. Take into account the first term of the series (3.11) as given in (3.18). As $x \to +\infty$, one shifts the integration contour over λ_1 into the lower complex half-plane, and the λ_2-contour into the upper half-plane:

$$\mathrm{Im}\,\lambda_1 \leqslant 0; \quad \mathrm{Im}\,\lambda_2 \geqslant 0. \qquad (5.5)$$

Remember that the qualitative behaviour of function $p_1(\lambda_1, \lambda_2)$ in (3.18) is just the same as of $-i(\lambda_1 - \lambda_2)$ (4.15). Function $p_1(\lambda_1, \lambda_2)$ permits to make the analytical continuation to the region (5.5) with no singularities arising (see (4.18)). The Fermi weight $\vartheta(\lambda)$ and the statistical weight $\omega(\lambda)$ are given as:

$$\vartheta(\lambda) = [1 + \exp\{\varepsilon(\lambda)/T\}]^{-1}; \quad \omega(\lambda) = \exp\{-\frac{1}{2\pi}\int\limits_{-\infty}^{\infty} K(\lambda,\mu)\vartheta(\mu)d\mu\}. \quad (5.6)$$

Analytical continuation of function $\varepsilon(\lambda)$ was discussed in detail in s.I.6. It follows from the arguments given there that the nearest to the real axis singularities which are obstacles for analitical continuation of the integrand in (3.18) are first-order poles of the Fermi weight $\vartheta(\lambda)$ situated at points $\alpha, -\alpha, \alpha^*, -\alpha^*$ so that

$$\mathcal{E}(\alpha) = i\pi T; \quad \mathrm{Im}\,\alpha > 0, \; \mathrm{Re}\,\alpha > 0. \tag{5.7}$$

The contribution of these four poles to the asymptotics is

$$\langle\!\langle j(x)\,j(0)\rangle\!\rangle \xrightarrow[x\to\infty]{}$$

$$\to 2T^2\left|\frac{\omega(\alpha)}{\mathcal{E}'(\alpha)}\right|^2\left(\frac{2\,\mathrm{Im}\,\alpha - c}{2\,\mathrm{Im}\,\alpha + c}\right)\left[\frac{p_1(\alpha^*,\alpha)}{2\,\mathrm{Im}\,\alpha}\right]^2\exp\{x\,p_1(\alpha,\alpha^*)\} + \tag{5.8}$$

$$+ 2T^2\,\mathrm{Re}\left\{\frac{\omega(\alpha)(2\alpha - ic)}{\mathcal{E}'(\alpha)(2\alpha + ic)}\left[\frac{p_1(-\alpha,\alpha)}{2\alpha}\right]^2\exp\{x\,p_1(-\alpha,\alpha)\}\right\}.$$

The remaining part of the integral decreases faster. It is to be mentioned that $p_1(\alpha^*,\alpha)$ is a negative real number (see (4.16)). It is convenient to introduce the following notation:

$$r_c \equiv -\left(p_1(\alpha^*,\alpha)\right)^{-1} > 0. \tag{5.9}$$

Quantity $p_1(-\alpha,\alpha)$ is a complex number and also

$$\mathrm{Re}\,p_1(-\alpha,\alpha) < p_1(\alpha^*,\alpha) < 0. \tag{5.10}$$

So again one sees damping of the oscillations as in s.2. The second term at the right hand side of (5.8) oscillates, but this term decreases faster than the first one. (As $c\to\infty$, $\mathrm{Re}\,p_1(-\alpha,\alpha) = p_1(\alpha^*,\alpha) - (8D/c^2)(\mathrm{Re}\,\alpha)^2$. But at $c\to\infty$ the both terms give the same contribution to the asymptotics recovering formula (5.4)). So one obtains

$$\langle\!\langle j(x)\,j(0)\rangle\!\rangle_T \xrightarrow[x\to\infty]{} \mathrm{Const}\,\exp\{-x/r_c\}, \quad c < \infty \tag{5.11}$$

The analysis of other terms of the series (3.11) leads to the conjecture that the form of the asymptotics (5.11) are exact.

Consider now different limiting cases.

(1) $T\to 0$. One has that $\mathrm{Im}\,\alpha \to 0$ (see (5.7)), and

$$\alpha = q_T + (i\pi T/\mathcal{E}'(q_1)); \quad \mathcal{E}(q_T) = 0. \tag{5.12}$$

The factor $S_1(t)$ entering (3.12) becomes equal to 1:

$$S_1(t) = \left(\frac{\alpha^* - t + ic}{\alpha^* - t - ic}\right)\left(\frac{\alpha - t - ic}{\alpha - t + ic}\right) \longrightarrow 1 , \qquad (5.13)$$

so that $P_1 \to 0$. Equation (3.12) is thus linearized and can be solved (see (I.4.30)):

$$P_1(t) = -T \overset{\circ}{F}(t|q)/\varepsilon'(q) . \qquad (5.14)$$

The correlation length $r_c \to \infty$ as

$$\frac{1}{r_c} = \frac{2\pi T}{\varepsilon'(q)}\left[1 + \frac{1}{2\pi}\int_{-0}^{q}\overset{\circ}{F}(t|q)dt\right], \qquad (5.15)$$

where q is the Fermi momentum at zero temperature (I.3.5). Equation (I.3.16) results in

$$r_c = v/(2\pi T) , \qquad (5.16)$$

where v is the sound velocity (I.4.28), (I.4.31). It should be noted that this formula agrees with that obtained in [35] by means of perturbation theory.

(2) Discuss now formula (5.9) at the strong coupling limit and arbitrary temperature. Using formulae of s.I.6 and (4.19), one obtains:

$$\alpha = \sqrt{h + 2c^{-1}Pr + i\pi T} \quad ; \quad \text{Im}\,\alpha > 0 , \ \text{Re}\,\alpha > 0; \quad (5.17)$$

$$r_c^{-1} = 2\,\text{Im}\,\alpha\,(1 + 2c^{-1}D) + 8D\bar{c}^{-2}(\text{Im}\,\alpha)^2 \qquad (5.18)$$

CONCLUSION

So the main ideas of the coordinate Bethe Ansatz as well as the algebraic Bethe Ansatz are demonstrated using as an example the model of the one-dimensional Bose gas. Our hope is that the reader is now convinced that the algebraic Bethe Ansatz lets to progress much further in calculation of correlation functions. In these lectures only the current correlator for the Bose gas was considered. The method of calculation, however, is quite general. It permits obtaining different correlators, e.g., correlator $\langle \psi^+(x)\psi(y)\rangle$ of the field operators in the NS model. The method can be also applied to any integrable model with the XXX or XXZ R-matrix. It can be thus applied to the spin correlation function in the XXZ Heisenberg model [36] and to the field correlator of the sine Gordon model, which is now in progress.

One of us (V.E.K.) would like to thank the Organising Committee of the Parchgani Winter School organized by Tata Institute of Fundamental Research, and especially the scientific secretary B.S. Shastry, for hospitality.

APPENDIX I

The equation (II.1.32) can be proved using induction in n. For $n=1$ it is valid. Supposing that (II.1.32) valid for $T(n,1|\lambda)$ (II.1.16):

$$T(n,1|\lambda) = \mathcal{L}(n|\lambda)\mathcal{L}(n-1|\lambda)\ldots\mathcal{L}(1|\lambda),$$

one proves that it is also valid for $T(n+1,1|\lambda) = \mathcal{L}(n+1|\lambda)T(n,1|\lambda)$:

$$\{T(n+1,1|\lambda) \overset{\otimes}{,} T(n+1,1|\mu)\} = \{\mathcal{L}(n+1|\lambda) \overset{\otimes}{,} \mathcal{L}(n+1|\mu)\}\times$$
$$\times(T(n,1|\lambda)\otimes T(n,1|\mu)) + (\mathcal{L}(n+1|\lambda)\otimes\mathcal{L}(n+1|\mu))\times$$
$$\times\{T(n,1|\lambda) \overset{\otimes}{,} T(n,1|\mu)\}.$$

Using now (II.1.31), (II.1.32) one can further write:

$$\{T(n+1,1|\lambda) \overset{\otimes}{,} T(n+1,1|\mu)\} = (\mathcal{L}(n+1|\lambda)\otimes\mathcal{L}(n+1|\mu))r(\lambda,\mu)\times$$
$$\times(T(n,1|\lambda)\otimes T(n,1|\mu)) - r(\lambda,\mu)(T(n+1,1|\lambda)\otimes T(n+1,1|\mu)) +$$
$$+ (T(n+1,1|\lambda)\otimes T(n+1,1|\mu))r(\lambda,\mu) -$$
$$- (\mathcal{L}(n+1|\lambda)\otimes\mathcal{L}(n+1|\mu))r(\lambda,\mu)(T(n,1|\lambda)\otimes T(n,1|\mu)),$$

so that (II.1.32) is also valid for $n \to n+1$. The proof is thus finished.

APPENDIX 2

The determinant of the exact $\mathcal{L}$ -operator (II.1.35) is:

$$\det \mathcal{L}(n|\lambda) = \Delta^2 (\lambda - \nu)(\lambda - \nu^*)/4 \; ; \quad \nu = -2i/\Delta . \tag{A.1}$$

The lattice monodromy matrix is constructed as

$$T(\lambda) = \mathcal{L}(N|\lambda) \ldots \mathcal{L}(1|\lambda) \; ; \tag{A.2}$$

$$\det T(\lambda) = \left[\det \mathcal{L}(1|\lambda) \right]^N . \tag{A.3}$$

At $\lambda = \nu$ the $\mathcal{L}$ -operator (as well as the monodromy matrix) becomes a one dimensional proector:

$$\mathcal{L}_{ik}(n|\nu) = \alpha_i(n)\beta_k(n) \qquad (i,k = 1,2). \tag{A.4}$$

The trace $\mathcal{T}(\lambda)$ of the monodromy matrix is thus factorized:

$$\tau(\lambda) = \mathrm{tr}\, T(\lambda) = \prod_{n=1}^{N} (\vec{\beta}(n+1)\vec{\alpha}(n)) \; ; \tag{A.5}$$

$$(\vec{\beta}\,\vec{\alpha}) \equiv \sum_{i=1}^{2} \beta_i \alpha_i \tag{A.6}$$

The logarithm of $\mathcal{T}(\lambda)$ is given as a local expression:

$$\ln \tau(\lambda) = \sum_{n=1}^{N} \tau_2(n,n+1); \quad \tau_2(n,n+1) = \ln(\vec{\beta}(n+1)\vec{\alpha}(n)), \tag{A.7}$$

describing the interaction of two nearest neighbours. It was also proved in paper $[25]$ that all the logarithmic derivatives $d^m \ln \tau(\lambda)/d\lambda^m \big|_{\lambda=\nu}$ are local, which permits construction of local lattice Hamiltonian. This method is valid also in the quantum case $[26]$.

APPENDIX 3

The proof is done again using induction in number n of the sites.

For $n=1$ the equation (II.2.14) is valid. Suppose that relation (II.2.16) is also valid for

$$T(x,0|\lambda) = \mathcal{L}(n|\lambda)\dots\mathcal{L}(1|\lambda) \equiv T(n,1|\lambda) \equiv T(n|\lambda)$$

Prove that then (II.2.16) is also valid for $T(n+1|\lambda) = \mathcal{L}(n+1|\lambda) \times T(n|\lambda)$. Write down the tensor product of T's as (we use (II.2.15)):

$$T(n+1|\lambda)\otimes T(n+1|\mu) = \big(\mathcal{L}(n+1|\lambda)\otimes\mathcal{L}(n+1|\mu)\big)\big(T(n|\lambda)\otimes T(n|\mu)\big).$$

Multiplying that by R from the left and by R^{-1} from the right and using (II.2.14), (II.2.16), one has

$$R(\lambda,\mu)\big(T(n+1|\lambda)\otimes T(n+1|\mu)\big)R^{-1}(\lambda,\mu) =$$

$$= R(\lambda,\mu)\big(\mathcal{L}(n+1|\lambda)\otimes\mathcal{L}(n+1|\mu)\big)R^{-1}(\lambda,\mu)R(\lambda,\mu)\times$$

$$\times\big(T(n|\lambda)\otimes T(n|\mu)\big)R^{-1}(\lambda,\mu) = T(n+1|\mu)\otimes T(n+1|\lambda),$$

which completes the proof.

REFERENCES

1. Faddeev, L.D., Sklyanin E.K.: Dokl.Akad.Nauk SSSR 243, 1430-1433 (1978)
2. Faddeev, L.D.: LOMI-preprint P-2-79, Leningrad, 1979 and Sov. Sci.Rev., Math.Phys.C1, 107-160 (1981)
3. Bethe, H.: Zeit.Phys. 71, 205-226 (1931)
4. Lieb, E.H., Liniger, W.: Phys.Rev. 130, 1605-1616 (1963)
5. Gaudin, M.: Preprint, Centre d'Etudes Nucleaires de Saclay, CEA-N-1559, (1), (1972)
6. Faddeev, L.D., Takhtajan, L.A.: Usp.Mat.Nauk. 34, 13-63 (1979)
7. Korepin, V.E.: Commun.Math.Phys. 86, 391-418 (1982)
8. Izergin, A.G., Korepin, V.E.: Commun.Math.Phys. 94, 67-92 (1984)
9. Korepin, V.E.: Commun.Math.Phys. 94, 93-113 (1984)
10. Bogoliubov, N.M., Korepin V.E.: Teor.i Mat.Fiz. 60, 262-269 (1984)
11. Lieb, E.H.: Phys.Rev. 130, 1616-1624 (1963)
12. Yang, C.N., Yang, C.P.: J.Math.Phys. 10, 1115-1122 (1969)
13. Berezin, F.A., Pokhil, C.P., Finkelberg, V.M.: Vest.Mos. Gos. Univ.Ser.1.1,21-28 (1964)

316

14. Mc-Guire, J.B.: J.Math.Phys. 5, 622-636 (1964)

15. Brezin, E., Zinn-Justin, J.: C.R.Acad.Sci.(Paris) 263, 670-673
    (1966)

16. Gaudin, M.: La fonction d'onde de Bethe pour les modèles exacts
    de la méchanique statistic. Paris, Commisariat à l'energie ato-
    mique 1983.

17. Korepin, V.E.: Teor.Mat.Fiz. 41, 169-189 (1979)

18. Kulish, P.P., Sklyanin, E.K.: Integrable quantum field theories.
    Lect.Notes in Phys. 151, 61-119 (1982)

19. Izergin, A.G., Korepin, V.E.: Fizika Ehlementarnykh Chastits I
    Atomnogo Yadra 13(3), 207-223 (1982)

20. Lenard, A.: J.Math.Phys.5, 930-943 (1964); J.Math.Phys.7, 1268-
    1272 (1966)

21. Jimbo, M., Miwa, T., Mori, Y., Sato, M.: Physica D1, 80-158
    (1980)

22. Zakharov, V.E., Manakov, S.V., Novikov, S.P., Pitajevskij, L.P.:
    Theory of Solitons. Moskva, Nauka, 1980.

23. Zakharov, V.E., Shabat, A.B.: Zh.Exp.Teor.Fiz. 61, 118-134 (1971)

24. Sklyanin, E.V.: LOMI-preprint E-3-1979, Leningrad, 1979.

25. Izergin, A.G., Korepin, V.E.: Dokl.Akad.Nauk SSSR 259, 76-79
    (1981)

26. Izergin, A.G., Korepin, V.E.: Nucl.Phys.B205 $\left[\text{FS5}\right]$, 401-413
    (1982)

27. Sklyanin, E.K.: Dokl.Akad.Nauk SSSR 244, 1337-1341 (1978); Zap.
    Nauchn.Seminarov LOMI 95,55-128 (1980)

28. Izergin, A.G., Korepin, V.E., Smirnov F.A.: Teor.Math.Fiz. 48,
    319- 323 (1981)

29. Smirnov, F.A.: Dokl.Akad.Nauk SSSR 262, 78-83 (1982)

30. Izergin, A.G., Korepin, V.E.: Lett.Math.Phys.6, 283-288 (1982)

31. Heisenberg, W.: Zeit.Phys.49, 619-636 (1928)

32. Korepin, V.E.: Dokl.Akad.Nauk SSSR 265, 1361-1364 (1982)

33. Izergin, A.G., Korepin, V.E.: Lett.Math.Phys. 8, 259-265 (1984)

34. Greamer, D.B., Thaker, H.B., Wilkinson, D.: Phys.Rev.D21, 1523-
    1528 (1980)

35. Popov, V.N.: Functional integrals in quantum field theory and
    statistical physics. D.Reidel Publishing Company, Dordrecht 1983

36. Izergin, A.G., Korepin, V.E.: Commun.Math.Phys. (1985)

37. Gelfand, I.M., Levitan, B.M.: Dokl.Akad.Nauk SSSR 88, 593-596
    (1953)

<u>LIST OF PARTICIPANTS</u>

S.No.	Name	Address
1.	J. Avan	Universite Pierre et Marie Curie, Lab. de Phys. Theor. et Hautes Energies, Paris, France
2.	M. Azam	Tata Institute of Fundamental Research, Bombay
3.	O. Babelon	Universite Pierre Marie Curie, Lab. de Phys. Theor. et Hautes Energies, Paris, France
4.	M. Barma	Tata Institute of Fundamental Research, Bombay
5.	S.N. Behera	Institute of Physics, Bhubaneswar
6.	G.R. Bhat	Bombay University, Bombay
7.	G. Bhattacharya	Saha Institute of Nuclear Physics, Calcutta
8.	P. Bhattacharyya	Tata Institute of Fundamental Research, Bombay
9.	I. Bose	Indian Association for the Cultivation of Sciences, Calcutta
10.	H. de Vega	Universite Pierre et Marie Curie, Lab. de Phys. Theor. et Hautes Energies, Paris, France
11.	A. Dhara	Bhabha Atomic Research Centre, Bombay
12.	P.P. Divakaran	Tata Institute of Fundamental Research, Bombay
13.	L. Faddeev	V.A. Steklov Mathematical Institute, Academy of Sciences of the USSR, Leningrad, USSR
14.	A.D. Gangal	University of Pune, Pune
15.	D.K. Ghosh	Indian Institute of Technology, Bombay
16.	V. Gupta	Tata Institute of Fundamental Research, Bombay
17.	N. Gupte	Central University of Hyderabad, Hyderabad
18.	S.S. Jha	Tata Institute of Fundamental Research, Bombay
19.	A. Khare	Institute of Physics, Bhubaneswar
20.	V. Korepin	V.A. Steklov Mathematical Institute, Academy of Sciences of the USSR, Leningrad, USSR
21.	H.R. Krishnamurthy	Indian Institute of Science, Bangalore
22.	D. Kumar	University of Roorkee, Roorkee
23.	A. Kundu	Birla Institute of Technology & Science, Pilani, Rajasthan
24.	C.K. Majumdar	Indian Association for the Cultivation of Sciences, Calcutta
25.	P. Majumdar	Saha Institute of Nuclear Physics, Calcutta
26.	S.D. Mathur	Tata Institute of Fundamental Research, Bombay
27.	J. Maharana	Institute of Physics, Bhubaneswar
28.	N. Meckenzie	Indian Association for the Cultivation of Sciences, Calcutta
29.	S.G. Mishra	Institute of Physics, Bhubaneswar
30.	P. Mitra	Saha Institute of Nuclear Physics, Calcutta

S.No.	Name	Address
31.	T. Miwa	Kyoto University, Kyoto, Japan
32.	S. Mukhi	Tata Institute of Fundamental Research, Bombay
33.	A. Mukherjee	Indian Institute of Technology, Kanpur
34.	S. Mukhopadhyay	Saha Institute of Nuclear Physics, Calcutta
35.	S. Naik	Institute of Physics, Bhubaneswar
36.	R. Nityananda	Raman Research Institute, Bangalore
37.	A.K. Raina	Tata Institute of Fundamental Research, Bombay
38.	T.R. Ramadas	Tata Institute of Fundamental Research, Bombay
39.	R. Ramaswamy	Tata Institute of Fundamental Research, Bombay
40.	B.S. Shastry	Tata Institute of Fundamental Research, Bombay
41.	L.P. Singh	Utkal University, Bhubaneswar
42.	B. Sutherland	University of Utah, Salt Lake City, U.S.A.
43.	L.A. Takhtajan	V.A. Steklov Mathematical Institute, Academy of Sciences of the USSR, Leningrad, USSR
44.	D.N. Verma	Tata Institute of Fundamental Research, Bombay
45.	S.R. Wadia	Tata Institute of Fundamental Research, Bombay
46.	G.L. Wiersma	University of Amsterdam, Amsterdam, The Netherlands

Administrative Staff

S.No.	Name	Address
47.	S.K. Bhonslay	Tata Institute of Fundamental Research, Bombay
48.	R. Ganesan	Tata Institute of Fundamental Research, Bombay

M. Toda, R. Kubo, N. Saitô

Statistical Physics I

Equilibrium Statistical Mechanics

1983. 90 figures. XVI, 249 pages. (Springer
Series in Solid-State Sciences, Volume 30)
ISBN 3-540-11460-2

Contents: General Preliminaries. – Outlines of
Statistical Mechanics. – Applications. – Phase
Transitions. – Ergodic Problems. – General
Bibliography. – References. – Subject Index.

R. Kubo, M. Toda, N. Hashitsume

Statistical Physics II

Nonequilibrium Statistical Mechanics

1985. 28 figures. XVI, 279 pages. (Springer
Series in Solid-State Sciences, Volume 31)
ISBN 3-540-11461-0

Contents: Brownian Motion. – Physical
Processes as Stochastic Processes. – Relaxation
and Resonance Absorption. – Statistical
Mechanics of Linear Response. – Quantum
Field Theoretical Methods in Statistical
Mechanics. – General Bibliography of Text-
books. – References. – Subject Index.

Springer-Verlag
Berlin Heidelberg
New York Tokyo

Lecture Notes in Physics

Vol. 214: H. Moraal, Classical, Discrete Spin Models. VII, 251 pages. 1984.

Vol. 215: Computing in Accelerator Design and Operation. Proceedings, 1983. Edited by W. Busse and R. Zelazny. XII, 574 pages. 1984.

Vol. 216: Applications of Field Theory to Statistical Mechanics. Proceedings, 1984. Edited by L. Garrido. VIII, 352 pages. 1985.

Vol. 217: Charge Density Waves in Solids. Proceedings, 1984. Edited by Gy. Hutiray and J. Sólyom. XIV, 541 pages. 1985.

Vol. 218: Ninth International Conference on Numerical Methods in Fluid Dynamics. Edited by Soubbaramayer and J. P. Boujot. X, 612 pages. 1985.

Vol. 219: Fusion Reactions Below the Coulomb Barrier. Proceedings, 1984. Edited by S. G. Steadman. VII, 351 pages. 1985.

Vol. 220: W. Dittrich, M. Reuter, Effective Lagrangians in Quantum Electrodynamics. V, 244 pages. 1985.

Vol. 221: Quark Matter '84. Proceedings, 1984. Edited by K. Kajantie. VI, 305 pages. 1985.

Vol. 222: A. García, P. Kielanowski, The Beta Decay of Hyperons. Edited by A. Bohm. VIII, 173 pages. 1985.

Vol. 223: H. Saller, Vereinheitlichte Feldtheorien der Elementarteilchen. IX, 157 Seiten. 1985.

Vol. 224: Supernovae as Distance Indicators. Proceedings, 1984. Edited by N. Bartel. VI, 226 pages. 1985.

Vol. 225: B. Müller, The Physics of the Quark-Gluon Plasma. VII, 142 pages. 1985.

Vol. 226: Non-Linear Equations in Classical and Quantum Field Theory. Proceedings, 1983/84. Edited by N. Sanchez. VII, 400 pages. 1985.

Vol. 227: J.-P. Eckmann, P. Wittwer, Computer Methods and Borel Summability Applied to Feigenbaum's Equation. XIV, 297 pages. 1985.

Vol. 228: Thermodynamics and Constitutive Equations. Proceedings, 1982. Edited by G. Grioli. V, 257 pages. 1985.

Vol. 229: Fundamentals of Laser Interactions. Proceedings, 1985. Edited by F. Ehlotzky. IX, 314 pages. 1985.

Vol. 230: Macroscopic Modelling of Turbulent FLows. Proceedings, 1984. Edited by U. Frisch, J. B. Keller, G. Papanicolaou and O. Pironneau. X, 360 pages. 1985.

Vol. 231: Hadrons and Heavy Ions. Proceedings, 1984. Edited by W. D. Heiss. VII, 458 pages. 1985.

Vol. 232: New Aspects of Galaxy Photometry. Proceedings, 1984. Edited by J.-L. Nieto. XIII, 350 pages. 1985.

Vol. 233: High Resolution in Solar Physics. Proceedings, 1984. Edited by R. Muller. VII, 320 pages. 1985.

Vol. 234: Electron and Photon Interactions at Intermediate Energies. Proceedings, 1984. Edited by D. Menze, W. Pfeil and W. J. Schwille. VII, 481 pages. 1985.

Vol. 235: G. E. A. Meier, F. Obermeier (Eds.), Flow of Real Fluids. VIII, 348 pages. 1985.

Vol. 236: Advanced Methods in the Evaluation of Nuclear Scattering Data. Proceedings, 1985. Edited by H. J. Krappe and R. Lipperheide. VI, 364 pages. 1985.

Vol. 237: Nearby Molecular Clouds. Proceedings, 1984. Edited by G. Serra. IX, 242 pages. 1985.

Vol. 238: The Free-Lagrange Method. Proceedings, 1985. Edited by M. J. Fritts, W. P. Crowley and H. Trease. IX, 313 pages. 1985.

Vol. 239: Geometrics Aspects of the Einstein Equations and Integrable Systems. Proceedings, 1984. Edited by R. Martini. V, 344 pages. 1985.

Vol. 240: Monte-Carlo Methods and Applications in Neutronics, Photonics and Statistical Physics. Proceedings, 1985. Edited by R. Alcouffe, R. Dautray, A. Forster, G. Ledanois and B. Mercier. VIII, 483 pages. 1985.

Vol. 241: Numerical Simulation of Combustion Phenomena. Proceedings, 1985. Edited by R. Glowinski, B. Larrouturou and R. Temam. IX, 404 pages. 1985.

Vol. 242: Exactly Solvable Problems in Condensed Matter and Relativistic Field Theory. Proceedings, 1985. Edited by B. S. Shastry, S. S. Jha and V. Singh. V, 318 pages. 1985.